MATHEMATICAL PROGRAMMING
AT OBERWOLFACH II

MATHEMATICAL PROGRAMMING STUDIES

Editor-in-Chief
R.W. COTTLE, Department of Operations Research, Stanford University, Stanford, CA 94305, U.S.A.

Co-Editors
L.C.W. DIXON, Numerical Optimisation Centre, The Hatfield Polytechnic, College Lane, Hatfield, Hertfordshire AL10 9AB, England
B. KORTE, Institut für Ökonometrie und Operations Research, Universität Bonn, Nassestrasse 2, D-5300 Bonn 1, W. Germany
M.J. TODD, School of Operations Research and Industrial Engineering, Upson Hall, Cornell University, Ithaca, NY 14853, U.S.A.

Associate Editors
E.L. ALLGOWER, Colorado State University, Fort Collins, CO, U.S.A.
W.H. CUNNINGHAM, Carleton University, Ottawa, Ontario, Canada
J.E. DENNIS, Jr., Rice University, Houston, TX, U.S.A.
B.C. EAVES, Stanford University, CA, U.S.A.
R. FLETCHER, University of Dundee, Dundee, Scotland
D. GOLDFARB, Columbia University, New York, USA
J.-B. HIRIART-URRUTY, Université Paul Sabatier, Toulouse, France
M. IRI, University of Tokyo, Tokyo, Japan
R.G. JEROSLOW, Georgia Institute of Technology, Atlanta, GA, U.S.A.
D.S. JOHNSON, Bell Telephone Laboratories, Murray Hill, NJ, U.S.A.
C. LEMARECHAL, INRIA-Laboria, Le Chesnay, France
L. LOVASZ, University of Szeged, Szeged, Hungary
L. MCLINDEN, University of Illinois, Urbana, IL, U.S.A.
M.J.D. POWELL, University of Cambridge, Cambridge, England
W.R. PULLEYBLANK, University of Waterloo, Waterloo, Ontario, Canada
A.H.G. RINNOOY KAN, Erasmus University, Rotterdam, The Netherlands
K. RITTER, Technische Universität München, München, W. Germany
R.W.H. SARGENT, Imperial College, London, England
D.F. SHANNO, University of California, Davis, CA, U.S.A.
L.E. TROTTER, Jr., Cornell University, Ithaca, NY, U.S.A.
H. TUY, Institute of Mathematics, Hanoi, Socialist Republic of Vietnam
R.J.B. WETS, University of Kentucky, Lexington, KY, U.S.A.

Senior Editors
E.M.L. BEALE, Scicon Computer Services Ltd., Milton Keynes, England
G.B. DANTZIG, Stanford University, Stanford, CA, U.S.A.
L.V. KANTOROVICH, Academy of Sciences, Moscow, U.S.S.R.
T.C. KOOPMANS, Yale University, New Haven, CT, U.S.A.
A.W. TUCKER, Princeton University, Princeton, NJ, U.S.A.
P. WOLFE, IBM Research Center, Yorktown Heights, NY, U.S.A.

NORTH-HOLLAND – AMSTERDAM · NEW YORK · OXFORD

MATHEMATICAL PROGRAMMING STUDY 22
A PUBLICATION OF THE MATHEMATICAL PROGRAMMING SOCIETY

Mathematical Programming at Oberwolfach II

Edited by B. KORTE and K. RITTER

M.L Balinski
F. Barahona
I. Barany
J.M. Bourjolly
W. Cook
W. Cunningham
J.E. Dennis
J. Fonlupt
A. Frank
S. Fujishige
B. Gollan
M.D. Grigoriadis
P.L. Hammer
T. Hsu
R.G. Jeroslow
P.V. Kamesam
L. Lovász
J. Lowe
O.L. Mangasarian
R.R. Meyer
M. Raco
S.M. Robinson
J.B. Rosen
A. Russakoff
A. Schrijver
A. Sebő
B. Simeone
E. Tardos
T.J. Van Roy
H.F. Walker
L.A. Wolsey
K. Zimmermann

1984

NORTH-HOLLAND – AMSTERDAM · NEW YORK · OXFORD

© THE MATHEMATICAL PROGRAMMING SOCIETY, INC. – 1984

All rights reserved. No part of this publication may be reproduced, stored in a retrieval system or transmitted, in any form or by any means, electronic, mechanical, photocopying, recording or otherwise, without the prior permission of the copyright owner.

This book is also available in journal format on subscription.

ISBN: 0 444 87691 X

Published by:
ELSEVIER SCIENCE PUBLISHERS B.V.
P.O. Box 1991
1000 BZ Amsterdam
The Netherlands

Sole distributors for the U.S.A. and Canada:
ELSEVIER SCIENCE PUBLISHING COMPANY, INC.
52 Vanderbilt Avenue
New York, N.Y. 10017
U.S.A.

Library of Congress Cataloging in Publication Data

Main entry under title:

Mathematical programming at Oberwolfach II.

(Mathematical programming study ; 22)
Proceedings of a conference on "Mathematische Optimierung," held Jan. 1-15, 1983, at the Mathematisches Forschungsinsitut, Oberwolfach.
1. Programming (Mathematics)--Congresses. I. Korte, B. H. (Bernard H.), 1938- . II. Ritter, K. (Klaus), 1936- . III. Balinski, M. L. IV. Title: Mathematical programming at Oberwolfach 2. V. Title: Mathematical programming at Oberwolfach two. VI. Series.
QA402.5.M35285 1984 519.7 84-28605
ISBN 0-444-87691-X (U.S.)

PRINTED IN THE NETHERLANDS

PREFACE

From January 9 to 15, 1983 the third conference on ''Mathematische Optimierung'' was held at the Mathematisches Forschungsinstitut Oberwolfach. The first Oberwolfach meeting devoted to the subject of mathematical programming took place in 1979. Part of its results has been documented in the Mathematical Programming Study No. 14 entitled ''Mathematical Programming at Oberwolfach''.

The second Oberwolfach conference in 1981 had no proceedings volume. We feel that mandatory proceedings announced in advance do not contribute to the spontaneuos workshop atmosphere of an Oberwolfach week.

After the third Oberwolfach conference we believed that many new results and ideas presented there would justify a collection of them which eventually led to this Mathematical Programming Study entitled ''Mathematical Programming at Oberwolfach II''.

This Oberwolfach meeting was attended by 63 scholars from 13 different countries. Not too surprisingly, almost all invitations were accepted. Thus, the total size was somewhat beyond the scope of the Oberwolfach Institute. However, 25 participants came from overseas and especially from North America. They made substantial travel expenditures from their own resources in order to spend one week in the stimulating environment of Oberwolfach. This indicates strongly that the mathematical programming community has developed a very positive attitude toward Oberwolfach and considers this meeting to be one of the highlights in the field.

The 18 papers of this volume reflect the different areas of mathematical programming as well as the interaction of its two main streams, namely continuous and nonlinear optimization on one side and discrete and combinatorial optimization on the other. It contains papers about special algorithms for nonlinear programming, polyhedral theory, submodularity, min–max and duality relations, complexity, quasi-Newton and approximation methods, network algorithms and other numerical aspects of optimization methods.

We appreciate very much the help of numerous referees. An acknowledgement of their work will appear with the next referee listing of Mathematical Programming.

Finally, we are very much indebted to the staff of the Oberwolfach Institute whose excellent support made this conference a success. Our most sincere thanks go to the director of the Mathematisches Forschungsinstitut, Professor Dr. M. Barner, for hosting the meeting at this unique conference center.

Bernhard Korte
Klaus Ritter

CONTENTS

Mathematical Programming Study 22 (1984) 1–8
North-Holland

FACES OF DUAL TRANSPORTATION POLYHEDRA

M.L. BALINSKI

C.N.R.S., Laboratoire d'Econométrie de l'Ecole Polytechnique, Paris, France

Andrew RUSSAKOFF

College of Business Administration, St John's University, Jamaica, NY, USA

Received 26 November 1982
Revised manuscript received 28 February 1984

The extreme points of any nondegenerate dual transportation polyhedron are characterized by the m-partitions (or n-partitions) of $m+n-1$. This is used to show that all such polyhedra have exactly the same number of r-dimensional faces, $0 \leqslant r \leqslant m+n-2$, and to count them.

Key words: Polyhedra, Transportation Problem, Faces.

Introduction

The *transportation polytope*[1] is

$$P_{m,n}(\boldsymbol{a}, \boldsymbol{b}) = \left\{ \boldsymbol{x} = (x_{ij});\ \sum_{j \in N} x_{ij} = a_i,\ \sum_{i \in M} x_{ij} = b_j,\ x_{ij} \geqslant 0,\ i \in M, j \in N \right\}$$

where $a_i > 0$ and $b_j > 0$ for all $i \in M = \{1, \ldots, m\}$ and $j \in N = \{1, \ldots, n\}$, and $\sum_M a_i = \sum_N b_j$. We assume throughout that $m \leqslant n$. Primal methods for solving the transportation problem travel on the extreme points and edges of $P_{m,n}$ so it is not surprising that $P_{m,n}$ for varying $\boldsymbol{a}, \boldsymbol{b}$ has been extensively studied [1, 2, 6, 7, 8]. The precise number of extreme points of P for certain choices of $\boldsymbol{a}, \boldsymbol{b}$ is known. It has been proven that for fixed m, n relatively prime, $\boldsymbol{a} = (n, \ldots, n)$ and $\boldsymbol{b} = (m, \ldots, m)$ gives the $P_{m,n}$ having the maximum number of extreme points over all choices of $\boldsymbol{a}, \boldsymbol{b}$ [7]. The number itself has simple expressions only for the cases $n = km+1$ when it is $n^{m-2} n!/(k!)^m$ and $n = km+m-1$ when it is $m^{n-2} n!/(k!)^m$ [2]. For the general case $n = km + c$, it is $(n!/(k!)^m) p(m, c, k)$ where $p(m, c, k)$ is a polynomial in k (with m and c fixed) with integer coefficients and highest term $m^{m-2} k^{m-c-1}$ [1]. For fixed m, n the minimum number of extreme points of $P_{m,n}$ is n^{m-1} when there is no degeneracy, and $n!/(n-m+1)!$ if degeneracy is allowed.

The *dual transportation polyhedron*[1] is

$$D_{m,n}(\boldsymbol{c}) = \{\boldsymbol{u} = (u_i),\ \boldsymbol{v} = (v_j);\ u_i + v_j \leqslant c_{ij},\ i \in M, j \in N,\ u_1 = 0\}.$$

[1] Arguments $\boldsymbol{a}$, $\boldsymbol{b}$, m, n and $\boldsymbol{c}$ will be dropped unless necessary in the sequel.

It is unbounded. Fixing $u_1 = 0$ is an arbitrary choice that rules out cylinders of solutions $(u_i + \delta)$, $(v_j - \delta)$. Dual methods for solving the transportation problem travel on the extreme points and bounded edges of $D_{m,n}$ and so it is surprising that relatively little is known about it, although folklore had it that dual methods required fewer iterations.

The fact is that the number of r-dimensional faces of *any* nondegenerate $D_{m,n}(\boldsymbol{c})$ is always the same, as is the number of bounded (and so also of unbounded) faces. It will be proved that the maximum number of extreme points of any $D_{m,n}$ is at most $\binom{m+n-2}{m-1}$, a number that is very much smaller than the minimum possible number for $P_{m,n}$. This last fact we subsequently learned was shown in a much earlier but somewhat obscure and involved paper of Zhu [9]. The principal tool for proving these results is a one-to-one correspondence between extreme points of nondegenerate $D_{m,n}$ and m-partitions of $m+n-1$.

The folklore had it that dual methods for the transportation problem were better. This may have something to do with the sheer number of extreme points of primal as versus dual polyhedra. The comparisons are dramatic. For example, for $m = 8$, $n = 9$, the primal polytope has a minimum of 181 440 extreme points allowing degeneracy and of 4 782 969 not allowing degeneracy, and a maximum number of $19\,285 \times 10^{11}$; in contrast, the dual polyhedron has a minimum of 1 allowing degneracy, and a maximum of 6435.

The characterization of the extreme points of nondegenerate dual transportation polyhedra by partitions turns out to be quite useful [4]. On the one hand, it may be used to develop an algorithm for the assignment problem requiring at most $(n-1)(n-2)/2$ pivots [3]; on the other hand it may be used to show that the diameter of $D_{m,n}$ is at most $(m-1)(n-1)$ [5].

1. Partitions

An extreme point of $D_{m,n}(\boldsymbol{c})$ is a $(\boldsymbol{u}, \boldsymbol{v})$ that is a unique solution to

$$u_1 = 0, \qquad u_i + v_j = c_{ij} \quad \text{for } (i,j) \in T \subset M \times N,$$
$$u_i + v_j \leq c_{ij} \quad \text{for } (i,j) \notin T$$

where T corresponds to some $m+n-1$ linearly independent equations $u_i + v_j = c_{ij}$. The extreme point is nondegenerate if $u_i + v_j < c_{ij}$ for all $(i,j) \notin T$, and D is nondegenerate if all of its extreme points are nondegenerate.

A natural model for extreme points and faces of D is a graph. Let $C(M, N)$ be the complete bi-partite graph having disjoint node sets M and N and all edges (i,j), $i \in M$, $j \in N$. To each $(\boldsymbol{u}, \boldsymbol{v}) \in D_{m,n}$ associate the subgraph $G(\boldsymbol{u}, \boldsymbol{v})$ of $C(M, N)$ defined as containing edge (i,j) if and only if $u_i + v_j = c_{ij}$.

Lemma 1. *$(\boldsymbol{u}, \boldsymbol{v}) \in D$ is an extreme point of D if and only if $G(\boldsymbol{u}, \boldsymbol{v})$ contains a spanning tree T.*

This is well known and easily established. If $(\boldsymbol{u}, \boldsymbol{v})$ is nondegenerate, $G(\boldsymbol{u}, \boldsymbol{v})$ *is* a spanning tree. A spanning tree of $C(M, N)$ must contain precisely $m+n-1$ edges. If $(\boldsymbol{u}, \boldsymbol{v})$ is degenerate then $G(\boldsymbol{u}, \boldsymbol{v})$ must contain a cyucle, say (see Figure 1), (i_1, j_1), (i_2, j_1), (i_2, j_2), ..., (i_k, j_k), (i_1, j_k). Call the edges of type (i_l, j_l) odd, the others even. Then $\sum_{\text{odd}} c_{ij} = \sum_1^k (u_{i_l} + v_{j_l}) = \sum_{\text{even}} c_{ij}$, characterizing how degeneracy occurs.

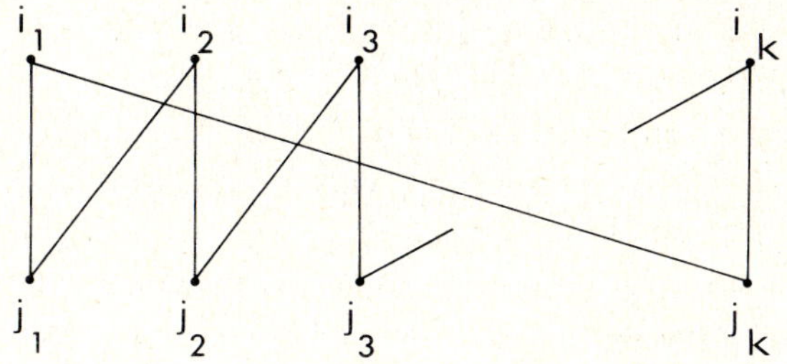

Fig. 1

A *row partition* is any $\boldsymbol{a} = (a_1, \ldots, a_m)$, $a_i \geqslant 1$, $\sum a_i = m+n-1$, corresponding to an assignment of valences to nodes of M; and a *column partition* is any $\boldsymbol{b} = (b_1, \ldots, b_n)$, $b_j \geqslant 1$, $\sum b_j = m+n-1$, corresponding to an assignment of valencies to nodes of N. To each spanning tree T contained in the graph $G(\boldsymbol{u}, \boldsymbol{v})$ of any extreme point $(\boldsymbol{u}, \boldsymbol{v})$ there corresponds a unique row partition which represents the valencies of the nodes M and a unique column partition which represents the valencies of the nodes N.

Lemma 2. *To one row (equivalently, column) partition there corresponds at most one extreme point of a nondegenerate* $D_{m,n}(\boldsymbol{c})$.

Proof. Suppose, instead, that the graphs $G(\boldsymbol{u}, \boldsymbol{v})$ and $G(\boldsymbol{u}', \boldsymbol{v}')$ of two *different* extreme points contained different spanning trees T and T' having the same row valencies $(a_1, \ldots a_m)$.

There is some arc $(i, j) \in T'$ but $\notin T$, call it (i_1, j_1). Since T is a spanning tree there exists a unique path joining j_1 and i_1 in T. Let $(i_2, j_1) \in T$ be the first edge on that path and $(i_2, j_2) \in T'$. There must be such an edge since the valence of i_2 in T is at least two and so the valence of i_2 in T' is also at least two. Now, take $(i_3, j_2) \in T$ to be the unique edge on the path joining j_2 to i_1 in T and continue. At some point a node previously visited must be encountered (see Fig. 2). But, then, calling edges $(i_l, j_l) \in T'$ for $l = s$ to t of the cycle odd and the others in the cycle which belong to T even, we have

$$\sum_{\text{odd}} c_{ij} = \sum_s^t (u'_{i_l} + v'_{j_l}) < \sum_{\text{even}} c_{ij}$$

and

$$\sum_{\text{even}} c_{ij} = \sum_s^t (u_{i_l} + v_{j_l}) < \sum_{\text{odd}} c_{ij}.$$

The inequalities are strict by nondegeneracy, giving a contradiction. □

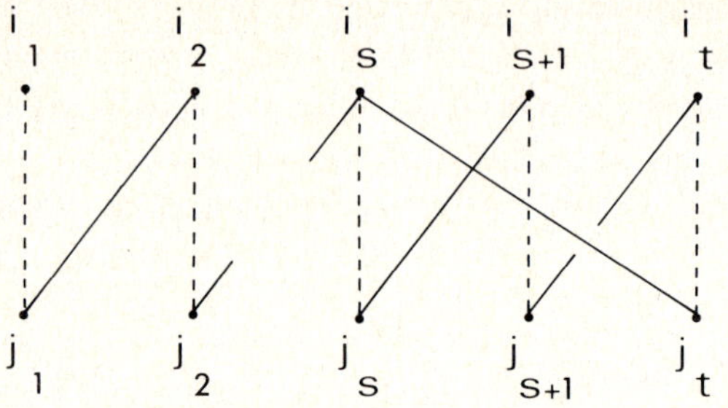

Fig. 2. Solid edges in T, dashed edges in T'.

In fact the lemma is true for all $D_{m,n}(\boldsymbol{c})$, degenerate or not [5]. To one degenerate extreme point there corresponds more than one partition, because degeneracy implies $G(\boldsymbol{u}, \boldsymbol{v})$ contains more than one spanning tree.

Theorem 1. *For nondegenerate $D_{m,n}(\boldsymbol{c})$ there exist one to one correspondences between row partitions, extreme points and column partitions.*

Proof. The proof is by induction on $m+n$. For $m+n=1, 2$ or 3 the result is obvious. Suppose it is true for all $m+n<k$.

Recall that $m \leq n$. Any column partition $(b_1, b_2, \ldots, b_n)$ must have at least one component equal to 1, say $b_1 = 1$. By the inductive hypothesis $(b_2, \ldots, b_n)$ determines a unique extreme point $\boldsymbol{u}^{(1)} = (u_1^{(1)}, \ldots, u_m^{(1)})$, $\boldsymbol{v}^{(1)} = (v_2^{(1)}, \ldots, v_n^{(2)})$ of $D_{m,n-1}(\boldsymbol{c}^{(1)})$, where $\boldsymbol{c}^{(1)}$ is $\boldsymbol{c}$ with the first column dropped. Define $u_i = u_i^{(1)}$, $v_j = v_j^{(1)}$ and $v_1 = \min_i(c_{i1} - u_i) = c_{k1} - u_k$. By nondegeneracy the k is unique. This is an extreme point of $D_{m,n}(\boldsymbol{c})$ since it is feasible and its graph forms a spanning tree. This shows that to each column partition $\boldsymbol{b}$ there corresponds an extreme point. These extreme points are all different, for otherwise, one would have two different partitions, implying degeneracy. There are $\binom{m+n-2}{n-1}$ column partitions, so the same number of extreme points. To each extreme point there corresponds a unique row partition, and since all must be different and there are $\binom{m+n-2}{m-1}$ of them, all are accounted for, which completes the proof. □

2. Examples

The three examples below show that polyhedra $D_{m,n}(\boldsymbol{c})$ for different $\boldsymbol{c}$ are not necessarily combinatorially equivalent (in the sense that there does not exist a one-to-one correspondence between faces that is inclusion preserving). Since $m=n=3$ in each, their dimensions are all 5; however, we draw only the extreme points and bounded extreme edges. (Since each is nondegenerate, every extreme point should have 5 adjacent extreme edges: the missing ones are unbounded).

Example 1

$$c' = \begin{bmatrix} 0 & 3 & 2 \\ 0 & 1 & 3 \\ 0 & 0 & 0 \end{bmatrix}, \qquad D_{3,3}(c'):$$

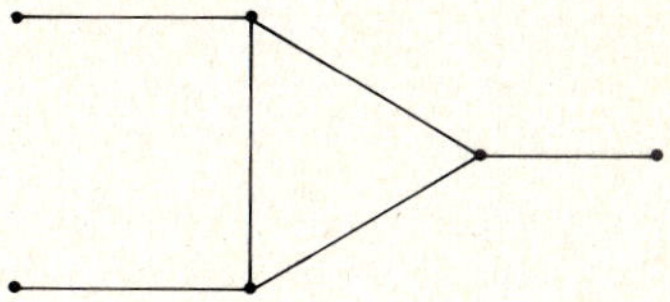

Example 2

$$c'' = \begin{bmatrix} 0 & 2 & 4 \\ 0 & -1 & 3 \\ 0 & 0 & 0 \end{bmatrix}, \qquad D_{3,3}(c''):$$

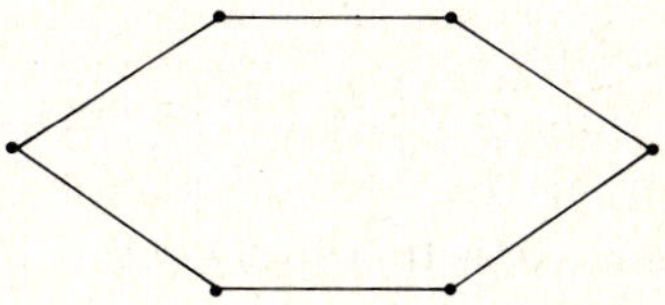

Example 3

$$d = \begin{bmatrix} 0 & 2 & 4 \\ 0 & 1 & 2 \\ 0 & 0 & 0 \end{bmatrix}, \qquad D_{3,3}(d):$$

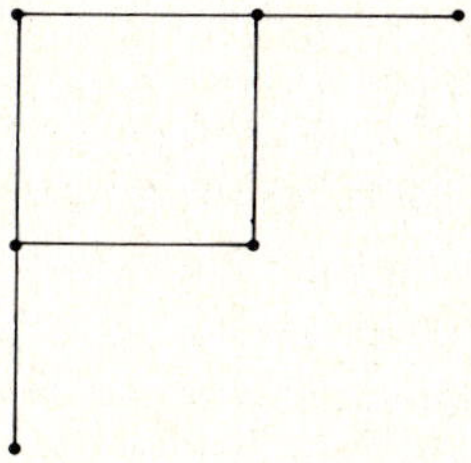

The last example belongs to a general class that is particularly easy to describe. Define $\boldsymbol{d}$ by $d_{ij} = (m-i)(j-1)$, and let $(\boldsymbol{u}, \boldsymbol{v})$ be any extreme point of $D_{m,n}(\boldsymbol{d})$ and $G(\boldsymbol{u}, \boldsymbol{v})$ its graph. Then, if $i_1 < i_2$ and $j_1 < j_2$ it is impossible for (i_1, j_2) and (i_2, j_1) to both belong to $G(\boldsymbol{u}, \boldsymbol{v})$. For if both did, then $u_{i_1} + v_{j_2} = (m-i_1)(j_2-1)$ and $u_{i_2} + v_{j_1} = (m-i_2)(j_1-1)$. But, $u_{i_1} + v_{j_1} \leq (m-i_1)(j_1-1)$ and $u_{i_2} + v_{j_2} \leq (m-i_2)(j_2-1)$ implying $(m-i_1)(j_2-1)+(m-i_2)(j_1-1) \leq (m-i_1)(j_1-1)+(m-i_2)(j_2-1)$, a contradiction. So $G(\boldsymbol{u}, \boldsymbol{v})$ can contain 'no crossing' edges. This immediately shows that the 'no crossing' polyhedron $D_{m,n}(\boldsymbol{d})$ is nondegenerate. Moreover, any spanning tree T of $C(M, N)$ that contains no crossing (see Figure 3) corresponds to an extreme point.

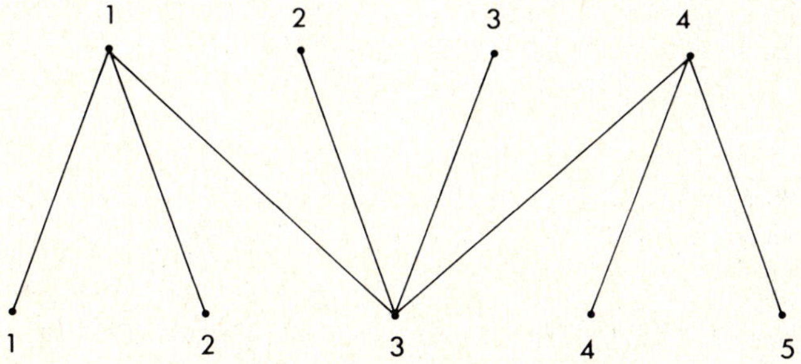

Fig. 3. A tree with no crossings.

3. Faces

Any $D_{m,n}(\boldsymbol{c})$ is of *dimension* $m+n-1$ because there always exist $\boldsymbol{u}, \boldsymbol{v}$ satisfying $u_1=0$, $u_i+v_j<c_{ij}$ for all i, j. Every $D_{m,n}$ has exactly *mn facets* (maximal proper faces) $F_{kl}=D\cap\{\boldsymbol{u}, \boldsymbol{v};\ u_k+v_l=c_{kl}\}$ because there exist $\boldsymbol{u}, \boldsymbol{v}$ satisfying $u_1=0$, $u_k+v_l=c_{kl}$ and $u_i+v_j<c_{ij}$ for $(i,j)\neq(k,l)$.

A *face* of $D_{m,n}(\boldsymbol{c})$ is the intersection of D with a set of solutions to $u_1=0$, $u_i+v_j=c_{ij}$ for $(i,j)\in R\subset M\times N$, where the equations corresponding to R are linearly independent. The edges that correspond to R in $C(M,N)$ must therefore form a forest: by an abuse of terminology we will refer to the forest R as well as the face R. R is of dimension r if the number of equations corresponding to $(i,j)\in R$ is $m+n-1-r$ and these admit a solution in D with $u_i+v_j<c_{ij}$ for $(i,j)\notin R$.

A face R of dimension r contains at least one extreme point $(\boldsymbol{u}, \boldsymbol{v})$: Therefore, the forest R belongs to a spanning tree $T\subset G(\boldsymbol{u}, \boldsymbol{v})$ or, what is the same thing, any such R may be obtained from some T by dropping exactly r edges. Note, finally, that a nonempty face R is bounded if and only if the forest R leaves no node of M or N 'exposed', that is with degree 0.

Theorem 2. *For any nondegenerate* $D_{m,n}(\boldsymbol{c})$:

(i) *the number of bounded r-dimensional faces is*

$$\beta(m,n;r)=\binom{n-1}{r}\binom{m+n-r-2}{n-1}=\binom{m-1}{r}\binom{m+n-r-2}{m-1}$$

(ii) *the total number of r-dimensional faces is*

$$\alpha(m,n;r)=\sum_{s,t}\binom{m}{s}\binom{n}{t}\beta(m,n;r).$$

Proof. We begin by counting the bounded r-dimensional faces. Each has a column partition $(b_1,\ldots,b_n)$ with $\sum b_j=m+n-1-r\geq n$, $b_j\geq 1$ the number of equations involving v_j. There must be some node, say node n, with $b_n=1$. Call it the root.

Choose a subset P of r nodes from among the $n-1$ column nodes excluding the root. To each one of the $\binom{n-1}{r}$ such choices there corresponds a different bounded r-dimensional face, and these are all the faces. The contruction below proves these statements. Given them it is immediate that

$$\beta(m,n;r)=\binom{n-1}{r}\binom{m+n-2-r}{n-1}.$$

Define $\boldsymbol{b}'=(b_j')$ by $b_j'=b_j+1$ for $j\in P$, $b_j'=b_j$ otherwise. $\boldsymbol{b}'$ determines a unique extreme point $\boldsymbol{u}, \boldsymbol{v}$ with spanning tree $T=G(\boldsymbol{u}, \boldsymbol{v})$. Therefore, there is a unique path in T from each $j\in P$ to the root. Drop from T the edge incident to j on this path for each $j\in P$ to obtain a bounded r-dimensional face.

Different choices of P yield different faces. For suppose two different P's gave the same R. Then different spanning trees $T = G(\boldsymbol{u}, \boldsymbol{v})$ and $T' = G(\boldsymbol{u}', \boldsymbol{v}')$ with different column partitions give the same R. This means there must be a column node j_1 with valence $b_{j_1}+1$ in T' and b_{j_1} in T. Consider the path from j_1 to the root in T'. The first edge (i_1, j_1) does not belong to R by construction; the second (i_1, j_2) must belong to R (it cannot be the first edge connecting j_2 to the root) and so to T; the third belongs to T'; the fourth to R and so to T; etc. Consider now the path from j_1 to the root in T. The first belongs to T but also to R; the second to R and so to T'; the third to T; the fourth to R and so to T'; etc. These paths are different but must meet at some first node, forming a cycle of edges $(i, j) \in C$ alternately in T and in T', with

$$\sum_{C \cap T} c_{ij} = \sum_{C} (u_i + v_j) \leq \sum_{C \cap T'} c_{ij} = \sum_{C} (u'_i + v'_j) \leq \sum_{C \cap T} c_{ij}$$

contradicting nondegeneracy.

Every bounded face with column partition $\boldsymbol{b}$ is obtained by the above construction. For, let R be any such face and consider the maximal connected component Q of R that contains the root. Increase u_i by $\varepsilon > 0$ and decrease v_j by ε for all nodes $i \in M \cap Q$, $j \in N \cap Q$ until some new equation $u_i + v_j = c_{ij}$ is satisfied for $i \in M \cap Q$, $j \in N \cap (\sim Q)$. Assign j to P. Repeat until a spanning tree obtains. The r edges introduced are precisely the edges dropped in the construction for this choice of P. This completes the proof of (i).

The proof of (ii) is immediate because unbounded faces have forests that span only a subset of the nodes: some nodes are exposed. □

Once it is known that all nondegenerate $D_{m,n}(\boldsymbol{c})$ have the same numbers of bounded and of unbounded r-dimensional faces, counts may be obtained from any one particular $\boldsymbol{c}$. A convenient choice is the 'no crossing' polyhedron $D_{m,n}(\boldsymbol{d})$: its faces are characterized by forests containing no crossings. This means that a bounded face of dimension r has row and column partitions $\boldsymbol{a}, \boldsymbol{b}$ (that sum to $m+n-1-r$) with (a_1, b_1) equal to either (1, 1), (1, 2 or greater), or (2 or greater, 1). Therefore,

$$\beta(m, n; r) = \beta(m-1, n-1; r-1) + \beta(m-1, n; r) + \beta(m, n-1; r),$$

and β may be found using a generating function. A similar approach may be taken to computing α.

References

[1] J.H. Ahrens, "A Conjecture of E.D. Bolker", *Journal of Combinatorial Theory* (*B*) 31 (1981), 1–8.
[2] M.L. Balinski, "On two special classes of itransportation polytopes", in M.L. Balinski, ed., *Pivoting and Extensions* (North-Holland, Amsterdam, 1974) pp. 43–58.
[3] M.L. Balinski, "Signature methods for the assignment problem", to appear in *Operations Research.*
[4] M.L. Balinski, "Signatures des points extrêmes du polyèdre dual du problème de transport", *Comptes Rendus de l'Académie des Sciences de Paris* 296 (1983), Série I, 457–459.

[5] M.L. Balinski, "The Hirsch conjecture for dual transportation polyhedra", to appear in *Mathematics of Operations Research.*
[6] M.L. Balinski and A. Russakoff, "On the assignment polytope", *SIAM Review* 16 (1974), 516–525.
[7] E.D. Bolker, "Transportation polytopes", *Journal of Combinatorial Theory* (B) 13 (1972), 251–262.
[8] V. Klee and C. Witzgall, "Facets and vertices of transportation polytopes", in: G.B. Dantzig and A.F. Veinott, Jr., eds., *Mathematics of the decision sciences*, vol. I (American Mathematical Society, Providence, R.I., 1968) pp. 257–282.
[9] Zhu Yeng-Jen, "Maximum number of iterations in the dual algorithm of the Kantorovic–Hitchcock transportation problem in linear programming", *Chinese Mathematics* 3 (1963) 307–313. Translation of paper in *Acta Mathematica Sinica* 12 (1962) 278–283.

Mathematical Programming Study 22 (1984) 9–31
North-Holland

A SUBMODULAR NETWORK SIMPLEX METHOD

Francisco BARAHONA* and William H. CUNNINGHAM**
Institut für Operations Research, Universität Bonn, FRG

Received 6 September 1983
Revised manuscript received 11 April 1984

We present a simplex method for the solution of the optimal submodular flow problem. Like the network simplex method for solving the minimum cost network flow problem, this algorithm is purely combinatorial. It requires an oracle which can minimize a submodular function. In general this oracle is available only via the ellipsoid algorithm, but in several applications it is provided by an efficient combinatorial procedure. We generalize the notion of strong feasibility in the network simplex method to give a finiteness proof for the new algorithm. Hence we obtain a constructive proof of the integrality theorem of Edmonds and Giles.

Key words: Network Flows, Submodular Functions, Simplex Method, Strong Feasibility.

1. Introduction

The submodular flow problem is a generalization of the network flow problem in which restrictions on net flows into vertices are generalized to restrictions on net flows into subsets of vertices. The subsets are members of a certain family of subsets, and the restrictions are given by the values of a submodular function on the family. This problem was introduced by Edmonds and Giles [7]. Related models can be found in [15, 18, 21]; see [25] for a survey. Its applications include, as well as ordinary network flow problems, the polymatroid intersection problem [6], the optimal dicut-covering problem [22], and the optimal m-strongly-connected reorientation problem [12]. The fundamental integrality result of [7] also generalizes important combinatorial minmax results in these applications.

On the algorithmic side Grötschel, Lovász and Schrijver [17] have proved that an optimal submodular flow can be computed in polynomial time using the ellipsoid method. This approach uses but does not prove the integrality result, and suffers from certain practical deficiencies. Frank [12] has given a generalization of the network flow primal–dual algorithm to the submodular flow problem, which does

* Research supported by the Alexander von Humboldt Stiftung. Present address: Dpto. de Matematices, U. de Chile, Casilla 5272, Santiago 3, Chile.
** Present address: Department of Mathematics and Statistics, Carleton University, Ottawa, Canada. Partially supported by a grant from N.S.E.R.C. of Canada, and by SFB 21 (DFG), Universität Bonn.

give a proof of the integrality theorem. Moreover, using an important labelling method for finding augmenting paths introduced by Lawler and Martel [21] and Schönsleben [24] (see also [3, 13]), together with the scaling technique [8], this approach has been refined to yield a polynomial-time algorithm [5]. (Other algorithmic approaches to optimizing linear functions for similar models can be found in [15, 18, 24, 26].) However, it should be noted that these primal-dual methods require an oracle which can minimize certain submodular functions associated with the problem. In general, the only polynomial-time algorithm for submodular function minimization also uses the ellipsoid method, but in most of the applications this oracle is available via an efficient combinatorial algorithm.

In spite of thirty years or more of research on network flow algorithms, the solution method most in use is still the simplex method, appropriately specialized and implemented. This 'network simplex method' is attractive, easy to explain, easy to implement, and empirically efficient. Its major deficiency is that little is known about its theoretical complexity [2], while primal-dual methods can be made to run in polynomial time [8]. (We remark that there are some recent improvements in theoretical running times for algorithms related to the network *dual* simplex method [19, 23]). The main result of this paper is a generalization of the network simplex method to solve the optimal submodular flow problem. It retains the attractive features of the network simplex method. In particular, the method of 'strongly feasible trees' is generalized to submodular flows, so that a simple pivoting rule ensures finiteness of the algorithm. As a consequence we obtain an algorithmic proof of the Edmonds-Giles integrality theorem. Like the primal-dual method, the simplex method needs an appropriate submodular oracle.

2. Preliminaries

Let G be a finite digraph (directed graph); we often denote its vertex-set $V(G)$ and its edge-set $E(G)$ by just V, E. For any $e \in E$ the head of e is denoted by $h(e) \in V$ and the tail by $t(e) \in V$. The *incidence matrix* of G is $(m_{ve} : v \in V, e \in E)$ given by: $m_{ve} = 1$ if $v = h(e)$, $m_{ve} = -1$ if $v = t(e)$, and $m_{ve} = 0$ otherwise, for any edge $e \in E$ which is not a loop. If e is a loop ($h(e) = t(e)$), then $m_{ve} = 0$ for all $v \in V$.

Two subsets P, Q of V *cross* if $P \cap Q$, $P \cap \bar{Q}$, $\bar{P} \cap Q$, $\bar{P} \cap \bar{Q}$ are all nonempty. (For $R \subseteq V$, $\bar{R}$ denotes $V \backslash R$.) A family $\mathcal{F}$ is *crossing* if, whenever P, $Q \in \mathcal{F}$ and P, Q cross, then $P \cap Q$, $P \cup Q \in \mathcal{F}$. A family $\mathcal{F}$ is *cross-free* if no two members of $\mathcal{F}$ cross. Obviously, every cross-free family is a crossing family. A function b on a crossing family $\mathcal{F}$ is *submodular* on $\mathcal{F}$ if, whenever P, $Q \in \mathcal{F}$ and P, Q cross, then

$$b(P) + b(Q) \geq b(P \cup Q) + b(P \cap Q).$$

For any set $F \subseteq V$ and any $j \in E$, we define a_j^F to be: 1 if j enters F, -1 if j leaves, F, 0 otherwise. We denote $(a_j^F : j \in E)$ by a^F, and think of a^F as a row vector; where $F = \{v\}$ for some $v \in V$, we may write a^F as a^v. Notice that, for a vector $(x_j : j \in E)$

of 'flows', $a^F x$ is the 'net flow' into F. Given digraph G, crossing family $\mathscr{F}$ of subsets of V, function b submodular on $\mathscr{F}$, real cost vector $(c_j: j \in E)$, and real bound vectors $(l_j: j \in E)$, $(u_j: j \in E)$ the (optimal) *submodular flow* problem is

$$\text{minimize} \quad cx = \sum (c_j x_j : j \in E)$$
$$\text{subject to } a^F x \leq b(F),\ F \in \mathscr{F}; \quad l \leq x \leq u. \tag{2.1}$$

We allow $l_j = -\infty$ and $u_j = \infty$. It is convenient to define a_j to be $(a_j^F: F \in \mathscr{F})$, and to think of a_j as a column vector. So a^F is a row and a_j a column of the constraint matrix A.

Edmonds and Giles [7] proved the following important result.

(2.2) Theorem. *Suppose that* (2.1) *has an optimal solution. If c is integer-valued, then the linear program dual to* (2.1) *has an optimal solution which is integer-valued. If b and (the finite components of) l, u are integer-valued, then* (2.1) *has an optimal solution which is integer-valued.*

Consider the special case of (2.1) in which $\mathscr{F} = \{\{v\}: v \in V\}$ and $\sum (b(\{v\}): v \in V) = 0$. Then the constraint matrix A is the incidence matrix of G. Adding the corresponding constraints yields the inequality $0 \leq 0$, and it follows that all of them must hold with equality. Therefore, in this case (2.1) is an ordinary minimum cost flow problem. (For this case, of course, (2.2) is a classical result.) The network simplex method for this problem will now be summarized; it provides motivation and useful terminology for the generalization. The basic theoretical fact underlying the algorithm is the following well-known result.

(2.3) Theorem. *Let $M = (m_j: j \in E(H))$ denote the incidence matrix of a connected digraph H. Then $(m_j: j \in T)$ is a column basis for M if and only if T is the edge-set of a spanning tree of H. Moreover, for any edge $k \notin T$, the (unique) expression for m_k in terms of this basis is $m_k = \sum (\alpha_{jk} m_j: j \in T)$, where*

$$\alpha_{jk} = \begin{cases} 1 & \text{if } j \text{ is a reverse edge of the path in } T \text{ from } h(k) \text{ to } t(k); \\ -1 & \text{if } j \text{ is a forward edge of the path in } T \text{ from } h(k) \text{ to } t(k); \\ 0 & \text{otherwise.} \end{cases}$$

Since elementary row operations on a matrix are reversible, we have the following useful consequence of (2.3)

(2.4) Corollary. *Let H be a connected digraph and let T be the edge-set of a spanning tree of H. Let M' be the matrix $(m'_{jk}: j \in T, k \in E(H))$ defined by: $m'_{jk} = \alpha_{jk}$ if $k \in E(H) \backslash T$; $m'_{jj} = 1$, $j \in T$; $m'_{jk} = 0$, else. Then M' is row-equivalent to the incidence matrix of H.*

We shall often abbreviate 'edge-set of a spanning tree' to 'tree'. The network simplex method maintains *feasible tree solutions*, feasible solutions x for which there exists a tree T such that $j \notin T$ implies $x_j = l_j$ or $x_j = u_j$. Throughout, we assume a fixed *root* vertex $r \in V$, and think of trees as being rooted at r. Given T with corresponding solution x and an edge $e \notin T$, we define a circuit $C = C(T, e)$ to be the unique circuit satisfying:

(i) The edge set of C is contained in $T \cup \{e\}$;
(ii) e is a forward edge of C if $x_e = l_e$ and otherwise is a reverse edge;
(iii) The initial vertex of C is the first common vertex s of the paths in T from $h(e)$ and $t(e)$ to r.

An iteration of the network simplex method begins by choosing edge $e \notin T$ such that C has negative cost. (The *cost* of a path is obtained by adding c_j on forward edges and subtracting c_j on reverse edges.) If no such e exists, the current solution is optimal, and the algorithm terminates. Otherwise, the new x is obtained by raising x_j on forward edges of C by $\theta \geq 0$ and lowering x_j on reverse edges by θ. The number θ is $\min(\theta_1, \theta_2)$, where $\theta_1 = \min(x_j - l_j\colon j$ reverse in $C)$ and $\theta_2 = \min(u_j - x_j\colon j$ forward in $C)$. (If $\theta = \infty$, then the problem is unbounded, and the algorithm terminates.) The new tree is obtained by adding e to T and deleting f, where f is some edge of C determining θ. We make one remark on implementation of the algorithm, which is also useful in its analysis. If π_v, $v \in V$, is defined to be the cost of the path in T from r to v, and $\bar{c}_j$, $j \in E$, to be $c_j + \pi_{t(j)} - \pi_{h(j)}$, then for $e \notin T$, the cost of C is $\bar{c}_e$ if $x_e = l_e$ and $-\bar{c}_e$ otherwise. Therefore the algorithm computes and updates $(\pi_v\colon v \in V)$ at each iteration.

We now describe the modification [1] of the above algorithm which guarantees finiteness. (Nontermination, due to 'cycling', can occur in the unrefined algorithm [2].) An edge of T is *directed away from* r in T if j is an edge of the path in T from r to $h(j)$; otherwise, j is *directed toward* r in T. A tree T is *strongly feasible* if its associated solution x is feasible and satisfies $x_j = l_j$, $j \in T$, only if j is directed away from r in T, and $x_j = u_j$, $j \in T$, only if j is directed toward r in T. The finite version of the network simplex method involves two refinements of the basic algorithm: (1) Begin with a strongly feasible tree; (2) Among the edges of C determining the minimum θ, choose f to be the first one.

In generalizing the above algorithm to one for solving (2.1), we need to assume the availability of an oracle to deal with the (possibly) exponentially-many constraints. Given $z = (z_v\colon v \in V)$ and $F \subseteq V$, we denote by $z(F)$ the sum $\sum (z_v\colon v \in F)$. We assume the availability of an upper bound M on $|b(F)|$ and an oracle which can solve

$$\text{minimize} \quad (b(F) - z(F)\colon F \in \mathscr{F}) \tag{2.5}$$

for any z. In fact, the vectors z for which our algorithm needs to solve (2.5) satisfy some special conditions. The most important is that there is an a priori bound on $|z_v|$, polynomial in M and $|V|$. (Thus, in particular, the ellipsoid algorithm [17] could in principle be used as the oracle; it could solve (2.5) in time polynomial in $|V|$

and log M.) Notice that, by making z_v sufficiently large (or small) we can use (2.5) to minimize over $F \in \mathcal{F}$ containing v (or not containing v). Thus for disjoint sets P, $Q \subseteq V$ we can minimize over $F \in \mathcal{F}$ with $P \subseteq F \subseteq \bar{Q}$. The simplex algorithm will need only the cases in which $|P| \leq 2 \geq |Q|$. In fact, most calls to the oracle will take the following form: Given a feasible solution x of (2.1) and vertices $w, v \in V$

$$\text{Minimize} \quad (b(F) - a^F x \colon w \in F \in \mathcal{F}, v \notin F). \tag{2.6}$$

(Where we define z_v to be $a^v x$ for $v \in V$, we have $z(F) = a^F x$, $F \subseteq V$.) Calling the oracle to solve (2.6) will be referred to as 'asking (x, w, v)'. The oracle answers with a minimizing F, and the value of the minimum, or the information that no such F exists.

Finally, it is convenient to recall some ideas from polymatroid theory. Two sets F, $F' \subseteq V$ are said to *intersect* if $F \cap F' \neq \emptyset$. A family $\mathcal{F}$ of subsets of V is an *intersecting* family if F, $F' \in \mathcal{F}$ and F, F' intersect, imply $F \cap F'$, $F \cup F' \in \mathcal{F}$. If b is a function on $\mathcal{F}$, submodular on intersecting members of $\mathcal{F}$, let $P = \{z = (z_v \colon v \in V) \colon z(F) \leq b(F) \text{ for } F \in \mathcal{F}\}$. It is known [6] that if $z' \in P$, then the polyhedron $P' = \{y \colon y \geq 0, y + z' \in P\}$ is a polymatroid. (Actually, [6] requires polymatroids to be bounded, so this statement is not quite true. However, if P' is not bounded, we can add V to $\mathcal{F}$ and make $b(V)$ large. The new P' will be bounded, and we can use the result we need from polymatroid theory.) Hence, given a number k, the problem of finding $z \in P$ with $z(V) = k$ is easily solved by the greedy algorithm, given a known $z' \in P$. Namely, beginning with $z = z'$, we successively choose $v \in V$ and raise z_v by $\min(k - z(V), \min(b(F) - z(F) \colon v \in F \in \mathcal{F}))$. Of course, to carry out this procedure we need to be able to perform the above minimization, and also to be able to obtain $z' \in P$. (For the latter, notice that it is enough to choose $z'_v \leq \min(0, \min(b(F) \colon F \in \mathcal{F}))$ for each $v \in V$, so it is enough to have a lower bound for $b(F)$.)

3. Cross-free bases

In this section we develop the essential ideas which underly the submodular network simplex algorithm. We remark that one of the techniques used in the analysis, looking at a partitioned form of the basis matrix, is standard in large scale linear programming. Moreover, the idea of solving exponentially-large, combinatorially-defined linear programs using the simplex method is far from new. Ford and Fulkerson [10] gave a simplex algorithm for a form of the multicommodity flow problem, which has exponentially many variables (at least). A shortest path problem is solved to find the column which enters the basis, but otherwise the method is an ordinary (revised) simplex method. Koch [20] (see also Edmonds and Koch [9]) developed a simplex method for the optimum b-matching problem, which (in its linear programming formulation) has an exponential number of constraints. His algorithm restricts attention to a certain class of well-behaved bases, so that the pivot steps can be made by a combinatorial algorithm, as in the network simplex method. Our algorithm is similar to Koch's in these respects. However, Koch did not prove his algorithm to be finite. (Whether it is remains an open problem.)

We begin the development of the simplex algorithm for (2.1) by adding slack variables:

$$\begin{aligned} &\text{minimize} \quad cx \\ &\text{subject to} \quad a^F x + s_F = b(F), \ F \in \mathcal{F}; \\ &\qquad\qquad l \leq x \leq u; \\ &\qquad\qquad s_F \geq 0, \ F \in \mathcal{F}. \end{aligned} \tag{3.1}$$

The constraint matrix $(A|I)$ of (3.1) has full row rank. Any basis of this matrix, $B = [(a_j: j \in T), (e_F; F \in \mathcal{B})]$ is determined, in principle, by a small amount of information: a set $T \subseteq E$ and a set $\mathcal{N} = \mathcal{F} \setminus \mathcal{B} \subseteq \mathcal{F}$. (Of course, e_F denotes the elementary column vector corresponding to $F \in \mathcal{F}$.) Notice that $|\mathcal{N}| = |T| = k$ (say), and that the condition that B be a basis is equivalent to the condition that the k by k submatrix $(a_j^F: F \in \mathcal{N}, j \in T)$ be non-singular. Moreover, $k \leq |V| - 1$, because the matrix A itself has rank at most $|V| - 1$. (This follows from the fact that the rows of A are linear combinations, in fact sums, of the rows of the incidence matrix of G.)

In order to devise a version of the simplex method for (3.1) having a simple combinatorial behaviour, we restrict attention to a particular class of bases B: those determined by $(T, \mathcal{N})$ where $\mathcal{N}$ is cross-free. This choice is motivated by the following result of [7], which was an essential step in the proof of Theorem (2.2). Notice that the dual solution of (2.1) corresponding to B, will be nonzero only on components corresponding to $\mathcal{N}$.

(3.2) Theorem. *If the linear program dual to* (2.1) *has an optimal solution* $(y_F: F \in \mathcal{F})$, *then it has one for which the subfamily* $(F \in \mathcal{F}: y_F \neq 0)$ *is cross-free.*

Of course, if a simplex method using only cross-free bases is to work, it is necessary (among other things) that every basic feasible solution of (2.1) be determined by a cross-free basis. This result implies (3.2) but is not equivalent to it, because (3.2) does not deal with basic solutions. We do not give a direct proof here of the stronger result; it is a consequence of the finiteness of our simplex algorithm.

Let $A(\mathcal{N})$ denote the constraint matrix of

$$a^F x + s_F = b(F), \quad F \in \mathcal{N}.$$

We shall show that $A(\mathcal{N})$ is really the constraint matrix of an ordinary network flow problem. (The idea for this is already contained in [7].) We know that $\mathcal{N}$ is *independent*, that is, that the vectors $(a^F: F \in \mathcal{N})$ are linearly independent, but we remark that the next result is true without this assumption.

(3.3) Theorem. *$A(\mathcal{N})$ is row-equivalent to the incidence matrix of a connected digraph $H(\mathcal{N})$.*

In fact, we shall prove (3.3) by explicitly constructing the digraph, because our algorithm will operate almost exclusively on this digraph. We shall define $H = H(\mathcal{N})$, and use Corollary (2.4) to prove that the incidence matrix of H is row-equivalent to $A(\mathcal{N})$. Two elements of V are *separated* by $F \subseteq V$ if and only if F contains exactly one of them. A (*pseudo*) *node* of $\mathcal{N}$ is a subset of V which is an equivalence class under the relation: u, v are equivalent if they are separated by no member of $\mathcal{N}$. The nodes of $\mathcal{N}$ form the set $V(H)$ of vertices of H. We denote by $N(v)$, for $v \in V$, the node which contains v. For an edge $j \in E$ we define the head (tail) of j in H to be $N(h(j))$ $(N(t(j)))$. (Where there is no danger of confusion we may abbreviate these to $h(j)$, $t(j)$.) Thus H, restricted to edges in E, is obtained by 'shrinking each node to a single vertex'. Notice that an edge may acquire the same head and tail in H; such an edge is *not* deleted from H. We have yet to define the heads and tails of the edges of H corresponding to the identity columns of $A(\mathcal{N})$, that is, those for the slack variables s_F, $F \in \mathcal{N}$. We need the following lemma.

(3.4) Lemma. *Let $\mathcal{N}$ be an independent cross-free subfamily of $\mathcal{F}$, let $C \in \mathcal{N}$, and let $N' = \mathcal{N} \backslash \{C\}$. There exist nodes P, Q of $\mathcal{N}$ such that the nodes of N' are those of $\mathcal{N}$ with P, Q replaced by $P \cup Q$.*

Proof. Let $\mathcal{N}_1 = \{F\colon F \text{ or } \bar{F} \in \mathcal{N},\ F \subset C,\ F \text{ maximal}\}$, and let $N_2 = \{F\colon F \text{ or } \bar{F} \in \mathcal{N},\ F \cap C = \emptyset,\ F \text{ maximal}\}$. Then the members of $\mathcal{N}_1$ are pairwise disjoint so that, where $S = \cup\{F\colon F \in \mathcal{N}_1\}$, we have $a^S = \sum (a^F\colon F \in \mathcal{N}_1)$. Since $\mathcal{N}$ is independent, $S \neq C$; since $S \subseteq C$, we have $P = C \backslash S \neq \emptyset$. Now each element of P is separated from each element of $C \backslash P$ by some member of $\mathcal{N}_1$ and from each element of $\bar{C}$ by C. But no two elements of P are separated by any member of $\mathcal{N}$, so P is a node of $\mathcal{N}$. Similarly, where $R = \cup\{F\colon F \in \mathcal{N}_2\}$, we have that $Q = \bar{C} \backslash R$ is a node of $\mathcal{N}$.

If a member F of $\mathcal{N}'$ separates $p \in P$ from $v \in \bar{C}$ then since no member of $\mathcal{N}_1$ separates p from v, it must be that p is separated from q by a member of $\mathcal{N}_2$, so $v \notin Q$. Applying a similar argument for $v \in C$, $q \in Q$, we conclude that $P \cup Q$ is a node of $\mathcal{N}'$. Finally, suppose that $p \in C$, $q \in \bar{C}$ are separated by no member of $\mathcal{N}'$. Then it is easy to see that $p \in P$, $q \in Q$, so the only nodes of $\mathcal{N}$ that are not nodes of $\mathcal{N}'$ are P and Q. □

We remark that (3.4) is true under a much weaker assumption than independence, as can be seen from the proof.

We are now ready to complete the definition of H. Let $C \in \mathcal{N}$, and choose P, Q of Lemma (3.4); we may assume $P \subseteq C$, $Q \subseteq \bar{C}$. Then the edge $C \in E(H)$ has head P and tail Q. We sometimes write $P = h(C)$, $Q = t(C)$, and we use C to refer both to the set and the edge. Figure 1 shows an example. From the digraph G and the cross-free family $\mathcal{N} = \{\{1, 2\}, \{1, 2, 3\}, \{5, 6\}\}$, we obtain the digraph H. The broken

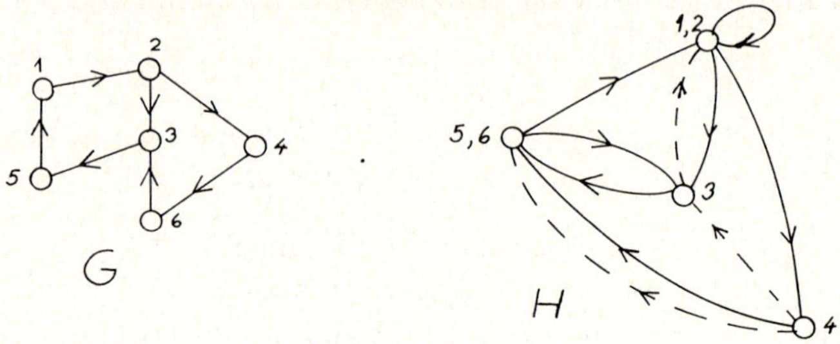

Fig. 1.

edges of H are the edges corresponding to members of $\mathcal{N}$. In general, we let $\mathcal{T}$ denote the spanning subgraph of H having only these edges. Some notation: For a tree T ($\mathcal{T}$) and vertices R, S of T ($\mathcal{T}$), $P(R, S)$ ($\mathcal{P}(R, S)$) denotes the simple path from R to S in T ($\mathcal{T}$).

(3.5) Lemma. *$\mathcal{T}$ is a tree. Moreover, for any two nodes, R, S of $\mathcal{N}$ and any $F \in \mathcal{N}$, F contains R and not S if and only if F is a reverse edge of $\mathcal{P}(R, S)$; F contains S and not R if and only if F is a forward edge of $\mathcal{P}(R, S)$.*

Proof. Suppose that nodes R, S of $\mathcal{N}$ are joined by a simple path $\mathcal{P}(R, S)$. Since the head and tail of an edge F are nodes which are separated precisely by F, it follows that $\mathcal{P}(R, S)$ has the properties claimed. Thus $\mathcal{P}(R, S)$ is unique, and so $\mathcal{T}$ is a forest. Since any subfamily of an independent, cross-free family is independent and cross-free, and since the empty family has one node, it follows that $|V(\mathcal{T})| = |\mathcal{N}| + 1 = |E(\mathcal{T})| + 1$. Thus $\mathcal{T}$ is a tree. □

We can now prove that $A(\mathcal{N})$ is row-equivalent to the incidence matrix of H, using (2.4). It will be enough to show, for any $j \in E$ ($= E(H) \backslash \mathcal{N}$) and any $F \in \mathcal{F}$, that

$$a_j^F = \begin{cases} 1, & \text{if } F \text{ is a reverse edge of } \mathcal{P}(h(j), t(j)); \\ -1, & \text{if } F \text{ is a forward edge of } \mathcal{P}(h(j), t(j)); \\ 0, & \text{otherwise.} \end{cases} \tag{3.6}$$

But by (3.5) F is a reverse (forward) edge of $\mathcal{P}(h(j), t(j))$ if and only if $h(j) \in F$ and $t(j) \in \bar{F}$ ($t(j) \in F$ and $h(j) \in \bar{F}$), which is true if and only if j enters F (j leaves F), that is, $a_j^F = 1$ (-1), as required. The analog of (2.3) for the submodular network simplex method is the following result.

(3.7) Theorem. *Where $\mathcal{N}$ is an independent cross-free subfamily of $\mathcal{F}$ and $T \subseteq E$, $(T, \mathcal{N})$ determines a basis of* (3.1) *if and only if T is a spanning tree of $H(\mathcal{N})$. Moreover, let k index a nonbasic column vector g. (Thus $g = a_k$, $k \in E \backslash T$,* or *$g = e_k$, $k \in \mathcal{N}$.) Then for any $j \in T$ the coefficient α_{jk} of a_j in the expression for g as a linear combination of basic* *columns is given by*:

$$\alpha_{jk}=\begin{cases}1, & \text{if } j \text{ is a reverse edge of } P(h(k), t(k));\\ -1, & \text{if } j \text{ is a forward edge of } P(h(k), t(k));\\ 0, & \text{otherwise.}\end{cases}$$

Proof. The first part is a direct consequence of (3.3) and (2.3). For the second part we have, for some numbers β_{Fk}, $F\in\mathcal{B}$, that $g=\sum(\alpha_{jk}a_j\colon j\in T)+\sum(\beta_{Fk}e_F\colon F\in\mathcal{B})$. Let $g^{\mathcal{N}}(a_j^{\mathcal{N}})$ denote the restriction of g (a_j) to the components indexed by elements of $\mathcal{N}$. Then $g^{\mathcal{N}}=\sum(\alpha_{jk}a_j^{\mathcal{N}}\colon j\in T)$. Since $A(\mathcal{N})$ is row-equivalent to the incidence matrix of $H(\mathcal{N})$, the result follows from (2.3). □

4. Pivoting

Suppose that we have a cross-free basis determined by $(T, \mathcal{N})$ which has an associated feasible solution x^0. Then x^0 satisfies

(i) $x_j^0=l_j$ or $x_j^0=u_j$ for all $j\in E\setminus T$;

(ii) F is x^0-tight for each $F\in\mathcal{N}$.

(For a set F and a vector $x=(x_j\colon j\in E)$, F is *x-tight*, or *tight*, if $a^Fx=b(F)$.) We could determine x^0 from $(T, \mathcal{N})$ and knowledge of which nonbasic edge variables are at their lower bounds. However, as in most simplex methods, successive solutions x^0 will be determined by updating.

A pivot step consists of choosing an index k of a nonbasic variable and moving its current value away from its bound, while holding the values of the other nonbasic variables fixed, and changing the values of the basic variables accordingly. It is usually adequate to discuss the case in which the value is increased from the lower bound, the other case being similar. (Of course, the lower bound of a slack variable is zero.) The amount θ by which the value is increased is as large as possible subject only to feasibility conditions. Because the values of the slack variables are determined by the values of the edge-variables (flows), and because they have zero cost, most of the discussion can be restricted to the flows. The index k is chosen so that cx is decreased if $\theta>0$. The *reduced cost* $\bar{c}_k$ is defined to be the increase in cx per unit increase in the value of the variable indexed by k. (For convenience, the reduced cost corresponding to a basic variable is defined to be zero.) Theorem (3.7) tells us precisely how the value of a basic edge variable changes, since the value of x_j, $j\in T$, will increase by $-\theta\alpha_{jk}$. Thus the new flow x^1 is given by

$$x_j^1=\begin{cases}x_j^0+\theta, & \text{if } j \text{ is a forward edge of } P(h(k), t(k)) \text{ or } j=k;\\ x_j^0-\theta, & \text{if } j \text{ is a reverse edge of } P(h(k), t(k));\\ x_j^0, & \text{otherwise.}\end{cases} \tag{4.1}$$

Hence, $\bar{c}_k$ is $c_k+$the cost of $P(h(k), t(k))$. (If $k=F\in\mathcal{N}$, then $c_F=0$, of course). Therefore, to evaluate the reduced costs conveniently, we can use the same trick as in the network simplex method. Let $r \in V$ be a fixed *root* vertex. Let R denote $N(r)$. For a node S, let π_S denote the cost of the path in T from R to S. For convenience, define π_v for $v \in V$ to be $\pi_{N(v)}$. Then, where k indexes a nonbasic variable, we have

$$\bar{c}_k = c_k + \pi_{t(k)} - \pi_{h(k)}. \tag{4.2}$$

(Notice that this formula also works for $k \in T$.) Therefore x^0 is optimal to (2.1) if

$$\begin{aligned} &c_e + \pi_{t(e)} - \pi_{h(e)} \geq 0 \quad \text{if } e \in E \backslash T \text{ with } x_e^0 = l_e; \\ &c_e + \pi_{t(e)} - \pi_{h(e)} \leq 0 \quad \text{if } e \in E \backslash T \text{ with } x_e^0 = u_e; \\ &\pi_{t(F)} \geq \pi_{h(F)} \quad \text{if } F \in \mathcal{N}. \end{aligned} \tag{4.3}$$

Our algorithm will work with π rather than with a dual solution of (2.1). (In the network flow case, the two are essentially the same.) But it is worthwhile to know that the dual solution corresponding to a particular cross-free basis, in particular an optimal one, can be constructed easily from π. Namely, $y_F = 0$, $F \in \mathcal{B}$, and for $F \in \mathcal{N}$, $y_F = ye_F = -\bar{c}_F = \pi_{h(F)} - \pi_{t(F)}$. (Notice that, in the dual of (2.1), the variables are required to be nonpositive.)

Since the variable which becomes basic or becomes non-basic can be of two types, edge-variable or set-variable, there are in principle four different types of pivot. However, we can use the following device to handle at the same time the two possibilities for the entering variable. Let k index the entering variable. If $k = e \in E \backslash T$, there is a unique circuit C formed by e together with $P(h(k), t(k))$. In this case put $\mathcal{N}^* = \mathcal{N}$, and $H^* = H$. If $k = F \in \mathcal{N}$, then put $\mathcal{N}^* = \mathcal{N} \backslash \{F\}$ and let H^* be the corresponding shrunken digraph. In H^*, $h(F)$ and $t(F)$ have been merged so that the edges of $P(h(F), t(F))$ form a circuit which we again call C. In either case, the change in flow resulting from the pivot amounts to 'sending flow around C': raising x^0 by θ on forward edges of C and lowering x^0 by θ on reverse edges. (Notice that the direction of C is thus well-defined.)

Let J denote the spanning subgraph of H^* having edge-set $T \cup \{k\}$ if $k \in E$ and T otherwise. Then J is connected and contains just one circuit, C. We assign C an initial node S_0; S_0 is the first node of C on the path in J from R to C. Let the node-edge sequence of C be: $S_0, e_1, S_1, e_2, \ldots, e_m, S_m$ where $S_0 = S_m$. For $1 \leq i \leq m$, let v_{i-1} be the end of e_i which is in S_{i-1} and let w_i be the end of e_i which is in S_i. Of course, it is possible that $w_i = v_i$, in particular, that $|S_i| = 1$. Figure 2 shows such a C, together with the path in j from C to R.

Let d be the incidence vector of C, so d_j is 1 (-1) if j is a forward (reverse) edge of C, $d_j = 0$ otherwise. Then the new flow x^1 will be $x^0 + \theta d$; θ is chosen as large as possible so that $x^0 + \theta d$ is feasible. (As usual, if no such θ exists, (2.1) is unbounded.) We have $\theta = \min(\theta_1, \theta_2, \theta_3)$, where θ_1, θ_2 are computed in the usual way based on the bounds for the edges of C, and θ_3, involves the submodular inequalities. The next result simplifies the calculation of θ_3, and also shows that the new basis can be chosen to be cross-free.

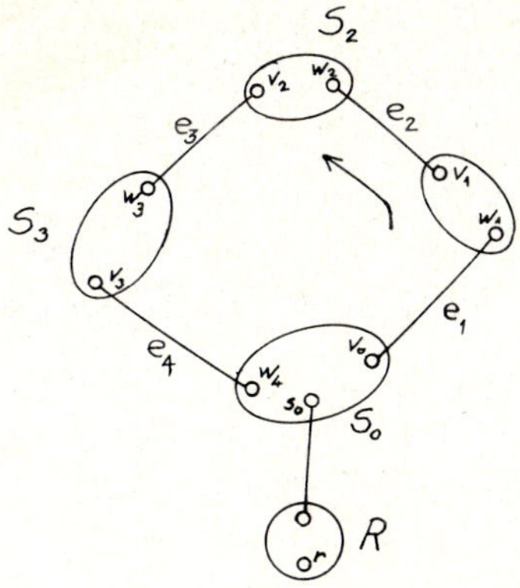

Fig. 2.

(4.4) Lemma. *Let $\mathcal{N}' \subseteq \mathcal{F}$ be independent and cross-free, let x' be a feasible flow such that F' is x'-tight for every $F' \in \mathcal{N}'$, and let $F \in \mathcal{F}$ be x'-tight such that, for a node S of $\mathcal{N}'$, $S \cap F \neq \emptyset \neq S \backslash F$. Then without changing $S \cap F$, we can require, in addition, that:*

(i) *$\mathcal{N}' \cup \{F\}$ is cross-free;*

(ii) *For every node S' of $\mathcal{N}'$ with $S' \neq S$, either $F \supseteq S'$ or $F \cap S' = \emptyset$.*

To prove (4.4) we use the following well-known result [6, 12]. We include its proof for completeness.

(4.5) Lemma. *Let x be a feasible solution of (2.1) and suppose that K, $L \in \mathcal{F}$ are crossing and x-tight. Then $K \cap L$, $K \cup L$ are also x-tight.*

Proof. $b(K \cap L) + b(K \cup L) \leq b(K) + b(L) = a^K x + a^L x$

$$= a^{K \cap L} x + a^{K \cup L} x \leq b(K \cap L) + b(K \cup L).$$

The result follows. □

Proof of (4.4). Consider the edges of the set-tree $\mathcal{T}'$ corresponding to $\mathcal{N}'$. If some $F' \in \mathcal{N}'$ has $h(F') = S$ and $F \backslash F'$, $F \cup F' \neq \emptyset$, then F, F' cross. By (4.5) $F \cap F' \in \mathcal{F}$ is also x'-tight, and it has $F \cap F' \cap S = F \cap S$, so we can replace F by $F \cap F'$. Similarly, if some $F' \in \mathcal{N}'$ with $t(F') = S$, has $F' \cap F$, $F' \backslash F \neq \emptyset$, then we can replace F by $F \cup F'$. It is easy to see that each such replacement of F decreases the number of violations of (ii), so (ii) is satisfied after at most $|\mathcal{N}'|$ such steps. Moreover, F does not cross F' for any $F' \in \mathcal{N}'$ incident with S in τ', so it crosses no $F' \in \mathcal{N}'$, and (i) is proved. □

Suppose that in (4.4) we take $\mathcal{N}' = \mathcal{N}^*$, $x' = x^1$, and $S = S_i$ for some i; we obtain F satisfying (i) and (ii). If F contains both or neither of v_i, w_i then it is easy to see that the number of times C enters F equals the number of times C leaves F, so $a^F d = 0$. Similarly, if $w_i \in F$, $v_i \notin F$, then $a^F d = 1$, and if $w_i \notin F$, $v_i \in F$, then $a^F d = -1$.

So to ensure that $a^F x^1 \leq b(F)$ for all $F \in \mathscr{F}$, it is sufficient that, for $0 \leq i \leq m-1$, and each $F \in \mathscr{F}$ with $w_i \in F$ and $v_i \notin F$, we have $\theta \leq b(F) - a^F x^0$.

We can now summarize the pivot operation. If (4.3) is satisfied, x^0 is optimal and the algorithm terminates. Otherwise choose an index k of a non-basic variable violating (4.3). Then form $\mathscr{N}^*$, H^*, $\mathscr{T}^*$, C. Compute $\theta_1 = \min(u_j - x_j^0 : j$ a forward edge of $C)$ and $\theta_2 = \min(x_j^0 - l_j : j$ a reverse edge of $C)$. For $0 \leq i \leq m-1$, let $\varepsilon_i = \min(b(F) - a^F x^0 : w_i \in F \in \mathscr{F}, v_i \notin F)$, and let $\theta_3 = \min(\varepsilon_i : 0 \leq i \leq m-1)$. Let $\theta = \min(\theta_1, \theta_2, \theta_3)$. If $\theta = \infty$, (2.1) is unbounded and the algorithm terminates. If $\theta = \theta_1$ or θ_2, choose an edge f of C giving the minimum; x_f is the leaving variable. Otherwise, beginning with $F \in \mathscr{F}$ giving the minimum $\varepsilon_i = \theta_3$, and $S = S_i$, use (4.4) to obtain such F with the additional properties (i), (ii), and make s_F the leaving variable. Figure 3 shows a typical pivot. In this case an edge-variable enters and a set-variable leaves.

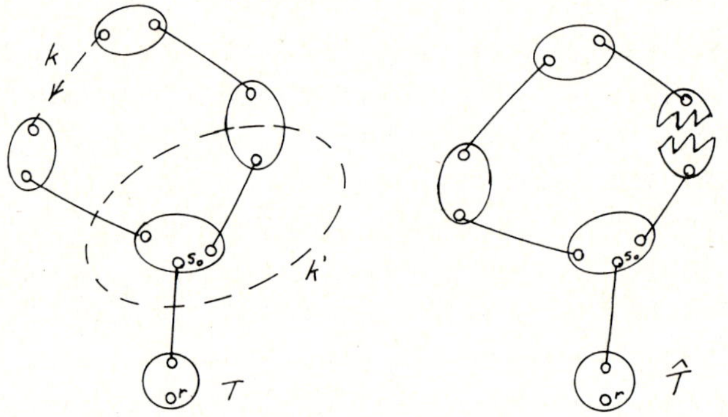

Fig. 3.

Let us consider some parts of the pivot step in a little more detail. To compute θ_3, and find a minimizing F, we use the oracle for (2.6). To perform a pivot, we need to call this oracle $m \leq |V|$ times. In Section 6 we discuss the availability of the oracle in special cases.

It is generally more useful to maintain $\mathscr{T}$ rather than $\mathscr{N}$, that is, to remember a cross-free family by its tree representation. In particular, it is quite easy to obtain $\hat{\mathscr{T}}$ from $\mathscr{T}^*$. (Of course, it is easy to obtain $\mathscr{T}^*$ from $\mathscr{T}$. If an edge-variable is entering, then $\mathscr{T}^* = \mathscr{T}$, and if a set-variable, indexed by $K \in \mathscr{F}$, is entering, $\mathscr{T}^*$ is obtained by contracting the edge K of $\mathscr{T}$.) If the leaving variable is indexed by $f \in E$, then $\hat{\mathscr{T}} = \mathscr{T}^*$. If the leaving variable is a set-variable, suppose $\varepsilon_i = \theta_3$ and F minimizes $(b(F) - a^F x^0 : w_i \in F \in \mathscr{F}, v_i \notin F)$. Then $\hat{\mathscr{T}}$ is obtained from $\mathscr{T}^*$ as follows. Split S_i into two nodes $S_i \cap F$, $S_i \backslash F$. For each $F' \in \mathscr{N}^*$ incident in $\mathscr{T}^*$ with S_i:

(a) If $t(F') = S_i$ and $F \cap F' \neq \emptyset \neq F' \backslash F$, put $t(F') = S_i \cap F$;
(b) If $t(F') = S_i$ and $F \supseteq F'$, put $t(F') = S_i \cap F$;
(c) If $t(F') = S_i$ and $F \cap F' = \emptyset$, put $t(F') = S_i \backslash F$;
(d) If $h(F') = S_i$, and $F \cup F' \neq V$ and $F \backslash F' \neq \emptyset$, put $h(F') = S_i \backslash F$; (4.6)
(e) If $h(f') = S_i$ and $F' \supseteq F$, put $h(F') = S_i \backslash F$;
(f) If $h(F') = S_i$ and $F \cup F' = V$, put $h(F') = S_i \cap F$.

All other tails, heads of sets $F' \in \mathcal{N}^*$ remain the same in $\hat{\mathcal{T}}$ as in $\mathcal{T}^*$. Finally, put $h(F) = S_i \cap F$, $t(F) = S_i \backslash F$. This construction is illustrated in Fig. 4, where F is indicated by the cross-hatched area, and certain subtrees are condensed to triangles.

As in the ordinary network simplex method, it is possible to compute $\hat{\pi}$, the node numbers corresponding to $(\hat{T}, \hat{\mathcal{N}})$, by updating. This updating technique is useful for implementation of the algorithm, but it is especially important for the finiteness analysis of the next section. Notice that each of T, $\hat{T}$ is obtained from J by an operation of one of two kinds: (i) deleting some e_i, or (ii) for some set F and node S_i of $\mathcal{N}^*$ as in (4.4), splitting S_i into $S_i \cap F$, $S_i \backslash F$. (Notice that in case (ii) F has an additional property not required in (4.4): F contains just one of v_i, w_i.) We refer to such edges e_i or sets F as *links*.

Fig. 4.

In order to give a formula for $\hat{\pi}$ in terms of π, we need some more terminology. Recall that in the ordinary network simplex method, the vertices v of C for which $\hat{\pi}_v \neq \pi_v$ are those between the entering edge and the leaving edge. First, we define s_0 to be r if $r \in S_0$, and otherwise to be the element of S_0 which is an end of the first edge in the path in J from S_0 to R. Given F as in (4.4.) and containing just one of v_i, w_i, we denote i by $i(F)$. However, we distinguish between the indices 0 and m, as follows. If F contains v_0 but not s_0 or contains w_0 and s_0, we put $i(F) = 0$; if F contains v_0 and s_0 or contains w_0 and not s_0, we put $i(F) = m$. In general, we denote by W_F the one of $S_{i(F)} \cap F$, $S_{i(F)} \backslash F$ which contains $w_{i(F)}$. For two links k, k' we define a set $Q(k, k') \subseteq \cup(S_i\colon 1 \leq i \leq m)$. Roughly speaking it consists of the vertices of C following k and preceding k'.

$$Q(e_i, e_j) = \cup(S_p\colon i \leq p < j);$$

$$Q(e_j, F) = \cup(S_p\colon j \leq p < i(F)) \cup W_F;$$

$$Q(F, e_j) = \cup(S_p\colon i(F) \leq p < j) \backslash W_F;$$

$$Q(F, F') = (\cup(S_p\colon i(F) \leq p \leq i(F')) \backslash W_F) \backslash (S_{i(F')} \backslash W_{F'}),$$

unless $i(F) = 0$, $i(F') = m$, when

$$Q(F, F') = \cup(S_p\colon 1 \leq p \leq m) \backslash (W_F \backslash W_{F'}) \backslash (W_{F'} \backslash W_F).$$

In the special case $i(F)=0$, $i(F')=m$, we make one more definition: $D(F, F')=W_{F'}\backslash W_{F}$. In all other cases $D(k, k')=\emptyset$.

Suppose that the entering variable is indexed by k and the leaving variable by k'. Normally, at most one of $Q(k, k')$, $Q(k', k)$ will be nonempty. An exception can occur when k, k' are both sets F, F' (say) and $i(F)=i(F')$. A pivot of this type is indicated in Figure 5. Here $p\in Q(F, F')$, $q\in Q(F', F)$. Normally, if $v\in Q(k, k')$ or

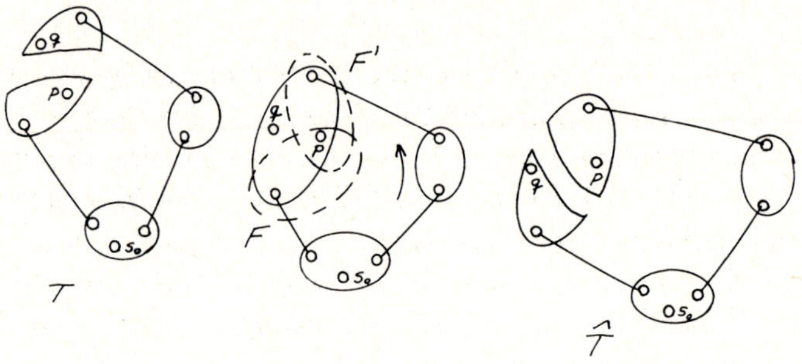

Fig. 5.

$Q(k', k)$, then the simple paths in T, $\hat{T}$ from r to v differ in that one traverses part of C in the forward direction, and the other traverses the complementary part of C in the reverse direction. Hence $\hat{\pi}_v=\pi_v\pm\bar{c}_k$. However, there is an exception here also. If k, k' are both sets F, F' (say), $i(F)$ is one of 0, m, $i(F')$ is the other, and $v\in D(F, F')$ or $D(F', F)$, then one of the paths traverses all edges of C in one direction and the other path traverses all edges of C in the other, so $\hat{\pi}_v=\pi_v\pm 2\bar{c}_k$. A pivot of this type is indicated in Figure 6. Here $v\in D(F', F)$. As in the network case, for a vertex not on C, $\hat{\pi}$ differs from π in the same way that it does for the 'nearest' vertex on C. We can state a formula for $\hat{\pi}$; the proof is straightforward, and is omitted.

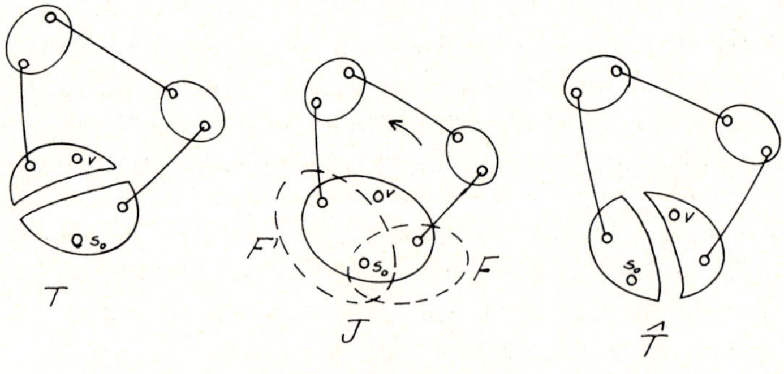

Fig. 6.

(4.7) Theorem. *Suppose that k indexes the entering variable, k' indexes the leaving variable, and $\bar{c}_k<0$. For $v\in\cup(S_i\colon 1\leqslant i\leqslant m)$, let $p(v)$ denote v, and for $v\in V\backslash\cup(S_i\colon 1\leqslant i\leqslant m)$, let $p(v)$ denote the first element of $\cup(S_i\colon 1\leqslant i\leqslant m)$ which is* *an end of the last edge of the simple path in J from v to C. Then*

(a) $\hat{\pi}_v = \pi_v + \bar{c}_k$, *if* $p(v) \in Q(k, k')$;
(b) $\hat{\pi}_v = \pi_v - \bar{c}_k$, *if* $p(v) \in Q(k', k)$;
(c) $\hat{\pi}_v = \pi_v + 2\bar{c}_k$, *if* $v \in D(k, k')$;
(d) $\hat{\pi}_v = \pi_v - 2\bar{c}_k$, if $v \in D(k', k)$;
(e) $\hat{\pi}_v = \pi_v$, *otherwise.*

If $\bar{c}_k > 0$, that is, if $k \in E$ and $x_k^0 = u_k$, then $\bar{c}_k$ is replaced by $-\bar{c}_k$ in the formula, because in this case $-\bar{c}_k$ is the cost of C. Of the five possibilities (a)-(e), three can occur on a given pivot (and normally only two will occur). This is because $D(k, k') \neq \emptyset$ implies $Q(k', k) = D(k', k) = \emptyset$, and similarly for $D(k', k) \neq \emptyset$. In fact, we encourage the reader to check that precisely these six combinations can occur on a given pivot: (e), (a); (e), (b); (e), (a), (b); (e), (a), (c); (e), (b), (d); (e). The last one is another surprising possibility. In this case T does not change, but of course $\mathcal{T}$ does. Such a pivot is illustrated in Fig. 7.

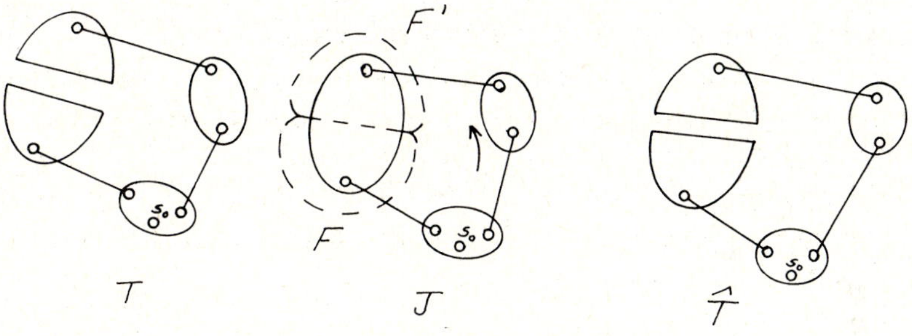

Fig. 7.

5. Strong feasibility

In this section we generalize the method of strongly feasible trees [1] to the submodular simplex method. By refining the choice of initial basis and the choice of leaving variable we obtain a version of the algorithm which is guaranteed to terminate. Thus we obtain a proof of the Edmonds–Giles Theorem (2.2) as well as characterizations of vertices and vertex-adjacency for the submodular flows polyhedron. (We remark that none of the known methods for refining the simplex method to prevent cycling in general linear programming can be used here, because of the restrictions we have made on the choice of leaving column.)

Let $(T, \mathcal{N})$ determine a cross-free basis of (3.1). For each node S of $\mathcal{N}$ we define the *root* of S as follows. If $r \in S$, then r is the root of S; otherwise, the root of S is the element of S which is an end of the last edge of the simple path from R to S in T. Let x^0 be the flow associated with $(T, \mathcal{N})$. We say that $(T, \mathcal{N})$ is *strongly feasible* (s.f.) if x^0 is feasible and

(i) If $j \in T$ and $x_j^0 = l_j$, then j is directed away from R in T;

(ii) If $j \in T$ and $x_j^0 = u_j$, then j is directed toward R in T;

(iii) If s is the root of a node S of $\mathcal{N}$, $P \in \mathcal{F}$ is x^0-tight, and $s \in P$, then $P \supseteq S$.

Notice that if $(T, \mathcal{N})$ determines a nondegenerate basic solution of (3.1), that is, if $l_j < x_j^0 < u_j$ for all $j \in T$, and $a^F x^0 < b(F)$ for all $F \in \mathcal{B}$, then $(T, \mathcal{N})$ is strongly feasible.

The modification of the simplex method which we will describe uses only s.f. cross-free bases. This is accomplished by a more specific choice of leaving variable which preserves strong feasibility. That choice, roughly speaking, is the same as the one in the network case, to choose the 'first' variable in C which gives the minimum θ. Among the candidates for leaving variable we choose one which is preceded by no other, where k *precedes* k' if and only if $Q(k, k') \neq \emptyset$. It is not obvious that such a candidate exists, since it is possible for k, k' to precede each other. This can happen only if k, k' are sets F, F' with $i(F) = i(F') = j$, say. Assume first that $0 < j < m$. Since $w_j \in F, F'$ and $v_j \notin F, F'$, we have either that one of F, F' contains the other or that F, F' cross. In either case $F \cap F' \in \mathcal{F}$ is x^1-tight too, and $F \cap F'$ is not preceded by F or F'. So the set $F'' = \cap(F\colon F \in \mathcal{F}, F\ x^1\text{-tight}, w_j \in F, v_j \notin F)$ is an x^1-tight set in $\mathcal{F}$ with $i(F'') = j$, and no other such set precedes F''. Similar arguments can be made for the cases $j = 0$, $j = m$. If $j = 0$, we choose $F'' = \cap(F\colon F \in \mathcal{F}, F\ x^1\text{-tight}, r, w_m \in F, v_0 \notin F)$; if $j = m$, we choose $F'' = \cap(F\colon F \in \mathcal{F}, F\ x^1\text{-tight}, w_m \in F, r, v_0 \notin F)$.

We must also show that such a set can be efficiently constructed using only the submodular oracle. Surprisingly, this seems to require the assumption of strong feasibility. Suppose we have $0 < j < m$ and $F \in \mathcal{F}$ with F x^1-tight, $w_j \in F$, $v_j \notin F$. We want such F with $F \cap S_j$ minimal. Take $v \in F \cap S_j$, $v \neq w_j$. We ask (x^1, w_j, v). If the answer is not an x^1-tight set, then v must be an element of the 'minimal' F we seek. If the answer is an x^1-tight set F' and F and F' cross or $F \subseteq F'$, then $F \cap F'$ is an x^1-tight set having the required properties and we can replace F by $F \cap F'$ and continue. Finally, suppose the answer is an x^1-tight set F' with $F' \not\subseteq F$ and F, F' not crossing; then $F \cup F' = V$, so F' contains both v_j and w_j. By (4.4) there exists another such F' with $F' \cap S = \emptyset$ or $F' \supseteq S$ for all nodes S of $\mathcal{N}^*$, $S \neq S_j$. It follows that $a^{F'} d = 0$, so F' is x^0-tight. But since $v \notin F'$, and one of v_j, w_j is the root of the node of $\mathcal{N}$ containing v, this contradicts strong feasibility of $(T, \mathcal{N})$.

In case $j = 0$ or m, the following observation will be useful: Any x^1-tight set $F \in \mathcal{F}$ with $s_0 \in F \not\supseteq S_0$, must contain exactly one of v_0, w_m. For suppose that F is such a set. By (4.4) F can be transformed, without changing $S_0 \cap F$, so that $F \cap S = \emptyset$ or $F \supseteq S$ for every node S of $\mathcal{N}^*$ different from S_0. If F contains both or neither of v_0, w_m, then F is x^0-tight. If $v_0, w_m \notin F$, then since the node of $\mathcal{N}$ having s_0 as root contains at least one of v_0, w_m, strong feasibility of $(T, \mathcal{N})$ is contradicted. If v_0, $w_m \in F$, then the node of $\mathcal{N}$ containing any $v \in S_0 \backslash F$ has one of v_0, w_m, s_0 as root, giving a contradiction. The claim is proved. If the answer to (x^1, s_0, v_0) is an x^1-tight set F, then $w_m \in F$, $j = 0$, and we can apply the above procedure to find such a tight set with $F \cap S_0$ minimal. Otherwise, $j \neq 0$ and, if $j = m$, we have to find an x^1-tight set F with $w_m \in F$, and s_0, $v_0 \notin F$ having $F \cap S_0$ minimal. We begin by asking (x^1, w_m, v_0) to get one such set, and then apply the procedure to obtain a minimal one.

The main step in the proof of finiteness of the modified algorithm is the following.

(5.1) Theorem. *If* $(T, \mathcal{N})$ *is s.f. and the leaving variable is chosen to be the first candidate on* C, *then* $(\hat{T}, \hat{\mathcal{N}})$ *is s.f. too.*

We let k index the entering variable and k' index the leaving variable.

(5.2) Lemma. *If* $\theta = 0$, *then* k' *does not precede* k.

Proof. If k' is an edge e_i, and k' precedes k, then k' is directed away from r in T if and only if k' is a forward edge of C. Since k' is the leaving edge and $\theta = 0$, it follows that k' is directed away from r if $x^0_{k'} = u_{k'}$ and is directed toward r if $x^0_{k'} = l_{k'}$. In either case, strong feasibility is contradicted.

Now suppose that k' is a set $F' \in \mathcal{F}$ and k' precedes k. Let $j = i(F')$. If $j > 0$, then there is a node S of $\mathcal{N}$ of which w_j is the root. (In most cases $S = S_{j}$.) Since there exists $v \in Q(k', k)$, we can choose such v to be in S. Then $v \notin F'$, $w_j \in F'$. Since $\theta = 0$, F' is x^0-tight, contradicting strong feasibility. In case $j = 0$ the same argument works, except that s_0 is the root of S. □

Combining (5.2) with (4.7) and the remark following (4.7), we obtain the next result, which will be useful later.

(5.3) Corollary. *If* $\theta = 0$, *then* $\hat{\pi}_v \leq \pi_v$ *for all* $v \in V$.

Proof of (5.1). Consider first the orientation requirements for edges $j \in \hat{T}$. It is easy to see that if j is not an edge of C, then j is directed away from (toward) r in $\hat{T}$ if and only if it is so directed in T. Since such edges satisfy $x^0_j = x^1_j$, they do not violate strong feasibility. Suppose that j is an edge of C satisfying $x^1_j = l_j$. If $\theta > 0$, then j is a reverse edge of C having $x^0_j = l_j$. Such an edge will be directed toward r in $\hat{T}$ only if it precedes k'. But this contradicts the choice of leaving variable. If $\theta = 0$, then j is a forward edge of C which precedes k, or $j = k$, or j is a reverse edge of C which is preceded by k. In the first two cases j is directed toward r in $\hat{T}$ only if k' precedes j, so that k' precedes k, contradicting (5.2). In the last case j is directed toward r in $\hat{T}$ only if j precedes k', which contradicts the choice of leaving variable. So j does not violate strong feasibility. The case when $x^1_j = u_j$ is similar.

Now suppose that $(\hat{T}, \hat{\mathcal{N}})$ is not s.f. because there exist a node S of $\hat{\mathcal{N}}$ having root s and a set $F \in \mathcal{F}$ such that F is x^1-tight and $s \in F \not\supseteq S$. Using (4.4) we may assume that $F \cap S' = \emptyset$ or $F \supseteq S'$ for every node S' of $\hat{\mathcal{N}}$ with $S' \neq S$. Notice that every node S' of $\mathcal{N}^*$, with a single exception, also satisfies $F \cap S' = \emptyset$ or $F \supseteq S'$. We divide the analysis into four cases.

Case 1. $S \cap (\cup S_j) = \emptyset$. Then S is a node of $\mathcal{N}$, and s is its root. Moreover, $a^F d = 0$, so F is x^0-tight. This contradicts the strong feasibility of $(T, \mathcal{N})$.

Case 2. $s = v_j$ for some j, $0 \le j < m$. If $w_j \in F$, then F is x^0-tight. Choose $v \in S \backslash F$, and let S' be the node of $\mathcal{N}$ containing v. If one of v_j, w_j is the root of S' (with respect to $(T, \mathcal{N})$), strong feasibility of $(T, \mathcal{N})$ is contradicted. Thus we may suppose that $j = 0$, and that s_0 is the root of S'. Then k, $k' \in \mathcal{F}$, $i(k') = 0$, and $i(k) = m$. Since $w_m \in S' \cap F$, $F \supseteq S'$, so $s_0 \in F$. But v is an element of the node of $\mathcal{N}$ containing s_0, so we have a contradiction. Now suppose that $w_j \notin F$. Then $a^F d = -1$, so $\theta = 0$ and F is x^0-tight. Again, choose $v \in S \backslash F$, and let S' be the node of $\mathcal{N}$ containing v. If the root of S' is v_j, then we have a contradiction. If the root of S' is w_j, and $0 < j < m$, then $v \in Q(k', k)$ so that k' precedes k, contradicting (5.2). If $j = 0$ and the root of S' is w_m, then $v \in D(k', k)$ which implies that k' precedes k, a contradiction. Finally, if the root of S' is s_0, then k' is a set F' which is x^0-tight and contains s_0 and not v, a contradiction.

Case 3. $s = w_j$ for some j, $1 \le j \le m$. When $v_j \in F$, the argument is exactly analagous to the corresponding part of Case 2, and so is omitted. Now suppose that $v_j \notin F$. Then $a^F d = 1$, so $b(F) - a^F x^0 = \theta$. It follows from the choice of leaving variable that $Q(k', F) = \emptyset$. But clearly, any $v \in S \backslash F$ is in $Q(k', F)$, a contradiction.

Case 4. $s = s_0$. If F contains neither of v_0, w_m, then F is x^0-tight. But the node of $\mathcal{N}$ containing s_0 contains at least one of v_0, w_m, so the strong feasibility of $(T, \mathcal{N})$ is contradicted. If F contains both v_0, w_m then F is x^0-tight. Choose $v \in S \backslash F$, and let S' be the node of $\mathcal{N}$ containing v. The root of S' is one of s_0, v_0, w_m, so strong feasibility of $(T, \mathcal{N})$ is contradicted. If F contains v_0 but not w_m, then $\theta = 0$ and F is x^0-tight. Choose $v \in S \backslash F$ and let S' be the node of $\mathcal{N}$ containing v. Then the root of S' must be w_m. (Otherwise, we have an immediate contradiction.) Thus $k \in \mathcal{F}$, $i(k) = m$, and $v \in Q(k', k)$. But this contradicts (5.2). Finally, suppose that $v_0 \notin F$, $w_m \in F$. Then, by the choice of leaving variable, $Q(F, k') = \emptyset$. But any $v \in S \backslash F$ is in $Q(F, k')$, a contradiction. □

We are now able to state and prove the main result.

(5.4) Theorem. *The submodular network simplex algorithm, beginning with an s.f. basis and always choosing the first candidate on C as the leaving variable, is finite.*

Proof. By (5.1) and (5.3) the only possibility for non-termination is a sequence of consecutive 'special' pivots: degenerate pivots in which $\hat{\pi} = \pi$. Such a pivot occurs precisely when the entering variable $F \in \mathcal{F}$ and the leaving variable $F' \in \mathcal{F}$ satisfy $i(F) = i(F')$ and $W_F = W_{F'}$, as in the example of Fig. 7. Thus the nodes of $\hat{\mathcal{N}}$ are the same as the nodes of $\mathcal{N}$, but the trees $\mathcal{T}$, $\hat{\mathcal{T}}$ are different. The tree $\hat{\mathcal{T}}$ has the same edges as $\mathcal{T}$ except that F' replaces F. Let $t(F) = U$, $h(F) = W$. Then in $\hat{\mathcal{T}}$, F' has head U and tail W. If the head (tail) of $F'' \neq F$ in $\mathcal{T}$ is $S \notin \{U, W\}$, then it remains the same in $\hat{\mathcal{T}}$. If the head (tail) of $F'' \neq F$ in $\mathcal{T}$ is one of U, W, then it will be one of U, W in $\hat{\mathcal{T}}$, but not necessarily the same one.

It is convenient to order the nodes of $\mathcal{N}$ as $S(1), \ldots, S(n)$, where $\pi_{S(1)} \leq \cdots \leq \pi_{S(n)}$. For each node S of $\mathcal{N}$, we define $g(S)(\hat{g}(S))$ to be the number of nodes in that component which includes S of the graph obtained from $\mathcal{T}$ $(\hat{\mathcal{T}})$ by deleting the edges having head S. It is easy to see that $\hat{g}(S) = g(S)$ for all $S \notin \{U, W\}$. Moreover, $\hat{g}(U) < g(U)$. Since $0 > \bar{c}_F = \pi_U - \pi_W$, we have $\pi_U < \pi_W$. It follows that the vector $(\hat{g}(S_1), \ldots, \hat{g}(S_n))$ is lexicographically smaller than $(g(S_1), \ldots, g(S_n))$, so a cycle of such pivots is impossible, and the algorithm terminates. □

We remark that the vector $(cx, \sum(\pi_v\colon v \in V), g(S_1), \ldots, g(S_k))$, lexicographically decreases with each pivot, although k can change.

From the finiteness of the algorithm, combined with some techniques for obtaining initial solutions to be described in the next section, we obtain a proof of the Edmonds-Giles theorem (2.2). (We remark that this constructive proof works directly with the original crossing family $\mathcal{F}$, whereas the proof of the validity of Frank's algorithm [12, 13, 5] requires dealing with the special case in which $\mathcal{F}$ is an intersecting family.) Similarly, we can derive the following polyhedral results.

(5.5) Theorem. *A vector is a vertex of the polyhedron of feasible solutions of* (2.1) *if and only if it is feasible to* (2.1) *and is determined by a cross-free basis.*

(5.6) Theorem. *Two vertices x^1, x^2 of the polyhedron of feasible solutions of* (2.1) *are adjacent if and only if there are cross-free bases $(T_1, \mathcal{N}_1)$, $(T_2, \mathcal{N}_2)$ determining them, and a single nondegenerate pivot in the simplex algorithm takes $(T_1, \mathcal{N}_1, x^1)$ to $(T_2, \mathcal{N}_2, x^2)$.*

It was proved in [1] that the strongly feasible bases of a network flow problem are precisely the feasible bases of another problem obtained from the first by a linear perturbation of the demand vector b, and moreover that the new linear program (in equality form) is nondegenerate. The first part of that result extends to the present context. However, the possibility of special pivots shows that the perturbed linear program is not nondegenerate. We state the result without proof. (Of course, it is possible that (2.1) has no s.f. cross-free bases, and hence the perturbed problem is infeasible.)

(5.7) Theorem. *Define b' on $\mathcal{F}$ as follows. If $r \notin F \in \mathcal{F}$, $b'(F) = b(F) + \varepsilon|F|$. If $r \in F \in \mathcal{F}$, $b'(F) = b(F) - \varepsilon|V \backslash F|$. Then b' is submodular on $\mathcal{F}$. For ε sufficiently small and positive, the feasible cross-free bases of the problem obtained from* (2.1) *by replacing b by b' are precisely the s.f. cross-free bases of* (2.1).

6. Initialization and separation

In this section we discuss two practical aspects of the algorithm which we have so far ignored: finding an initial solution, and implementing the separation oracle.

First, we discuss these issues in the context of the general model, where it seems to be necessary to use a two-phase approach to the initialization problem, that is, to solve another submodular flow problem to obtain an initial (artificial) solution. Then we discuss several important special cases. In each of these there are direct methods for solving the initialization problem, and often the necessary separation oracle is available via an efficient combinatorial procedure.

As for the network simplex method, the basic idea for initializing the algorithm is to use artificial edges which are given sufficiently large cost. Suppose that we have a vector $z=(z_v\colon v\in v)$ satisfying

$$z(V)=0; \qquad z(F)\leq b(F), \quad F\in\mathcal{F}. \tag{6.1}$$

Of course, in the network flow case, taking $z_v=b_v$, $v\in V$, satisfies (6.1). More generally, it is easy to see that solvability of (6.1) is a necessary condition for feasibility of (2.1); namely, given a feasible solution x of (2.1), put $z_v=a^v x$, $v\in V$. One more remark: If it happens that $\mathcal{F}$ is an intersecting family and b is submodular on intersecting pairs of $\mathcal{F}$, then z satisfying (6.1) can be found using the greedy algorithm, beginning with $z_v=\min(0,\min(b(F)\colon F\in\mathcal{F}))$ for all $v\in V$, which can easily be found using the oracle.

On the other hand, given z satisfying (6.1), we can add artificial edges to G as follows. For each vertex $v\neq r$ with $z_v\geq 0$, add an edge $e(v)$ having tail r, head v, and flow $x^0_{e(v)}=z_v$. For each vertex $v\neq r$ with $z_v<0$, add an edge $e(v)$ with tail v, head r, and flow $x^0_{e(v)}=-z_v$. Put $l_{e(v)}=0$ for all v, put $u_{e(v)}=\infty$ if $x^0_{e(v)}=0$, and put $u_{e(v)}=x^0_{e(v)}$ if $x^0_{e(v)}\neq 0$. Put $x^0_j=0$ for each original edge j. Then with respect to this enlarged graph, x^0 satisfies $a^F x^0\leq b(F)$, $F\in\mathcal{F}$. As in the network flow case we can put $c_{e(v)}=|V|\max(|c_j|\colon j\in E)+1$; it is not too hard to show that, if (2.1) has a feasible solution, then every optimal solution x of the resulting problem has $x_{e(v)}=0$ for all v and hence gives an optimal solution to (2.1).

In order to apply the simplex method, we need to construct a cross-free s.f. basis corresponding to the artificial solution x^0. Suppose that we have a cross-free basis $(T,\mathcal{N})$ corresponding to x^0, with $T\subseteq A$, the set of artificial edges. Initially, $T=\emptyset$, $\mathcal{N}=\emptyset$ will do. If $(T,\mathcal{N})$ is not s.f., then there is a node S of $H(\mathcal{N})$ having root s and an x^0-tight set $F\in\mathcal{F}$ such that $s\in F\not\supseteq S$. We can find, by the usual methods, such an F so that $\mathcal{N}\cup\{F\}$ is cross-free. Then we add F to $\mathcal{N}$ and add an edge $e(v)$, $v\in S\backslash F$, to T, and change $u_{e(v)}$ to ∞. Continuing this procedure, we obtain the required cross-free s.f. basis.

Now we must show how to obtain z of (6.1). Frank [13] has shown how this can be done by solving another submodular flow problem. Let G^* be a digraph having vertex-set $V_1\cup V_2$, where $V_i=\{v_i\colon v\in V\backslash\{r\}\}$ for $i=1$ and 2, and edge-set $V\backslash\{r\}$. For any $F\subseteq V\backslash\{r\}$ and $i=1$ or 2, F_i denotes $\{v_i\colon v\in F\}$. Each $v\in V\backslash\{r\}$ has head v_1 and tail v_2. We define the family $\mathcal{F}^*$ and function b^* as follows. For each $F\in\mathcal{F}$ with $r\notin F$, put $F_1\in\mathcal{F}^*$ with $b^*(F_1)=b(F)$. For each $F\in\mathcal{F}$ with $r\in F$, put $(\bar{F})_2\in\mathcal{F}^*$ with $b^*((\bar{F})_2)=b(F)$. It is easy to check that b^* is submodular on the crossing family

$\mathscr{F}^*$, and the resulting flow constraints, using flows z_v with $l_v = -\infty$, $u_v = \infty$, are:

$$z(F) \leqslant b(F), \quad r \notin F \in \mathscr{F}; \qquad -z(\bar{F}) \leqslant b(F), \quad r \in F \in \mathscr{F}. \tag{6.2}$$

If we define z_r to be $-z(V \backslash \{r\})$, then $z(V) = 0$, and so $-z(\bar{F}) = z(F)$. Thus (6.2) is equivalent to (6.1). But in fact, $\mathscr{F}^*$ is an intersecting family and b^* is submodular on intersecting pairs, so the analogue of (6.1) for this new submodular flow problem, can be solved with the greedy algorithm. Thus we can use the simplex method to solve (6.1). One must check that the oracles needed for the greedy algorithm and for the 'phase-one' simplex method are available from the oracle we are assuming. This is not difficult to do.

We now turn to the discussion of some interesting applications of submodular flows. These are not new, so we will not justify the formulations, but only discuss the application of the simplex method to them.

Dicut Covers [22, 7]. Given a digraph $G = (V, E)$, and $F \subseteq V$, let $\rho(F)$ denote $\{e \in E: h(e) \in F, t(e) \notin F\}$ and let $\delta(F)$ denote $\rho(\bar{F})$. A *dicut* is a set $D \subseteq E$ of the form $\delta(F) \neq \emptyset$ for which $\rho(F) = \emptyset$. A (*dicut*) *cover* is a set $C \subseteq E$ such that $C \cap D \neq \emptyset$ for every dicut D. Hence, assuming that G is weakly connected, a cover is a set whose contraction makes G strongly connected. The problem of finding a minimum cost dicut cover can be formulated as a submodular flow problem as follows. (Notice that it is obvious what to do with edges having negative cost, so we may assume that $c \geqslant 0$.) Let $\mathscr{F} = \{F: F \subseteq V, \delta(F) \neq \emptyset, \rho(F) = \emptyset\}$, let $b(F) = -1$ for each $F \in \mathscr{F}$, and $l_e = 0$ and $u_e = \infty$ for each $e \in E$.

To solve (6.1) for this instance is easy: Put $z_v = |\rho(\{v\})| - |\delta(\{v\})|$ for each $v \in V$. Then for all $F \in \mathscr{F}$, $z(F) = |\rho(F)| - |\delta(F)| \leqslant -1$, as required. Moreover, equality holds if and only if $\delta(F) = \{e\}$, say, in which case e is in every dicut cover, and the problem can be decomposed. So obtaining an initial $\mathcal{N}$ to go with the artificial x determined by this z is easy: $\mathcal{N} = \emptyset$ will work. The problem (2.6) for this application is easily solved by a minimum cut calculation [12]. (In fact, in implementing the simplex method for this application, there are some additional simplifications, and minimum cut calculations can be replaced by a more efficient combinatorial algorithm. We believe that the simplex method is the most practical algorithm available for this problem.) Finally, we remark that the dual problem, finding a maximum packing of dicuts, is also of interest, and the simplex method also solves it. Frank's algorithm [11] also has this property, but some other approaches [14, 16] seem to solve only the primal problem.

Reorientations [12]. Given a digraph $G = (V, E)$, a positive integer m, and a cost c_e of reversing the direction of e, for each $e \in E$, we want to find a reorientation of G which is m-strongly-edge-connected, and has minimum cost. (A digraph is *m-strongly-edge-connected* if, for every proper nonempty subset F of V, $|\delta(F)| \geqslant m$.) This problem can be formulated as a submodular flow problem as follows. Let $\mathscr{F} = \{F: \emptyset \subset F \subset V\}$, for each $F \in \mathscr{F}$, let $b(F) = |\rho(F)| - m$, and for each $e \in E$, let $l_e = 0$, $u_e = 1$. A feasible integral solution x^0 to the resulting instance of (2.1) corresponds to the desired reorientation by: e is redirected if and only if $x_e^0 = 1$.

To solve (6.1) in this case, we define z by: $z_v = \frac{1}{2}(|\rho(\{v\})| - |\delta(\{v\})|)$, $v \in V$. Clearly, $z(V) = 0$. For any $F \in \mathscr{F}$, $z(F) = \frac{1}{2}(|\rho(F)| - |\delta(F)|)$, which is $\leqslant b(F)$ if and only if $|\rho(F)| + |\delta(F)| \geqslant 2m$. That the latter condition hold for every $F \in \mathscr{F}$, is an obvious necessary condition for the existence of an m-strong-connected reorientation. Moreover, it is easily checked by network flow methods. It is interesting to observe that, as a result of the simple solution to (6.1) we have found, the simplex algorithm will maintain flows which are guaranteed to be only half-integral, so long as any artificial variable remains positive. This does not seem to be a serious disadvantage, and it does not seem to be trivial to find an integral solution to (6.1). Finally, it is not difficult to see that the desired oracle for (2.6) is available via minimum-cut calculations. One oracle call of the kind used in the simplex method requires solving one minimum-cut problem. Alternative techniques for the reorientation problem include the specialization of the primal-dual method [12], as well as a method based on a primal matroid intersection algorithm [14]. This latter approach requires an initial feasible solution, but a refinement [4] overcomes this difficulty. Finally we remark that, from the primal viewpoint, the dicut covering application is equivalent to the case $m = 1$ of the reorientation application [16].

Polymatroid Intersection [6, 7]. For $i = 1$ and 2 suppose that $\mathscr{F}_i$ is an intersecting family of subsets of V, and that b_i is submodular on intersecting pairs of members of $\mathscr{F}_i$. Let $P_i = \{x: x(F) \leqslant b_i(F)$ for all $F \in \mathscr{F}_i\}$, $i = 1$ and 2. Let k be a fixed number and let $P = P_1 \cap P_2 \cap \{x: x(V) = k, l \leqslant x \leqslant u\}$. (Here, as usual, we can allow l, u to have infinite values.) We want to solve the problem $\min(cx: x \in P)$. To formulate this as a submodular flow problem, we make a digraph G having edge-set V and vertex-set $V_1 \cup V_2$ where $V_i = \{v_i: v \in V\}$, with v having head v_1, and tail v_2 for each $v \in V$. For each $F \in \mathscr{F}_1$ we put $F_1 \in \mathscr{F}$ with $b(F_1) = b_1(F)$. For each $F \in \mathscr{F}_2$, we put $V_1 \cup (V_2 \backslash F_2) \in \mathscr{F}$ with $b(V_1 \cup (V_2 \backslash F_2)) = b_2(F)$. Finally, we put $V_2 \in \mathscr{F}$ with $b(V_2) = -k$, and we put $V_1 \in \mathscr{F}$ with $b(V_1) = \min(b_1(V), k)$. (If we wish to drop the constraint $x(V) = k$ from the definition of P, we simply drop these last two constraints from the submodular flow problem.)

To solve (6.1) in this instance, we proceed as follows. Apply the greedy algorithm to find, if possible, a point $z^i \in P_i$ with $z^i(V) = k$ for $i = 1$ and 2. Define $z = (z_v: v \in V_1 \cup V_2)$ by: $z_v = z_v^1$ for $v \in V_1$ and $z_v = -z_v^2$ for $v \in V_2$. The separation problem in this rather general setting is essentially equivalent to submodular function minimization. To consider a more concrete special case, suppose that $\mathscr{F}_1 = \mathscr{F}_2 = \{F: F \subseteq V\}$, $l = 0$, $u = \infty$, k is a positive integer, and b_i is the rank function of a matroid M_i on V for $i = 1$ and 2. Then we have a submodular flow formulation for the optimal k-element common independent set problem for two matroids. For this formulation the oracle (2.5) is actually available via an efficient combinatorial algorithm [3]. However, this algorithm is not needed; because the simplex method will keep (0, 1)-valued flows, the required oracle can be reduced to the problem of testing independence in each of the two matroids. Hence the simplex method provides an alternative to the usual primal–dual algorithms for the matroid intersection problem.

References

[1] W. H. Cunningham, "A network simplex method", *Mathematical Programming* 11 (1976) 105–116.

[2] W.H. Cunningham, "Theoretical properties of the network simplex method", *Mathematics of Operations Research* 4 (1979) 196–208.

[3] W.H. Cunningham, "Testing membership in matroid polyhedra", *Journal of Combinatorial Theory B* 36 (1984) 161–188.

[4] W.H. Cunningham, "Minimum cuts, modular functions, and matroid polyhedra", *Networks*, to appear.

[5] W.H. Cunningham and A. Frank, "A primal-dual algorithm for submodular flows", *Mathematics of Operations Research*, to appear.

[6] J. Edmonds, "Submodular functions, matroids, and certain polyhedra", in: R.K. Guy et al, eds., *Combinatorial structures and applications* (Gordon and Breach, New York, 1970) pp. 69–87.

[7] J. Edmonds and R. Giles, "A min max relation for submodular functions on graphs", *Annals of Discrete Mathematics* 1 (1977) 185–204.

[8] J. Edmonds and R.M. Karp, "Theoretical improvements in algorithmic efficiency for network flow problems", *Journal of ACM* 19 (1972) 248–264.

[9] J. Edmonds and E. Koch, "Simplex methods for optimum matching problems", Technical Report 109, IBM Israel Scientific Center (Haifa, 1981).

[10] L.R. Ford Jr. and D.R. Fulkerson, "A suggested computation for maximal multicommodity flows", *Management Science* 5 (1958) 97–101.

[11] A. Frank, "How to make a digraph strongly connected", *Combinatorica* 1 (1981) 145–153.

[12] A. Frank, "An algorithm for submodular functions on graphs", *Annals of Discrete Mathematics* 16 (1982) 97–120.

[13] A Frank, "Finding feasible vectors of Edmonds-Giles polyhedra", *Journal of Combinatorial Theory B* 36 (1984) 221–239.

[14] A. Frank and E. Tardos, "Matroids from crossing families", Proceedings of Sixth Hungarian Combinatorial Colloquium (Eger, 1981), to appear.

[15] S. Fujishige, "Algorithms for solving the independent flow problems", *Journal of the Operations Research Society of Japan* 21 (1978) 189–203.

[16] S. Fujishige and N. Tomizawa, "An algorithm for finding a minimum-cost strongly connected reorientation of a directed graph", to appear.

[17] M. Grötschel, L. Lovász and A. Schrijver, "The ellipsoid method and its consequences in combinatorial optimization", *Combinatorica* 1 (1981) 169–197.

[18] R. Hassin, "Minimum cost flow with set constraints", *Networks* 12 (1982) 1–21.

[19] Y. Ikura and G. L. Nemhauser, "A polynomial-time dual simplex algorithm for the transportation problem", Technical Report 602, School of OR and IE, Cornell University (Ithaca, N.Y., 1983).

[20] E. Koch, "Ramifications of matching theory", Ph.D. Thesis, University of Waterloo (Waterloo, Ontario, 1979).

[21] E.L. Lawler and C.U. Martel, "Computing maximal polymatroidal network flows", *Mathematics of Operations Research* 7 (1982) 334–347.

[22] C. Lucchesi and D.H. Younger, "A minimax theorem for directed graphs", *Journal of the London Mathematical Society* 17 (1978) 369–374.

[23] J.B. Orlin, "A polynomial-time parametric simplex algorithm for the minimum cost network flow problem", Sloan Working Paper 1484–83, M.I.T. (Cambridge, MA, 1983).

[24] P. Schönsleben, "Ganzzahlige Polymatroid-Intersektions-Algorithmen", Ph.D. Thesis, ETH (Zurich, 1980).

[25] A. Schrijver, "Total dual integrality from directed graphs, crossing families, and sub- and supermodular functions", in: W.R. Pulleyblank, ed., *Progress in combinatorial optimization*, Academic Press (New York, 1984) pp. 315–361.

[26] U. Zimmermann, "Minimization on submodular flows", *Discrete Applied Mathematics* 4 (1982) 303–323.

Mathematical Programming Study 22 (1984) 32-43
North-Holland

UNCAPACITATED LOT-SIZING: THE CONVEX HULL OF SOLUTIONS

Imre BARANY
Mathematical Institute, Budapest, Hungary

Tony Van ROY
CORE, Université Catholique de Louvain, Louvain-la-Neuve, Belgium

Laurence A. WOLSEY
CORE, Université Catholique de Louvain, Louvain-la-Neuve, Belgium

Received 23 April 1983
Revised manuscript received 14 March 1984

It is well-known that the economic lot-sizing model is well-solved by dynamic programming. On the other hand, the standard mixed integer programming formulation of this problem leads to a very large duality gap. Here the convex hull of the solutions of the economic lot-sizing model is given. In addition, an alternative formulation as a simple plant location problem is examined, and here too the convex hull of solutions is obtained.

Key words: Mixed Integer Programming, Lot-Sizing, Polyhedral Theory.

Introduction

The economic lot-sizing problem can be stated like this:

$$\min \sum_{i=1}^{n} (f_i y_i + c'_i x_i + p_i s_i),$$

$$x_i + s_{i-1} = d_i + s_i. \quad i = 1, \dots, n,$$

$$s_0 = 0, \quad s_n = 0,$$

$$x_i, s_i \geq 0,$$

$$y_i \in \{0, 1\}, \quad y_i = 1 \text{ if } x_i > 0, \quad i = 1, \dots n,$$

where x_i, s_i represent production level and stock in period i, y_i indicates whether a set-up cost must be incurred in period i, $d_i \geq 0$, p_i, c'_i and f_i are the demand, storage, production and set-up costs, respectively.

In order to avoid uninteresting special cases we assume throughout the paper that $d_i > 0$ $(i = 1, \dots, n)$.

The variables s_i can clearly be eliminated from the above formulation. Using the notation $N=\{1,\ldots,n\}$ and $d_{st}=\sum_{i=s}^{t} d_i$, we define the set

$$X=\{(x,y)\in\mathbb{R}^{2n}: \sum_1^t x_i\geq d_{1t}\ (t\in N), \sum_1^n x_i=d_{1n},$$
$$x_i\geq 0\ (i\in N),\ y_i\in\{0,1\},\ y_i=1 \text{ if } x_i>0\ (i\in N)\}.$$

Then our problem becomes

$$\min\left\{\sum_1^n (f_i y_i+c_i x_i): (x,y)\in X\right\}$$

with the c_i different from the above c_i'. It is well known [6] that this problem can be solved by a dynamic programming $O(n^2)$ algorithm. Moreover, it is also well known that for any objective function, there is an optimal solution $(x,y)\in X$ having a special structure which can be described as follows: Let $I=\{i_1=1<i_2<\cdots<i_k\leq n\}$ and $I\subseteq J\subseteq N$. We associate with each such pair a point $(x,y)=(x,y)_{I,J}\in X$ by

$$x_i=\begin{cases}0 & \text{if } i\notin I,\\ d_{i_j}+d_{i_j+1}+\cdots+d_{i_{j+1}-1} & \text{if } i=i_j\in I,\end{cases}$$

where $i_{k+1}-1=n$ by definition, and

$$y_i=\begin{cases}1 & \text{if } i\in J,\\ 0 & \text{otherwise.}\end{cases}$$

Write X^* for the set of these specially structured points of X. The aim of this paper is to find a description of conv X^* by a set of linear inequalities. The practical usefulness of such a description is to give stronger mixed integer programming fomulations of classes of problems with embedded lot sizing. In a companion paper [1] computational results for the class of multi-item capacitated lot-sizing problems are given and a simple separation algorithm for conv X^* is presented.

It is clear that any point $(x,y)\in X^*$ satisfies the relations

$$x_i\geq 0\quad (i\in N), \tag{1a}$$

$$1\geq y_i\geq 0\quad (i\in N), \tag{1b}$$

$$y_1=1, \tag{1c}$$

$$\sum_{i=1}^n x_i=d_{1n}. \tag{1d}$$

Let $l\in N$ and $S\subseteq L=\{1,2,\ldots,l\}$. As has been shown in [1], the so-called (S,l) inequality

$$\sum_{i\in S} x_i+\sum_{i\in L\setminus S} d_{il}y_i\geq d_{1l} \tag{2}$$

is a valid inequality for X^*, and under mild assumptions defines a facet of conv X^*.

Now we can formulate the main result of this paper.

Theorem 1. *The* (S, l) *inequalities plus the inequalities* (1a), ..., (1d) *describe the convex hull of* X^*.

2. Further results

In this section we indicate in somewhat more detail the contents of the paper. In particular, we outline the steps in our proof of Theorem 1, as this shows the close connection to an alternative formulation of the economic lot-sizing model, namely to a simple plant location problem:

$$\begin{aligned} &\min \left(\sum_{i=1}^{n} f_i y_i + \sum_{i=1}^{n} \sum_{l=i}^{n} c_i d_l x_{il} \right), \\ &\sum_{i=1}^{l} x_{il} = 1 \quad (l \in N), \\ &y_i - x_{il} \geq 0 \quad (i, l \in N, i \leq l), \\ &-y_i \geq -1 \quad (i \in N), \\ &x_{il}, y_i \geq 0 \quad (i, l \in N, i \leq l) \end{aligned} \tag{PL}$$

where x_{il} is the fraction of the demand of period l produced in period i. This simple plant location problem (PL) has integral solutions in x and y because of the special structure of the objective function. At the end of this section we state a theorem that shows what inequalities should be added to the constraints of (PL) so that each vertex of the corresponding feasible set be integral.

Using Theorem 1 we can give the set of facets of conv X^*. First we mention that (1d) and the (S, l) inequality with $l = n$ and $S = N \backslash \{i\}$ imply

$$-x_i + d_{in} y_i \geq 0 \quad (i \in N), \tag{3}$$

which, in turn, implies each (S, l) inequality with $l = n$ (using (1d) again). They also imply the inequalities $y_i \geq 0$ $(i \in N)$. Moreover, the (S, l) inequality with $S = L$ is a consequence of the (S, l) inequality $\sum_1^l x_i + d_{l+1} y_{l+1} \geq d_{1l+1}$ and $1 \geq y_{l+1}$. Finally, the (S, l) inequality with $1 \notin S$ follows from (1a), (1b) and (1c).

Theorem 2. *The* (S, l) *inequalities with* $l < n$ *and* $1 \in S \subsetneqq L$, *the inequalities* (3), *plus the inequalities* $x_i \geq 0$, $1 \geq y_i$ $(i = 2, \ldots, n)$ *determine distinct facets of* conv X^*. *Moreover, each facet of* conv X^* *is determined by one of these inequalities.*

Our proof of Theorem 1 is based on the following mixed integer programming reformulation of the original problem:

$$\begin{aligned}
&\min \sum_1^n (f_i y_i + c_i x_i), \\
&\sum_{i=1}^{l} q_{il} = d_l \quad (l \in N), \\
&x_i - \sum_{l=i}^{n} q_{il} = 0 \quad (i \in N), \\
&d_l y_i - q_{il} \geq 0 \quad (i, l \in N, i \leq l), \\
&y, x, q \geq 0, \\
&y_i \in \{0, 1\} \quad (i \in N).
\end{aligned} \tag{R}$$

Here q_{il} can be thought of as production in period i to satisfy demand in period l, x_i, y_i are as before. Each point $(x, y) \in X^*$ gives rise to a canonically determined point (x, y, q) where

$$q_{il} = \begin{cases} 0 & \text{if } i \notin I \text{ or } i = i_j \in I \text{ and } l \geq i_{j+1} \\ d_l & \text{if } i = i_j \in I \text{ and } l < i_{j+1}. \end{cases}$$

Let $Q = \{(x, y, q)\colon (x, y, q)$ is feasible for (R)$\}$ and let Q^{LP} and $(\mathrm{R}^{\mathrm{LP}})$ denote the feasible set and linear program respectively obtained by replacing the integrality constraint on y_i by $1 \geq y_i \geq 0$ $(i \in N)$.

The importance of the reformulation (R) lies in the following result:

Theorem 3. *For all choices of c and f,* R^{LP} *has an optimal solution* (x^*, y^*, q^*) *with* $(x^*, y^*) \in X^*$.

This result was essentially shown by Krarup and Bilde [5] when looking at the closely related simple plant reformulation (PL). It can also be proved by writing the canonical covering problem associated with this simple plant reformulation and observing that the underlying matrix is totally balanced, see Kolen [4]. Our proof of Theorem 3 is based on finding the explicit primal and dual solutions to R^{LP}.

We then prove Theorem 1 by decomposing the dual solution.

We mention here that, having proved Theorem 3 one can easily find a set of inequalities whose solutions form the set conv X^*. This can be done using the following observations:

Proposition 4. conv $X^* = \{(x, y) \in \mathbb{R}^{2n}\colon \exists q$ *such that* $(x, y, q) \in Q^{\mathrm{LP}}\}$.

Proposition 5. $(x, y) \in \{(x, y) \in \mathbb{R}^{2n}\colon \exists q$ *such that* $(x, y, q) \in Q^{\mathrm{LP}}\}$ *if and only if* $x_i \geq 0$, $1 \geq y_i \geq 0$ $(i \in N)$ *and there exists a feasible flow in the following transportation network (see Fig. 1) with inflows equal to* x_i, *outflows equal to* d_i *and capacities* $d_l y_i$ *on the* (i, l) *arc* $(i, l \in N, i \leq l)$.

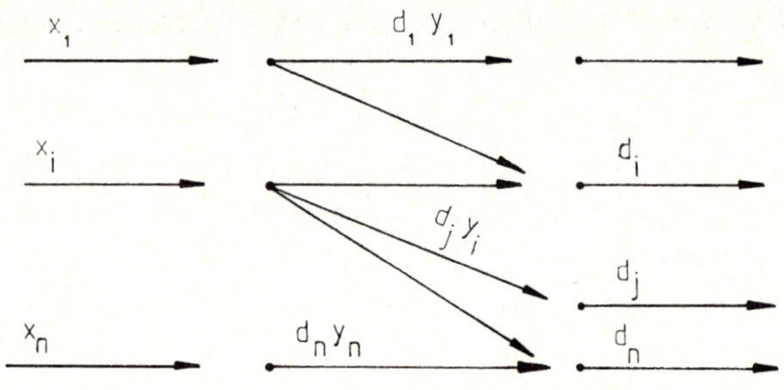

Fig. 1.

The necessary and sufficient conditions for such a flow to exist are well known (see Ford and Fulkerson [3]): There is a feasible flow if and only if each cut has capacity at least d_{1n}. So these cut-inequalities give another description of conv X^*.

It is easily checked that the (S, l) inequalities correspond to some cuts. But there are many more cuts than (S, l) inequalities. In this sense, the main content of Theorem 1 is that the essential cuts are just those corresponding to the (S, l) inequalities.

We indicate another proof of Theorem 1 here. This proof-method may be useful in other cases as well. For any fixed cut C consider the set of those points $X_C^* \subset X^*$ that satisfy the inequality corresponding to cut C with equality. If we exhibit an (S, l) inequality (where $l = l(C)$ and $S = S(C)$), which is satisfied as equality by all points in X_C^*, then we are done; this particular (S, l) inequality is stronger than the cut-inequality corresponding to C. We state without proof that it is indeed possible to find this (S, l) inequality using the special structure of X^*. A third unusual proof is presented in the original version of this paper [2].

There is one more point here which seems to be worth mentioning. Theorem 3 holds because of the special structure of the objective function (the coefficient of each q_{ij} is zero). This means that some vertices of Q^{LP} have fractional y-coordinates. By adding some extra constraints ('new cuts') one can get rid of these bad vertices. To be more precise, consider the simple plant location problem (PL), which is obtained from (R) by substituting $x_{il} = (1/d_l)q_{il}$. Again, the feasible set of this linear program has vertices fractional in y. The extra constraints needed are just $x_{il} - x_{il+1} \geq 0$ $(i, l \in N,\ i \leq l < n)$.

We state this theorem more generally. Let (V, E) be a rooted tree with root v_0 directed so that each node of V can be reached from v_0 by a directed path. We write $v < u$ $(v, u \in V)$ if v lies on the unique path between v_0 and u. In particular $u < u$, and $(v, u) \in E$ implies that $v < u$. Consider the following inequalities: with variables $x(v, u)$ for each pair $v, u \in V$, $v < u$

$$\sum_{v: v < u} x(v, u) = 1 \quad (\forall u \in V),$$

$$x(v, u) - x(v, u') \geq 0 \quad (\forall v, u, u' \in V,\ v < u < u',\ (u, u') \in E), \tag{T}$$

$$x(v, u) \geq 0 \quad (\forall u, v \in V,\ v < u)$$

Theorem 6. *Each vertex of the polytope defined by inequalities* (T) *is integral.*

3. Proofs

Before proving Theorem 3 we need to derive certain properties of the optimal solution values for the model. Let $G(l)$ be the optimal solution value of the problem

$$\min\left\{\sum_{i=1}(f_i^+ y_i + c_i x_i): (x, y) \in X_l\right\},$$

where

$$X_l = \left\{(x, y) \in \mathbb{R}^{2l}, \sum_{i=1}^{t} x_i \geq d_{1t}, t \in L, \sum_{i \in L} x_i = d_{1l},\right.$$

$$\left. x_i \geq 0,\ y_i \in \{0, 1\},\ y_i = 1 \text{ if } x_i > 0,\ i \in L\right\}.$$

and we use the notation $z^+ = \max(z, 0)$, $z^- = \min(z, 0)$. Wagner and Whitin [6] gave an $O(n^2)$ algorithm to calculate $G(1), \ldots, G(n)$ by dynamic programming. We set $G(0) = 0$.

Proposition 7. (i) *For* $i \leq k$, $G(k) \leq G(i-1) + f_i^+ + c_i d_{ik}$,

$$\text{(ii)} \qquad \frac{G(l) - G(l-1)}{d_l} \geq \frac{G(l+1 - G(l)}{d_{l+1}} \qquad (l = 1, \ldots, n-1).$$

Proof. (i) Given the optimal solution for the first $(i-1)$ periods, add to it $y_i = 1$, $x_i = d_{ik}$, $y_j = x_j = 0$ $(j = i+1, \ldots, k)$. This is a feasible solution in X_k with value $G(i-1) + f_i^+ + c_i d_{ik}$, and the result follows.

(ii) Suppose the demand d_l for period l is produced in period $i \leq l$. Then a feasible solution for $G(l+1)$ is to use the same solution and also produce the demand for period $l+1$ in period i. Hence:

$$G(l+1) \leq G(l) + c_i d_{l+1}.$$

Now consider the optimal solution for the first $(l-1)$ periods. If $i < l$, the optimal solution for l periods provides a feasible solution and hence $G(l-1) \leq G(l) - c_i d_l$. If $i = l$, $G(l) - G(l-1) = f_i^+ + c_i d_l \geq c_i d_l$. Therefore, in all cases

$$\frac{G(l) - G(l-1)}{d_l} \geq c_i \geq \frac{G(l+1) - G(l)}{d_{l+1}}$$

for some i, and the result follows. □

After these preparations we are able to prove Theorem 3.

Proof of Theorem 3. To construct a primal feasible solution to $(\mathrm{R^{LP}})$, we can use the Wagner–Whitin dynamic programming algorithm to obtain a solution $(x^*, y^*) \in X^*$ with value $z = G(n) + \sum_1^n f_i^-$. We then calculate the canonical q^* and take (x^*, y^*, q^*) as our primal feasible solution. Below we exhibit a feasible solution to the dual of $(\mathrm{R^{LP}})$ having value z. This shows that the solutions are optimal, and proves the theorem. The dual to (R^{LP}) is

$$\begin{aligned} \max \sum_{i\in N} d_i v_i \qquad & - \sum_{i\in N} z_i, \\ \sum_{l\geq i} d_l w_{il} - \quad & z_i \leq f_i \quad (i\in N), \\ u_i \quad & \leq c_i \quad (i \in N), \\ v_l - u_i - w_{il} \quad & \leq 0 \quad (i, l \in N, i \leq l), \\ & w, z \geq 0. \end{aligned}$$

We claim that

$$v_l = \frac{G(l)-G(l-1)}{d_l}, \quad u_i = c_i, \; w_{il} = |v_l - c_i|^+, \; z_i = -f_i^-$$

is an optimal dual solution. Clearly, the objective value

$$\sum_1^n d_i v_i - \sum_1^n z_i = G(n) + \sum_1^n f_i^- = z.$$

To check dual feasibility we just need to show that

$$\sum_{l\geq i} d_l w_{il} \leq f_i + z_i = f_i^+.$$

Now by Proposition 7(ii), $(G(l)-G(l-1))/d_l$ is nonincreasing and hence $w_{il}=0$ implies $w_{it}=0$ for $t \geq l$. Thus for a given i either $w_{il}=0$ for $l \geq i$ or there exists $k \geq i$ such that $w_{il} > 0$ for $l = i, \ldots, k$ and $w_{il} = 0$ for $l > k$. In the first case $\sum_{l\geq i} d_l w_{il} = 0 \leq f_i^+$. In the second

$$\sum_{l\geq i} d_l w_{il} = \sum_{l=i}^{k} (G(l) - G(l-1) - c_i d_l) = G(k) - G(i-1) - c_i d_{ik},$$

and this is not greater than f_i^+ by Proposition 7(i). □.

Now we use this proof to obtain the main result.

Proof of Theorem 1. Consider the following linear program:

$$\begin{aligned} & \min \sum_{i\in N} (f_i y_i + c_i x_i) \\ & \quad -y_i \geq -1 \quad (i \in N), \\ & \sum_{i\in N} x_i = d_{1n}, \qquad \text{(P)} \\ & \sum_{i\in S} x_i + \sum_{i\in L\setminus S} d_{il} y_i \geq d_{1l} \quad (S \subseteq L, l \in N), \\ & y_i \geq 0, \; x_i \geq 0 \quad (i \in N). \end{aligned}$$

For the proof it clearly suffices to find an optimal solution (x^*, y^*) to (P) with $(x^*, y^*) \in X^*$. This is found by the Wagner–Whitin algorithm. Its optimality for (P) will be shown by exhibiting an optimal dual solution to (P).

We suppose, without loss of generality, that $c_i \geqslant 0$ $(i \in N)$, because by the equation $\sum_{i\in N} x_i = d_{1n}$, replacing c_i by $c_i + M$ in the objective function does not change the set of optimal solutions to (P).

Writing the dual of (P) gives

$$\max - \sum_{i\in N} z_i + u_0 d_{1n} + \sum_{l} \sum_{S\subseteq L} d_{1l} u(S, l),$$

$$u_0 + \sum_{l} \sum_{i\in S\subseteq L} u(S, l) \leqslant c_i \quad (i \in N)$$

$$-z_i + \sum_{l} \sum_{i\in L\setminus S} d_{il} u(S, l) \leqslant f_i \quad (i \in N).$$

$$z_i \geqslant 0 \quad (i \in N),$$

$$u(s, l) \geqslant 0 \quad (l \in N, S \subseteq l).$$

In finding the optimal dual variables we use the optimal dual solution to $(\mathrm{R}^{\mathrm{LP}})$ from the proof of Theorem 3. Namely, we set $z_i = -f_i^-$ $(i \in N)$, and we are looking for feasible u_0 and $u(S, l)$ such that

$$\sum_{S\subseteq L} u(S, l) = v_l - v_{l+1} \quad (l = 1, \ldots, n-1), \qquad u_0 + \sum_{S\subseteq N} u(S, n) = v_n \tag{4}$$

where $v_l = (G(l) - G(l-1))/d_l$ is again taken from Theorem 3. If this holds, then the value of the objective function is

$$-\sum_{i\in N} z_i + \sum_{1}^{n-1} d_{1l}(v_l - v_{l+1}) + d_{1n} v_n = -\sum_{i\in N} z_i + \sum_{i\in N} d_i v_i = G(n) + \sum_{i\in N} f_i^- = z,$$

and hence it coincides with the value of the primal solution.

So it suffices to find feasible u_0 and $u(S, l)$ satisfying (4). We distinguish two cases:

Case 1. $v_n \geqslant 0$. Take v_l as above. Define

$$v_{il} = |v_l - c_i|^+ - |v_{l+1} - c_i|^+ \quad \text{for } 1 \leqslant i \leqslant l < n,$$

$$v_{in} = |v_n - c_i|^+ \quad \text{for } i \in N,$$

$$u_{il} = v_l - v_{l+1} - v_{il} \quad \text{for } 1 \leqslant i \leqslant l < n,$$

$$u_{in} = v_n - v_{in} \quad \text{for } i \in N,$$

$$u_0 = 0.$$

We are going to find $u(S, l)$ satisfying

$$\sum_{i\in S\subseteq L} u(S, l) = u_{il} \quad \text{for } 1 \leqslant i \leqslant l \leqslant n,$$

$$\sum_{i\in L\setminus S} u(S, l) = v_{il} \quad \text{for } 1 \leqslant i \leqslant l \leqslant n. \tag{5}$$

If we can do so, then – as it is easily checked – (4) will hold. Moreover, z_i, u_0, $u(S, l)$ is then feasible for the dual because

$$u_0+\sum_l \sum_{i\in S\subseteq L} u(S,l) = \sum_{l\geqslant i} u_{il}$$

$$= \sum_{l=i}^{n-1} (v_l - v_{l+1} - [|v_l - c_i|^+ - |v_{l+1} - c_i|^+]) + (v_n - |v_n - c_i|^+)$$

$$= v_i - |v_i - c_i|^+ \leqslant c_i,$$

and

$$-z_i + \sum_l \sum_{i\in L\setminus S} d_{il} u(S,l) = -z_i + \sum_{l=i}^{n} d_{il} v_{il}$$

$$= -z_i + \sum_{l=i}^{n-1} d_{il}(|v_l - c_i|^+ - |v_{l+1} - c_i|^+) + d_{in}|v_n - c_i|^+$$

$$= -z_i + \sum_{l=i}^{n} d_l |v_l - c_i|^+ \leqslant f_i$$

where the last inequality is justified at the end of the proof of Theorem 3.

So it suffices to find $u(S, l) \geqslant 0$ satisfying (5). Notice that u_{il}, $v_{il} \geqslant 0$ and $u_{il} + v_{il} = v_l - v_{l+1} \geqslant 0$ (for $i \leqslant l$). Hence it is enough to find them for fixed l. This is done by the following

Algorithm to find $u(S, l)$ for fixed $l \in N$ satisfying (5). Initially set $u(S, l) \leftarrow 0$ $(\forall S \subseteq L)$.

Step 1. If $u_{1l} + v_{1l} = 0$, then STOP, else set

$$\Delta \leftarrow \min[\{\min u_{il}\colon u_{il} > 0\ i \in L\}, \{\min v_{il}\colon v_{il} > 0\ i\in L\}]$$

Case 1. $\Delta = u_{il}$ for some $i \leqslant l$.
Set $T = \{i\colon u_{il} \geqslant \Delta\}$.
Set

$$u_{il} \leftarrow u_{il} - \Delta, \quad i \in T, \qquad v_{il} \leftarrow v_{il} - \Delta, \quad i \in L\setminus T.$$

Set $u(T, l) \leftarrow \Delta$, and go to Step 1.
Case 2. $\Delta = v_{il}$ for some $i \leqslant l$.
Set $T = \{i\colon v_{il} \geqslant \Delta\}$.
Set

$$v_{il} \leftarrow v_{il} - \Delta, \quad i \in T, \qquad u_{il} \leftarrow u_{il} - \Delta, \quad i \in L\setminus T.$$

Set $u(L\setminus T, l) \leftarrow \Delta$, and go to Step 1.

This algorithm produces the desired $u(S, l)$ after at most $(l+1)$ iterations. This follows because the set $\{i \in L\colon u_{il} = 0$ or $v_{il} = 0\}$ increases by at least one at each iteration. Hence after l iterations, the remaining nonzero entries are all equal, and the algorithm terminates after the next iteration.

Case 2. $v_n < 0$. Define v_{il}, u_{il} for $1 \le i \le l < n$ in the same way as before and

$$v_{in} = u_{in} = 0 \quad (i \in N), \qquad u_0 = v_n.$$

Again, we are looking for $u(S, l)$ satisfying (5) for $1 \le i \le l \le n$. What has to be checked is feasibility. In this case:

$$u_0 + \sum_l \sum_{i \in S \subseteq L} u(S, l) = v_n + (v_i - |v_i - c_i|^+) - (v_n - |v_n - c_i|^+)$$

$$= v_i - |v_i - c_i|^+ \le c_i$$

because $v_n < 0$ and $c_i \ge 0$ $(i \in N)$ by our assumption.

The rest is the same as in Case 1. □

For the Proof of Theorem 2 we mention that its first part was proved in the companion paper [1] and the second part is justified by the discussion preceding Theorem 2.

Proof of Theorem 4. Suppose $x = (x(u, v), u, v \in V, u < v)$ is a fractional point of T where T denotes the polytope determined by (T). To prove the theorem we are going to find two distinct points x^1, $x^2 \in T$ with $x = \frac{1}{2}x^1 + \frac{1}{2}x^2$.

For each $u \in V$ we define the set S_u in the following way:

$$u \in S_u \quad \text{if } 0 < x(u, u) < 1,$$

$$v \in S_u \quad \text{if } v > u,\ v \ne u,\ x(u, v) > 0 \text{ and } x(u, v) < x(u, v')$$

where v' is the predecessor of v on the tree.

For each pair $u \in V$ and $v \in S_u$ we define the set C_{uv} by

$$w \in C_{uv} \text{ if } w > v \text{ and } x(u, v) = x(u, w).$$

So if $w \in C_{uv}$, then, for each node $z \in V$ on the unique path connecting v and w, $0 < x(u, v) = x(u, z) = x(u, w) < 1$. This shows that C_{uv} is a subtree with smallest (with respect to $<$) node v. Furthermore, $\bigcup_{v \in S_u} C_{uv}$ is the set of those $w \in V$, $w > u$ with $0 < x(u, w) < 1$.

Now let us define new variables $\xi(u, v)$ for each pair u, v with $v \in S_u$ and consider the following system of equations:

$$\sum \{\xi(u, v) \colon w \in C_{uv}\} = 0 \qquad (E(w)).$$

Claim. *This system of equations has a nontrivial solution.*

Suppose for a moment that the claim has been proved. Then the theorem follows. Let ξ be a small enough solution and set

$$x^1(u, w) = x^2(u, w) = x(u, w) \quad \text{if } x(u, w) = 0 \text{ or } 1,$$

$$x^1(u, w) = x(u, w) + \xi(u, v) \text{if } w \in C_{uv},$$

$$x^2(u, w) = x(u, w) - \xi(u, v) \text{if } w \in C_{uv}.$$

Clearly $x = \frac{1}{2}x^1 + \frac{1}{2}x^2$ and $x^i \in T$ $(i = 1, 2)$ because $\sum_{u<w} x^i(u, w) = 1$ and, for $u < w_1 < w_2$,

$$x^i(u, w_1) = x^i(u, w_2) \quad \text{if } w_1, w_2 \in C_{uv},$$

$$x^i(u, w_1) > x^i(u, w_2) \quad \text{if } w_1 \in C_{uv} \text{ and } w_2 \notin C_{uv}.$$

Proof of the Claim. Equation $E(w)$ may be trivial (i.e., saying that $0 = 0$) and this is the case iff $w \notin C_{uv}$ for any u and $v \in S_u$. (This simply means that $x(u, w) = 0$ or 1 for each $u \in V$, $u < w$.) Deleting these trivial equations we have a set $V' \subset V$ such that $E(w)$ $(w \in V')$ is nontrivial. This means that for each $w \in V'$ $x(u, w)$ is fractional for some $u \in V$.

Define now an ordering $v_1, v_2, \ldots, v_n$ of V' in such a way that $i < j$ implies $v_i \not< v_j$. Such an ordering clearly exists.

Consider $v_i \in V'$. Then $x(u, v_i)$ is fractional for some $u \in V$. There are two cases:

Case 1. $v_i \in S_u$ for some $u \in V$

Case 2. $v_i \notin S_u$ for any $u \in V$.

In Case 1 $\xi(u, v_i)$ appears in equation $E(v_i)$ but does not appear in any equation $E(v_j)$, $j > i$.

In Case 2, consider the predecessor on the tree of v_i. Clearly this is a $v_j \in V'$ with $j > i$. We claim that equations $E(v_i)$ and $E(v_j)$ are the same. To see this consider u, v with $v \in S_u$ and $v_i \in C_{uv}$. Then $v_j \in C_{uv}$ by definition of C_{uv} and $v_i \notin S_u$. Consider the two equations from (T)

$$\sum_{u<v_i} x(u, v_i) = 1 \quad \text{and} \quad \sum_{u<v_j} x(u, v_j) = 1.$$

Clearly each variable here is less than 1 (because there are fractional variables). So the implication $v_i \in C_{uv} \Rightarrow v_j \in C_{uv}$ shows that

$$\sum_{u<v_i} x(u, v_i) \leq \sum_{u<v_j} x(u, v_j).$$

But equality must hold here. This shows, in turn, that $v_j \in C_{uv}$ implies $v_i \in C_{uv}$. Hence equations $E(v_i)$ and $E(v_j)$ are identical.

Now we can prove the Claim. Consider the equations $E(v_1), E(v_2), \ldots, E(v_n)$ one-by-one in this order. If $v_i \in S_u$ for some u, then $E(v_i)$ contains a variable ($\xi(u, v_i)$) which will not come up again. If not, then equation $E(v_i)$ can be dropped from the system because it comes up again as some $E(v_j)$, $j > i$. So each remaining equation contains 'its own variable'. We prove finally that $E(v_n)$ contains at least two variables (they will clearly be different from the previously fixed variables $\xi(u, v_i)$ $(i < n)$). This is so because otherwise the equation $\sum_{u<v_n} x(u, v_n) = 1$ could contain only one fractional variable and so could not hold. □

References

[1] I. Barany, T.J. Van Roy and L.A. Wolsey, "Strong formulations for multi-item capacitated lot-sizing", CORE Discussion Paper No. 8312, Université Catholique de Louvain, Louvain-la-Neuve, Belgium, March 1983.

[2] I. Barany, T.J. Van Roy and L.A. Wolsey, "Uncapacitated lot-sizing: The convex hull of solutions: CORE Discussion Paper No. 8314, Université Catholique de Louvain, Louvain-la-Neuve, Belgium, March 1983.

[3] L.R. Ford and D.R. Fulkerson, *Flows in networks* (Princeton University Press, Princeton, NJ, 1962).

[4] A. Kolen, "A polynomial time algorithm for solving the set covering problem on a totally balanced matrix", Report BW 147/81, Mathematisch Centrum, Amsterdam, 1981.

[5] J. Krarup and O. Bilde, "Plant location, set covering an economic lot size: An O(*mn*) algorithm for structural problems", in: *Numerische Methoden bei Optimierungsaufgaben, Band* 3: *Optimierung bei grapentheoretische und ganzzahligen Problemen*, International Series of Numerical Mathematics (ISNM), Vol. 36 (Birkhäuser-Verlag, Basel und Stuttgart, 1977).

[6] H.M. Wagner and T.M. Whitin, "Dynamic version of the economic lot size model", *Management Science* 5 (1958) 89–96.

[7] W.I. Zangwill, "A deterministic multi-period production scheduling model with backlogging", *Management Science* 13 (1966) 105–119.

Mathematical Programming Study 22 (1984) 44–63
North-Holland

NODE-WEIGHTED GRAPHS HAVING THE KÖNIG–EGERVÁRY PROPERTY

J.-M. BOURJOLLY
University of Waterloo, Ontario, Canada

P.L. HAMMER
Rutgers University, New Brunswick, NJ, USA

B. SIMEONE
University of Rome, Italy

Received 5 July 1983
Revised manuscript received 7 May 1984

A graph G with positive integral node-weights b_i is said to be a b-König–Egerváry graph (b-KEG for short) if there exist a transversal T and a b-matching λ such that the weight of T is equal to the value of λ; the usual König–Egerváry graphs correspond to $b = (1, 1, \ldots, 1)$.

Several characterizations and two polynomial recognition algorithms for b-KEGs are presented. Algorithm 1 either recognizes (G, b) as being a b-KEG, or else exhibits an 'obstruction'. Algorithm 2 associates a quadratic boolean function f with (G, b) and reduces the problem of recognizing whether (G, b) is a b-KEG to that of solving the equation $f = 0$. In case (G, b) is a b-KEG, both algorithms provide a minimum weight transversal.

Some of our results generalize previous results obtained in the unweighted case by Deming, Gavril and Sterboul.

A companion paper will analyze applications of the weighted König–Egerváry property to quadratic 0–1 optimization.

Key words: Graphs, b-Matchings, Transversals, Stable Sets, Quadratic Boolean Equations, Algorithms.

1. Introduction

Let $G = (V, E)$ denote a finite, undirected graph without loops and multiple edges, which is assumed to be connected. V is the set of vertices (or nodes) of G and E is the set of edges. Let $|V|$, the cardinality of V, be n and $|E|$ be m. Let b denote an n-vector of positive integral weights associated with the vertices of G. When all the b_j's are equal, the graph is said to be *unweighted.* A *stable set* $S \subseteq V$ is a set of pairwise nonadjacent vertices of G. A *transversal* $T \subseteq V$ is a set of vertices such that each edge has at least one endpoint in T. Finding a stable set of maximum weight or, equivalently, a transversal of minimum weight are known to be NP-complete problems [4]. They, however, can be solved in polynomial time in some particular cases. The purpose of this paper is to study a class of weighted graphs that can be recognized in polynomial time and for which the aforementioned problems can also be solved in polynomial time.

It is well known that the above problems can be expressed as linear binary programs.

The maximum weight stable set problem (ISP)

$$\begin{aligned} \max \quad & \sum_{i=1}^{n} b_i x_i \\ \text{subject to} \quad & x_i + x_j \leq 1 \text{ for every edge } (i, j), \\ & x_i \in \{0, 1\} \text{ for } i = 1, 2, \ldots, n. \end{aligned} \tag{1.1}$$

The minimum weight transversal problem (ITP)

$$\begin{aligned} \min \quad & \sum_{i=1}^{n} b_i x_i \\ \text{subject to} \quad & x_i + x_j \geq 1 \text{ for every edge } (i, j), \\ & x_i \in \{0, 1\} \text{ for } i = 1, 2, \ldots, n. \end{aligned} \tag{1.2}$$

Replacing the constraints $x_i \in \{0, 1\}$ by $x_i \geq 0$, one obtains the continuous relaxation of ISP (denoted by CSP) and of ITP (denoted by CTP). Note that the constraints $x_i \leq 1$ are certainly satisfied by all feasible solutions to CSP and by all optimal solutions to CTP. The feasible solutions to CSP and CTP are called *fractional stable sets* and *fractional transversals* of G, respectively.

The *b-matching* problem, denoted by WMP, is

$$\begin{aligned} \max \quad & \sum_{(i,j) \in E} \lambda_{ij} \\ \text{such that} \quad & \sum_{j \in N(i)} \lambda_{ij} \leq b_i \text{ for } i = 1, 2, \ldots, n, \\ & \lambda_{ij} \geq 0 \text{ and integral for every edge } (i, j). \end{aligned} \tag{1.3}$$

where $N(i)$ denotes the set of vertices that are adjacent to vertex i.

Note. We regard each component λ_{ij} as being indexed by the *unordered* pair $(i, j) \in E$.

The continuous relaxation CWMP of WMP (obtained by dropping the integrality constraints from (1.3)) is precisely the dual linear program of CTP. The feasible solutions to WMP and CWMP are called *b-matchings* and *fractional b-matchings*, respectively. The *value* of a fractional b-matching λ is $\sum_{(i,j) \in E} \lambda_{ij}$. When $\sum_{j \in N(i)} \lambda_{ij} = b_i$, node i is said to be *saturated* by λ. When all nodes are saturated by λ, λ is said to be *perfect*. Finally, we shall call an edge (i, j) *active* (in λ) if $\lambda_{ij} > 0$; else (i, j) is called *passive*.

Let us denote by $v(\text{P})$ the optimal value of an optimization problem P. One has

$$v(\text{WMP}) \leq v(\text{CWMP}) = v(\text{CTP}) \leq v(\text{ITP}). \tag{1.4}$$

The following theorem, whose proof relies on the 'persistency' results of [7], yields a useful identity for the 'integrality gap' $v(\text{ITP}) - v(\text{CTP}) = v(\text{ITP}) - v(\text{CWMP})$.

Theorem 1.1. *Let (G, b) be a weighted graph, λ any maximum fractional b-matching, T any minimum weight transversal of G. They every node of T is saturated by λ, and*

$$v(\text{ITP}) - v(\text{CWMP}) = \sum_{i \text{ and } j \in T} \lambda_{ij}. \tag{1.5}$$

Proof. CWMP is the dual of CTP and by the complementary slackness theorem any pair (x, λ) of optimal solutions is such that

$$x_i > 0 \Leftrightarrow \sum_{j \in N(i)} \lambda_{ij} = b_i.$$

Let λ be any maximum fractional b-matching and T any minimum weight (integral) transversal. Let

$$\tilde{x}_i = \begin{cases} 1 & \text{if } i \in T, \\ 0 & \text{if } i \notin T. \end{cases}$$

Let $i \in T$. Since $\tilde{x}_i = 1$ there is at least one optimal solution x of CTP such that $x_i > 0$, for otherwise one would have $x_i^* = 0$ for all optimal solutions x^* of ITP [7]. By complementary slackness, node i must be saturated by λ. Identity (1.5) follows easily. □

In case G is unweighted and bipartite, the König–Egerváry Theorem states that there are an (integral) matching and a transversal in G having the same cardinality (see e.g. [2, Section 7.3]). This result extends easily to the bipartite weighted case. Harary and Plummer gave a structural characterization of the unweighted graphs that have the *König-Egerváry property*, i.e. the graphs for which there exist an (integral) matching and a transversal having the same cardinality (see [8, Section 10]). Deming [3] found an $O(nm)$ time algorithm for testing whether or not an unweighted graph has the König–Egerváry property. The algorithm also finds a minimum transversal and a maximum stable set in polynomial time. An $O(m+n)$ algorithm has been given by Gavril [5]. In [14] it is shown that an unweighted graph has the König–Egerváry property if and only if a specific quadratic boolean equation is consistent. In the present paper the weighted case is studied. Section 2.1 provides several alternative characterizations of the so-called b-König–Egerváry graphs (b-KEGs, for short), i.e. those graphs for which the optimal values of WMP and ITP coincide. In Section 2.2 it is shown that a graph does not have the b-K–E property if and only if there exists a certain 'obstruction' (namely an 'alternating bicycle'). Our proof makes an essential use of Algorithm 1, which is described in Section 3. Section 2.4 provides a spanning forest characterization of b-KEGs. Algorithm 2 in Section 3 is based on the reduction of the recognition of a b-KEG to the recognition of a KEG (i.e. an unweighted graph having the K–E property) or, equivalently, to the solution of a quadratic boolean equation.

A generalization of b-KEGs has been studied in [13].

2. Study of the *b*-KEGs

2.1. Definitions and general properties

In the present paragraph we collect a few elementary facts about weighted König–Egerváry graphs. Some of these results will be exploited in later paragraphs.

Definition 2.0. A weighted graph (G, b) is said to have the *König–Egerváry property* (or to be a b-KEG) when it has a b-matching λ and a transversal T such that the value of λ is equal to the weight of T, i.e. when $v(\text{WMP}) = v(\text{ITP})$. A $(1, \ldots, 1)$-KEG is just called a KEG.

Thus when (G, b) is a b-KEG the inequalities in (1.4) become equalities

$$v(\text{WMP}) = v(\text{CWMP}) = v(\text{CTP}) = v(\text{ITP}). \tag{2.1}$$

We shall see later (Theorem 2.1(ii)) that the third equality implies the first one (this was proved in the unweighted case by Lovász [11]).

The converse is not true in general: Fig. 1 shows an example for which $v(\text{WMP}) = v(\text{CWMP}) = 3$, but $v(\text{ITP}) = 4 > v(\text{CTP}) = 3$.

A useful tool for studying b-KEGs is the following 'blowing-up' construction, introduced by Tutte [16]. The *blow-up* of the weighted graph (G, b) is the unweighted graph G^* obtained from G by replacing each vertex i by a set B_i of b_i vertices, and by linking every vertex of B_i to every vertex of B_j whenever i and j are linked by an edge in G.

A set Y of vertices of G^* will be called *consistent* if for all $i = 1, \ldots, n$ either $B_i \subseteq Y$ or $B_i \cap Y = \emptyset$. Equivalently, Y is consistent if $Y = \bigcup_{i \in X} B_i$ for some $X \subseteq V$. It is easy to see that each transversal of G^* contains a consistent transversal. It follows that the minimum cardinality of a transversal of G^* is equal to the minimum weight of a transversal of G. On the other hand, it is also easily shown that the maximum cardinality of a matching of G^* is equal to the maximum value of a b-matching of G. Hence G is a b-KEG iff G^* is a KEG.

The above construction reduces the recognition of b-KEGs to the recognition of KEGs. However, the resulting recognition algorithm for b-KEGs is pseudo-poly-

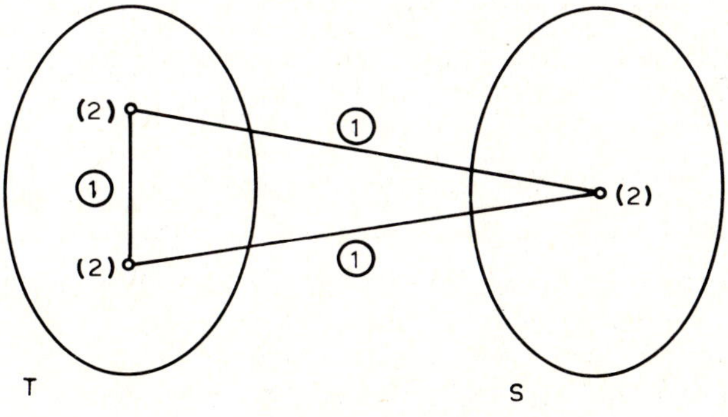

Fig. 1. $(\cdot)$ node weight; $\odot$ λ_{ij}.

nomial (in the sense of [4]), since G^* has $\sum_{i=1}^{n} b_i$ vertices. Berge ([2, Section 8.1]) has described a different reduction of the b-matching problem to the ordinary matching problem, but this reduction too, for similar reasons, fails to yield a polynomial recognition algorithm. On the other hand, it will be shown in Section 3 that, starting from (G, b), one can build in $O(n^3)$ time an unweighted graph $\tilde{G}$ with *fewer* vertices than G and such that (G, b) is a b-KEG iff $\tilde{G}$ is a KEG. In this way one eventually obtains an $O(n^3)$ recognition algorithm for b-KEGs.

Let $N(A)$ denote the *neighbourhood* of a subset A of V, i.e. $N(A)=\{j \notin A$: j is adjacent to some vertex of $A\}$. Let $G(A)$ denote the subgraph of G induced by the set A. If G' is an induced subgraph of G which is a b'-KEG, where b' is the restriction of b to the vertex-set of G', by a slight abuse of language we shall say that G' is a b-KEG. Similarly for terms such as b-matching, etc. If $X \subseteq V$, let $b(X)=\sum_{i\in X} b_i$.

Theorem 2.1. *The following are equivalent.*

(i) *(G, b) is a b-KEG.*

(ii) *CTP has an integral optimal solution.*

(iii) *CSP has an integral optimal solution.*

(iv) *For every minimum weight transversal T and for every maximum fractional b-matching λ, every edge whose endpoints are both in T is passive in λ.*

(v) *There exist a minimum weight transversal T and a maximum fractional b-matching λ such that every edge whose endpoints are both in T is passive in λ.*

(vi) *G has a stable set S such that, for every subset X of $T = V\setminus S$, $b(X) \leq b(N(X)\cap S)$.*

(vii) *Given a maximum fractional b-matching λ, there is a $T \subseteq V$ such that*

(a) *every node of T is saturated by λ;*

(b) *every edge which is active in λ has exactly one endpoint in T;*

(c) *every edge which is passive in λ has at least one endpoint in T.*

Proof. (i) *implies* (ii) follows from (1.4).

(ii) *is equivalent* to (iii) follows from the easily verifiable identities

$$v(\mathrm{ITP}) + v(\mathrm{ISP}) = v(\mathrm{CTP}) + v(\mathrm{CSP}) = \sum_{i=1}^{n} b_i.$$

(ii) *is equivalent* to (iv) follows from (1.4) and (1.5).

(iv) *implies* (v) is obvious.

(v) *implies* (vi): Assume that the pair (T, λ) satisfies the conditions (v), and let $S = V\setminus T$. Then, for every $X \subseteq T$,

$$\begin{aligned} b(X) &= \sum_{i\in X} \sum_{j\in N(i)} \lambda_{ij} \quad \text{(by Theorem 1.1)} \\ &= \sum_{i\in X} \sum_{j\in N(i)\cap S} \lambda_{ij} \quad \text{(by (v))} \\ &\leq b(N(X)\cap S). \end{aligned}$$

(vi) *implies* (i): Let S be a stable set of G satisfying (vi), let $T = V \setminus S$, and let $G^* = (V^*, E^*)$ be the blow-up of (G, b). Then $T^* = \bigcup_{i \in T} B_i$ is a transversal of G^*, and $S^* = V^* \setminus T^*$ a stable set of G^*. Condition (vi) implies that $|Y| \leq |N(Y) \cap S^*|$ for all consistent subsets Y of T^*.

Let Z be an arbitrary subset of T^*. The set $Y = \bigcup\{B_i: i = 1, \ldots, n;\ B_i \cap Z \neq \emptyset\}$ is consistent and includes Z. Moreover one has $N(Y) = N(Z)$. Hence

$$|N(Z) \cap S^*| - |Z| \geq |N(Y) \cap S^*| - |Y| \geq 0.$$

But then by P. Hall's Theorem ([2, Section 7.3]) T^* can be matched into S^*. Klee (quoted in [10, Section 5, Exercise 3.4]) has observed that KEGs are precisely the graphs having a transversal that can be matched into its complement. Hence G^* is a KEG. But then, as remarked above, (G, b) is a b-KEG.

(ii) *is equivalent* to (vii):

(ii) $\Leftrightarrow$ "Given an arbitrary optimal solution λ to CWMP, there exists an integral feasible solution x to CTP (i.e. the dual of CWMP) such that the pair (x, λ) satisfies the complementary slackness conditions".

$\Leftrightarrow$ "Given a maximum fractional b-matching λ, there exists a transversal T such that (a) and (b) hold".

$\Leftrightarrow$ (vii). □

Corollary 2.2. *Let λ be any maximum fractional b-matching. A necessary condition for (G, b) to be a b-KEG is that the subgraph G' spanned by the active edges be bipartite.*

Proof. If G is a b-KEG then there is a $T \subseteq V$ satisfying the conditions of Theorem 2.1(vii). In particular, condition (b) implies that every cycle of G' is even. □

In deriving necessary and sufficient conditions for a graph to be a b-KEG, we shall occasionally assume that the graph has a perfect fractional b-matching. Although not necessary, this assumption makes the presentation much simpler. On the other hand, Theorem 2.3 below shows that we do not lose in generality by making such an assumption.

In fact, let (G, b) be an arbitrary weighted graph with vertex-set V. Let V_1 (resp. V_0) be the set of all $j \in V$ such that $x_j = 1$ (resp. $x_j = 0$) for all optimal solutions of CSP. Let $V_{1/2} = V \setminus (V_0 \cup V_1)$. The following 'persistency' result was proven in [7]: V_1 is the set of nodes which are saturated by some maximum fractional b-matching; $V_0 = N(V_1)$ and for all optimal solutions x of ISP one must have $x_j = 1$ for all $j \in V_1$ and $x_j = 0$ for all $j \in V_0$. Otherwise stated: Given any minimum weight transversal T, let $S = V \setminus T$; then S is a maximum weight stable set and $V_1 \subseteq S$, $V_0 \subseteq T$.

For this reason, the set $V_0 \cup V_1$ was called the *permanent* of (G, b); by contrast, the subgraph $G(V_{1/2})$ induced by $V_{1/2}$ will be called the *fickle subgraph* of (G, b).

It was shown in [7] that the permanent – and hence the fickle subgraph – can be determined in polynomial time and that $G(V_{1/2})$ has a perfect fractional b-matching. Furthermore, we do not lose anything by restricting ourselves to $G(V_{1/2})$, as shown by Theorem 2.3.

Theorem 2.3. *G is a b-KEG if and only if its fickle subgraph is such.*

Proof. Since $V_0 = N(V_1)$, a vertex in $V_{1/2}$ is never adjacent to a vertex in V_1. Therefore there is a natural one-to-one correspondence between the feasible solutions to CSP for $G(V_{1/2})$ and those for G. The thesis then follows from Theorem 2.1(iii). □

2.2. Combinatorial structures related to b-KEGs

As mentioned in the previous sections, a maximum fractional b-matching of (G, b) induces a bipartition B_V of V into *saturated* and *unsaturated* vertices, and a bipartition B_E of E into *active* and *passive* edges. The results of this section imply that the b-König–Egerváry property for G depends only on the triple (G, B_V, B_E). Hence the problem of recognizing b-KEGs becomes a purely combinatorial one, *once the maximum fractional b-matching problem CWMP has been solved.* (As we shall see in Section 3, CWMP can be solved in $O(n^3)$ time.)

Given (G, B_V, B_E), we build a 2-graph (in the sense of [2]) $H = (V, D)$ as follows. The vertex set of H is V. Two vertices of H are linked by a *weak* edge if they were linked by a passive edge in G; two vertices of H are linked by a pair of edges, one being *strong* and the other one *weak*, if they were linked by an active edge in G.

Let B_D be the bipartition of D into strong and weak edges. A vertex of H is saturated if and only if it was such in G. In conclusion, starting from the triple (G, B_V, B_E), we have built a new triple (H, B_V, B_D). The introduction of the 2-graph H is a technical necessity and will be justified later on (see Remark 2.8).

Alternating paths, cycles and bicycles

Let Γ be an odd cycle of H which is *simple*, i.e. each vertex of Γ is incident to exactly two edges of Γ.

Definition 2.4. (i) a path of H is *alternating* if each weak edge is followed by a strong edge, and vice-versa.

(ii) Γ is an *alternating cycle* if it is simple and if every node of Γ but one is incident to two edges of Γ having different nature.

The unique vertex which is incident to two edges of the same nature is called the *tip* of the alternating cycle. A tip is called *weak* or *strong*, depending on the nature of its incident edges.

(iii) An *alternating bicycle* consists of two (odd) alternating cycles, together with an alternating path connecting their tips, in such a way that a weak (resp. strong) tip is incident to a strong (resp. weak) edge of the path. The connecting path is allowed to have zero length, provided that the two tips have different nature. Note that the cycles and the path are allowed to overlap (i.e. to have common edges).

(iv) *A degenerate alternating bicycle* is defined exactly as above, except for the fact that one of the two odd cycles is replaced by a single unsaturated vertex, which is regarded as a strong tip. A (possibly degenerate) alternating bicycle is also called an *obstruction* for reasons which will be apparent in the statement of Theorem 2.5.

With each alternating cycle Γ in H we can naturally associate a cycle Δ in G, having the same vertices as Γ. The cycle Δ will be called the *support* of Γ. The support of an alternating path, of an alternating bicycle, etc., is similarly defined.

Theorem 2.5. *(G, b) is a b-KEG if and only if H has no (possibly degenerate) alternating bicycle.*

Proof. (*'Only if' part.*) Suppose that (G, b) is a b-KEG and that H has a (possibly degenerate) alternating bicycle C. There exists a subset $T \subseteq V$ satisfying the conditions (a), (b), (c) of Theorem 2.1(vii).

In view of the definition of strong and weak edges, conditions (b) and (c) imply:
(b′) every strong edge has exactly one endpoint in T;
(c′) every weak edge has at least one endpoint in T.

Claim 1. *Let Γ be an alternating odd cycle of H with tip v.*
If v is a weak tip, then $v \in T$; if v is a strong tip, then $v \in S$.

Proof of Claim 1. If v is weak and belongs to S, its neighbours on Γ belong to T and are incident to a strong edge, and so on. Eventually, we find either two nodes in S linked by a weak edge, contradicting (c′), or two nodes in T linked by a strong edge, contradicting (b′).

Hence $v \in T$. A similar proof holds for the second half of the claim.

Claim 2. *Let P be an alternating path, let v be an endpoint of P and let e be the edge of P incident to v. Assume that one of the following two cases occurs:*
(i) *$v \in S$ and e is weak;*
(ii) *$v \in T$ and e is strong.*
Then the nodes along P alternate between S and T.

The proof of Claim 2 immediately follows from (b′) and (c′).

Consider the alternating bicycle C. By Claim 1 a weak tip belongs to T and a strong tip belongs to S. Therefore, the bicycle can only be completed by one of three types of alternating paths:
- the first and the last nodes belong to T; the first and the last edges are strong.
- the first and the last nodes belong to S; the first and the last edges are weak.
- the extremal nodes are in T and S; the corresponding extremal edges are strong and weak, respectively.

Taking into account Claim 2, each of these patterns contradicts (b′) or (c′).

(*'If' part.*) This will follow from Algorithm 1 which, for a given weighted graph (G, b) and a maximum fractional b-matching of G, either identifies the graph as being a b-KEG, or else exhibits an obstruction. □

Example 2.6. Figure 2 shows an obstruction. The tips are weak and forced to be in T. But they are joined by an active edge, which contradicts the condition (b′) in the proof of Theorem 2.5.

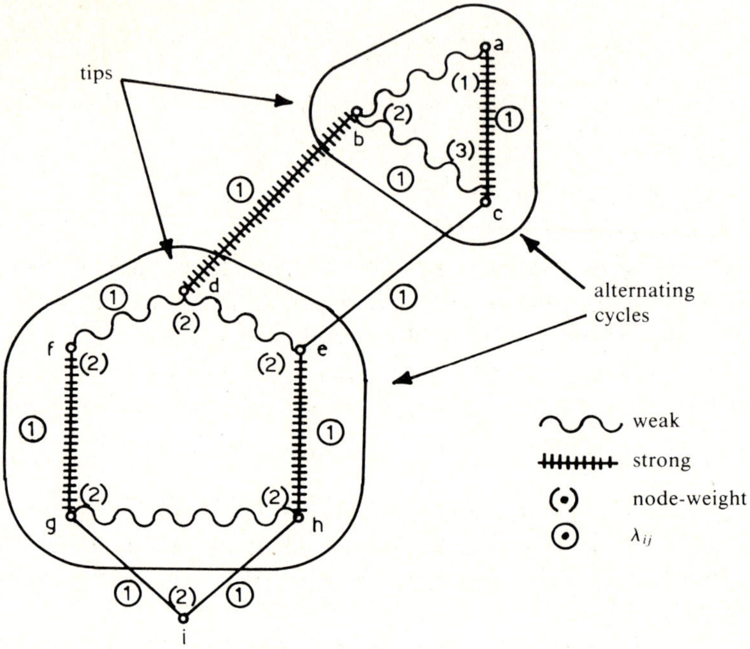

Fig. 2. The graph is not a b-KEG: $\sum_{(i,j)\in E} \lambda_{ij} = 9$ and $b(T) = 10$. A minimum weight transversal is $T = (a, c, d, g, h)$.

Remark 2.7. In view of the complexity results of next section, the above Theorem 2.5, which generalizes to weighted graphs previous results of Deming [3] and Sterboul [15], provides a 'good characterization' (in the sense of Edmonds) of b-KEGs.

Remark 2.8. We have introduced the 2-graph H in order to capture the ambiguity of the rôle played by active edges, which might be strong in one cycle and weak in the other cycle of an alternating bicycle (remember that the two cycles may overlap). This complication does not arise in the unweighted case, since then the active edges form a matching.

2.3. *Structure of alternating (bi)cycles*

The purpose of this paragraph is to give an intrinsic description of alternating (bi)cycles in terms of passive and active edges.

(G, b) is given as well as a maximum fractional b-matching λ. Thus the 2-graph H introduced in the previous section is well-defined. Let Δ be an odd simple cycle in G. If the edges of Δ are all active then by Corollary 2.2 G cannot be a b-KEG. Actually in this case H has a trivial alternating bicycle consisting of two alternating cycles C and C' both having Δ as a support, and such that each edge of Δ corresponds to a strong edge in C and to a weak edge in C', or vice-versa.

Now suppose there is at least one passive edge in Δ. Choose such an edge and label it 0. Choose an orientation of the cycle and starting from edge 0 label the edges as $1, 2, \ldots,$ in a circular fashion according to the chosen orientation.

Proposition 2.9. *H has an alternating cycle with support Δ if and only if there exists a choice of edge 0 for which the labels of all passive edges but 0 have the same parity.*

The proof is obvious.

Definition 2.10. A path in G is *uniform* if its edges are all passive or all active. A uniform path is *passive* or *active*, according to the nature of its edges. A passive (resp. active) uniform path is *maximal* if it cannot be made larger by addition of a passive (resp. active) edge. A 2-*path* is a path with two edges.

Obviously an odd cycle whose edges are all active contains no uniform path which is maximal and even.

Theorem 2.11. (i) *In an alternating cycle there is a unique uniform maximal even path. More precisely, one of the following (mutually exclusive) cases must occur:*

(a) *The cycle has a unique maximal passive 2-path.*

(b) *The cycle has a unique active even path which is maximal.*

(ii) *In case of a passive 2-path there is only one possible tip, namely the central node of the 2-path.*

(iii) *In case of an active even path there are three potential tips: the endpoints and the centre of the path.*

(iv) *If the tip is the centre of an active even path, it is weak if the length of the path is a multiple of 4; otherwise, the tip is strong.*

Proof. (i) Label the edges as before. Assume edges 0 and 1 to be passive; they form a passive 2-path. It is impossible to have a passive path longer than 2 because of alternation. It is impossible to have two passive even paths since the passive edges – except 0 – have the same parity. Again, alternation precludes the existence of both a passive 2-path and a maximal active even path.

Assume now that edge 1 is active. Let $2i+1$ be the smallest positive label of a passive edge. Edges $1, 2, \ldots, 2i$ define an active even path which is maximal. A passive even path is impossible as well as a second maximal active even path, because of alternation.

(ii), (iii) and (iv) follow easily. □

2.4. A spanning forest characterization

Let (G, b) be a weighted graph. We assume the existence of a perfect fractional b-matching λ for G throughout this section. Let G' be the *subgraph* of G *spanned by the active edges in* λ: the edges of G' are the active edges; the nodes of G' are the nodes of G which are the endpoints of some active edge. In case λ is a basic optimal solution (to CWMP), the basic edges form a *spanning pseudo-forest* Q of G, which means that Q is a partial graph [2] of G and each connected component of Q is a tree with an extra edge that creates one odd cycle (if G is bipartite, and only then, the basic edges form a spanning tree of G). This is a well-known result (see for instance [9, Section 5.2]).

Lemma 2.12. *If* (G, b) *is a b-KEG and* λ *is a basic fractional perfect b-matching of* G *then the subgraph* G' *spanned by the edges active in* λ *is a spanning forest of* G.

Proof. Assuming, as we do, that $b_i > 0$ at node i, at least one λ_{ij} around i must be >0. Therefore G' spans G. Furthermore, G' is a partial graph of Q and is bipartite by Corollary 2.2. □

Let F be a spanning forest of G.

Definition 2.13. F is *supporting* if and only if there exists a transversal T such that

$$(i, j) \text{ is an edge of } F,\ i \in T \quad \Rightarrow \quad j \in S = V \setminus T.$$

Let $(\mathscr{C}, t)$ be a node-weighted tree with node-weights t_i. Since $\mathscr{C}$ is bipartite there is a unique partition of its vertex-set into two stable sets B and W. The vertices in B and W will be called *black* and *white*, respectively. If e is any edge of $\mathscr{C}$, e has a black endpoint x_B and a white one x_W. By deleting e one obtains two subtrees called *branches. Let* $\mathscr{C}(B, e)$ be the branch containing x_B and $\mathscr{C}(W, e)$ the one containing x_W. Denote by $W(W, e)$ (resp. $W(B, e)$) the set of white vertices of $\mathscr{C}(W, e)$ (resp. $\mathscr{C}(B, e)$) and by $B(W, e)$ (resp. $B(B, e)$) the set of black vertices of $\mathscr{C}(W, e)$ (resp. $\mathscr{C}(B, e)$).

Definition 2.14. $(\mathscr{C}, t)$ is *t-partitionable* if
 (i) for every edge e, $t(W(W, e)) > t(B(W, e))$,
 (ii) $t(B) = t(W)$.

Remark 2.15. Assuming (ii), (i) can be replaced by the equivalent condition:
 (i′) for every edge e, $t(B(B, e)) > t(W(B, e))$.

Definition 2.16. If (G, b) is a weighted graph, a spanning forest F of G is *b-partitionable* if each of its connected components is b-partitionable.

The main result of this section is the following.

Theorem 2.17. (G, b) *is a b-KEG if and only if* G *has a spanning forest* F *which is supporting and b-partitionable.*

We need a preliminary result in order to prove this theorem.

Theorem 2.18. *A node-weighted tree* ($\mathscr{C}, t$) *has a positive perfect t-matching if and only if* $\mathscr{C}$ *is t-partitionable.*

Proof. We first claim that the system of linear equations

$$\sum_{j\in N(i)} \lambda_{ij} = t_i \quad \text{for every vertex } i \tag{2.2}$$

has a solution – possibly with some negative λ_{ij}'s – if and only if $t(B)=t(W)$. In fact, if λ is a solution of 2.5 then, $\mathscr{C}$ being bipartite,

$$t(W) = \sum_{i\in W}\sum_{j\in N(i)} \lambda_{ij} = \sum_{j\in B}\sum_{k\in N(j)} \lambda_{kj} = t(B).$$

Conversely, suppose that $t(B)=t(W)$. Let $e=(i,j)$ be an arbitrary edge where i is white and j is black. Let $\mathscr{C}_i$ be the branch of $\mathscr{C}\setminus e$ containing i and $\mathscr{C}_j$ the branch containing j. If $\mathscr{C}'$ is any subtree of C, denote by $W(\mathscr{C}')$ and $B(\mathscr{C}')$ the set of white vertices and that of black vertices of $\mathscr{C}'$, respectively. Set, for every edge (i,j),

$$\lambda_{ij} = t(W(\mathscr{C}_i)) - t(B(\mathscr{C}_i))(= t(B(\mathscr{C}_j)) - t(W(\mathscr{C}_j)) \text{ because } t(B)=t(W)). \tag{2.3}$$

We claim that λ is a solution of (2.2). Indeed, consider any white vertex i; let $j_1, j_2, \ldots, j_r$ be the (necessarily black) neighbours of i; let $\mathscr{C}_{j_1}, \mathscr{C}_{j_2}, \ldots, \mathscr{C}_{j_r}$ be defined as above with respect to $e_1=(i,j_1), \ldots, e_r=(i,j_r)$. Then

$$\sum_{j\in N(i)} \lambda_{ij} = \sum_{k=1}^{r} (t(B(\mathscr{C}_{j_k})) - t(W(\mathscr{C}_{j_k})))= t(B)-(t(W)-b_i)=b_i.$$

The computation is analogous in case i is a black vertex. This proves the claim. We observe next that the system (2.2) has $n-1$ variables, n equations and rank $n-1$; hence, when a solution to (2.2) exists it is necessarily unique and integral, and is given by (2.3). Condition (i) of Definition 2.14 merely expresses the fact that such a solution is positive. This completes the proof. □

(We are indebted to Professor Uri Peled for useful suggestions concerning the above proof).

Proof of Theorem 2.17. Assume G is a b-KEG. Let λ be a basic fractional perfect b-matching and let F be the subgraph spanned by the active edges in λ. By Lemma 2.12 F is a spanning forest. By Theorem 2.18, F is b-partitionable. F is supporting because of Theorem 2.1(vii).

Now assume that F is b-partitionable and supporting. There exists a perfect b-matching λ for G which is also a perfect b-matching for F (the active edges of λ are actually the edges of F).

There exists a transversal T such that

$$(i,j)\in F, \quad i\in T \quad \Rightarrow \quad j\in S=V\setminus T.$$

By Theorem 2.1(v), (G, b) is a b-KEG. □

Theorem 2.19. *Every graph has a supporting spanning forest.*

Proof. Let S_0 be a maximal stable set of G and $T_0 = V \setminus S_0$. Let F be the set of edges that join a vertex of S_0 with a vertex of T_0. Every vertex of S_0 is incident to an edge of F because G is connected. Every vertex of T_0 is incident to an edge of F since S_0 is maximal. The graph spanned by F is a forest. It spans G and is supporting. □

We can use the above theorem for proving the following result.

Theorem 2.20. *For any graph G there exists an n-vector b of positive integers such that (G, b) is a b-KEG.*

Proof. Find a maximal stable set S_0 and a supporting spanning forest F as in the proof of Theorem 2.19. Assign arbitrary positive values to the edges of F; assign value zero to the remaining edges of G. Define the weight b_i of node i as the sum of the values of all edges incident to it. Then, clearly, (G, b) is a b-KEG and S_0 is a maximum weight stable set. □

Finally we remark that, using an algorithm similar to the labelling procedure of Algorithm 1 (see next section), one can check in polynomial time whether a given spanning forest is supporting.

3. Recognizing b-KEGs

In this section we present two recognition algorithms based on the results obtained in Section 2. The maximum fractional b-matching problem plays a central rôle in the present study and the fact that it can be reduced to a maximum flow problem, and hence solved in polynomial-time, is crucial. The first algorithm is based on the material developed in Sections 2.2 and 2.3 and also on a scheme devised by Deming [3] to check whether a graph is a KEG. Given a maximum fractional b-matching, the algorithm will attempt to label the nodes either S or T so that condition (vii) of Theorem 2.1 is satisfied. The labelling can be successfully completed if and only if the graph is a b-KEG. In case of success the labels indicate a bipartition of the node-set into a maximum weight stable set and a minimum weight transversal. When the graph is not a b-KEG it is possible to trace back an alternating bicycle, using two predecessor functions $p_1(\cdot)$ and $p_2(\cdot)$.

Algorithm 1: Node-labelling and obstructions

Step 1. Find a maximum fractional b-matching λ of G. If all nodes are saturated, go to Step 2. Else for every node i which is unsaturated in λ label it S. Set $k=2$ and go to Step 3.

Step 2. Choose an unlabelled vertex, v. Label it T. Let $k=1$, and go to Step 3.

Step 3. Node Labelling. Choose a labelled vertex i, and *scan* it by using the following rules to label its neighbours:

If i is labelled T and if j is an unlabelled neighbour of i for which (i,j) is active, j is labelled S; if i is labelled S, every unlabelled neighbour j of i receives label T. (3.1)

Let $p_k(j)=i$ for every node j which is labelled when i is scanned. Repeat Step 3 until one of the following occurs:

(i) a conflict arises. This is the case when two T-nodes are connected by an active edge or when two S-nodes are connected by any edge. Record the two nodes in question. If $k=2$, go to Step 5. If $k=1$ then erase all the labels; assign to v the label S; let $k=2$ and go to Step 3.

(ii) the process goes stuck. It is impossible to label any new node by (3.1). If all the nodes of G have been labelled, go to Step 4; else, discard all the labelled nodes and restart at Step 2 with the subgraph induced by the remaining nodes.

Step 4. The graph is a b-KEG. The set of T-nodes (resp. S-nodes) is a minimum weight transversal (resp. maximum weight stable set).

Step 5. Exhibit an obstruction. The graph is not a b-KEG and an obstruction can be identified. There are two cases:

Case a. k was set to both values 1 and 2. Let i_k and j_k $(k=1,2)$ be the two nodes where a conflict arises. Take $p_k(i_k)$, $p_k(j_k)$ and then $p_k(p_k(i_k))$, $p_k(p_k(j_k))$, and so on. There exists a node v_k and integers l_k and l'_k such that $v_k=p_k^{l_k}(i_k)=p_k^{l'_k}(j_k)$. The nodes $i_k, p_k(i_k), p_k^2(i_k), \ldots, p_k^{l_k}(i_k)\equiv v_k\equiv p_k^{l'_k}(j_k), \ldots, p_k^2(j_k), p_k(j_k), j_k$ are the vertices of an odd cycle. Then, take $p_k(v_k)$, $p_k^2(v_k), \ldots,$ until the paths from v_1 to v and from v_2 to v meet at a node v'. The two odd cycles, together with the concatenation of the two paths v_1v' and $v'v_2$ are the support of an alternating bicycle of H, defined as follows.

Because of (3.1) we can say that

if a node i is in T, its predecessor $p_k(i)$ must be in S and vice-versa. (3.2)

Hence along the backpaths from i_k to v' and from j_k to v' weak or strong edges are selected according to the following rule:

if $p_k(i)$ is in S, select the weak edge linking i and $p_k(i)$,
if $p_k(i)$ is in T, select the strong edge linking i and $p_k(i)$. (3.3)

Finally, if i_k and j_k both belong to T (resp. to S) select the strong (resp. weak) edge linking them. The selected edges constitute the required alternating bicycle, whose tips are precisely v_1 and v_2.

Case b. k was never set to 1. (This can only happen when there are unsaturated vertices.) In this case only the predecessor index $p_2(\cdot)$ is available. Using only $p_2(\cdot)$ and repeating the procedure described in case (a), one obtains a degenerate alternating bicycle whose tips are v_2 and the unsaturated vertex v. In both cases, the two cycles (or the single cycle) in the alternating bicycle are seen to be simple.

Theorem 3.1. (i) *When applied to graphs which are not b-KEGs, Algorithm* 1 *identifies an obstruction.* (ii) *Algorithm* 1 *recognizes the b-KEGs.* (iii) *Algorithm* 1 *has an* $O(n^3)$ *time-complexity.*

Proof. (i) Obviously, the last two edges on the paths from i_k to v_k and from j_k to v_k are of the same nature and, because of (3.2), these paths, together with the edge (i_k, j_k) define an alternating odd cycle (see Figure 3(a)).

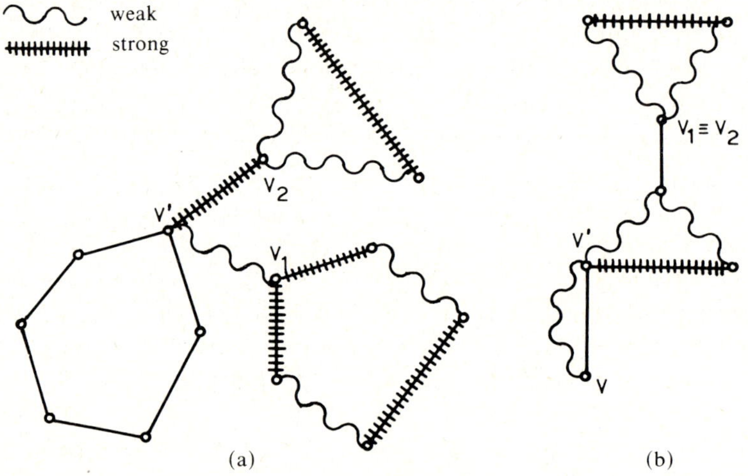

Fig. 3.

Assume that case (a) holds. If the alternating cycles contain exactly the same edges and if the common tip is weak in one alternating cycle and strong in the other, then the edges are all active and this defines an alternating bicycle. If the cycles contain the same edges and if the common tip is weak or strong in both of them, then $v_1 = v_2$. Starting at v_1 and v_2, backtrack to v along the paths defined by p_k $(k = 1, 2)$. If v_1 and v_2 are weak (resp. strong) the first edge of each path is strong (resp. weak). There is alternation; v' is reached the first time a weak edge of one path meets a strong edge of the other (see Figure 3(b)). This must occur eventually since v is first set T and then S. When $v_1 \neq v_2$ backtrack along the paths until they meet in v'. The edges that meet in v' cannot be both weak and strong because otherwise the conflict arising in Step 2(i) of the algorithm would be produced by the same pair of nodes i and j, for $k = 1$ and 2; p_1 and p_2 would be identical along the paths from i and j to v'. Therefore we would identify a single alternating cycle.

In case (b), we obtain a single alternating cycle with tip v_2. This cycle, together with the alternating path connecting v_2 and the unsaturated vertex v, forms a degenerate alternating bicycle.

(ii) Clearly Algorithm 1 embodies two parts: it attempts to partition the vertices by labelling them S or T, and when it fails to complete the labelling it points out an obstruction. If the labelling is successfully completed, the set of all nodes labelled T, denoted again T, satisfies the conditions (vii) of Theorem 2.1. Hence (G, b) is a b-KEG. If the labelling is aborted before completion, Algorithm 1 has been shown in (i) to identify an obstruction in H. By the "only if" part of Theorem 2.5, (G, b) cannot be a b-KEG.

(iii) Step 5 obviously can be executed in $O(n)$ time. Step 2 is executed at most n times and the number of elementary operations performed in Step 3 between any two consecutive executions of Step 2 is $O(m)$. Hence the overall complexity of the labelling procedure is $O(m \cdot n)$. Note that when the labelling process goes stuck, every edge connecting a labelled vertex i and an unlabelled one j must be passive, and i must be labelled T. Hence no backtracking is needed. Finally, CWMP is known to be reducible to a maximum flow problem (see e.g. [6, Section 7.5, Exercise 16]). Using this reduction and, say, the Malhotra–Kumar–Maheshwari maximum flow algorithm [12] one can find a maximum fractional b-matching in $O(n^3)$ time. Therefore the overall complexity of Algorithm 1 is $O(n^3)$. □

Remark 3.2. Gavril [5] has described a recognition algorithm for KEGs whose complexity (assuming that a maximum matching is available) is $O(m+n)$, as opposed to the $O(m \cdot n)$ complexity of Deming's algorithm. It is possible to design a variant of Algorithm 1 based on Gavril's scheme rather than on Deming's one. We have chosen the version based on Deming's scheme for two reasons: (1) Deming's scheme is conceptually simpler. (2) Some computational experiments performed by the third author seem to suggest that the average complexity of Deming's algorithm is lower than Gavril's one.

The second algorithm is quite different in spirit. Given a maximum fractional b-matching of G, it first focuses on the subgraph spanned by the active edges and breaks it into 'parts'. G is a b-KEG if and only if a certain quadratic boolean equation expressing a compatibility condition for the 'parts' is consistent. An alternative is to build an auxiliary graph $\tilde{G}$ whose nodes are the 'parts' in question and in which each edge indicates the incompatibility of two 'parts'. G is a b-KEG if and only if $\tilde{G}$ is a KEG (cf. Theorem 3.7).

Algorithm 2 (Break-down and build-up)

Step 1. Find a fractional maximum b-matching λ.

Step 2. Let Y_0 be the set of unsaturated vertices, and $X_0 = N(Y_0)$ (Y_0 is stable). Let G_0 be the subgraph induced by $V \setminus (X_0 \cup Y_0)$.

Step 3 (Break-down). Let G' be the partial graph of G_0 spanned by the active edges. If G' is not bipartite, STOP: (G, b) is not a b-KEG. Else go to Step 4.

Step 4 (Build-up). (Every component of G' is bipartite.) Let (X_i, Y_i), $i = 1, \ldots, r$, be the unique partition of the vertex set of the i-th component of G' into two stable sets. If neither X_i nor Y_i is stable in G, STOP: (G, b) is not a b-KEG. Else, assume that Y_i is stable in G for every i. Call each set X_i or Y_i, $i = 0, 1, \ldots, r$ a *cell*. Associate binary variables x_i with X_i and $\bar{x}_i$ with Y_i, $i=0, 1, \ldots, r$. Create a linear term x_0; Create a linear term x_i for each set X_i which is either empty or not stable in G $(i=1, \ldots, r)$; for every pair of cells Z, Z' which are adjacent in G, create a quadratic term zz' involving the binary variables z and z' associated with Z and Z', respectively. Finally, let f be the quadratic boolean expression formed by the union of all the (linear or quadratic) terms created so far. (G, b) is a b-KEG if and only if the equation $f=0$ is consistent. If this is the case and if x is a solution of the equation, a maximum weight stable set of G is given by

$$S = \bigcup_{i:x_i=1} X_i \cup \bigcup_{i:x_i=0} Y_i$$

END.

Theorem 3.3. *Algorithm 2 is correct and its time-complexity is* $O(n^3)$.

Proof. In G', every edge (i, j) is active. According to Corollary 2.2, G' must be bipartite in order that G be a b-KEG. In Step 3 we collect the pieces in order to build T and S. The partition of the node-set of a bipartite connected component is unique and in every bipartite component (X_i, Y_i, E_i), at most one of the sets X_i, Y_i is allowed to be non-stable in G, in order for this graph to be a b-KEG. One of these sets must be included in S and the other in T. This condition is met by requiring that the binary variable associated with Y_i be the complement of the variable associated with X_i (we adopt the convention that a variable is equal to 1 if and only if the corresponding set is included in S). If two cells Z and Z' are adjacent in G at most one of them may be included in S, i.e. zz' must be 0. Clearly $Y_0 \subseteq S$ by Theorem 1.1. Hence x_0 must be 0. If X_i is not stable in G then it must be in T, i.e. x_i must be 0. This is why the linear term x_i is introduced. Similarly, if $X_i = \emptyset$, then Y_i must be in S, i.e. x_i must be 0. Finally $f=0$ has a binary solution if and only if all the terms of f are simultaneously equal to zero for some (0, 1)-assignment to the variables; $f=0$ is a quadratic boolean equation and can be solved (see [1, 14]) in time proportional to the number or terms, that is in $O(m+n)$ time. Finding the connected components of a graph and testing a graph for bipartiteness both require a time proportional to the number of edges of the graph [6].

Checking whether Z is a stable set obviously requires $O(|Z|^2)$ elementary operations. Finally, one can solve CWMP in $O(n^3)$ time, as previously remarked. Hence the overall complexity of Algorithm 2 is $O(n^3)$. □

Remark 3.4. When $b=(1, 1, \ldots, 1)$ every connected component found in Step 3 is K_2 (two nodes linked by an edge of the maximum (integral) matching defined in Step 1). Then the process described in Step 3 is similar to the treatment of [14].

Example 3.5 (cf. Fig. 4(a)). $Y_1 \subseteq S$; $X_1 \subseteq T$.

$$f = \bar{x}_1 x_2 \vee x_2 \bar{x}_3 \vee \bar{x}_3 x_4 \vee \bar{x}_3 \bar{x}_4 \vee x_1.$$

$$f=0 \Rightarrow x_1=0; \quad x_2=0; \quad x_3=1; \quad x_4=0 \text{ or } 1.$$

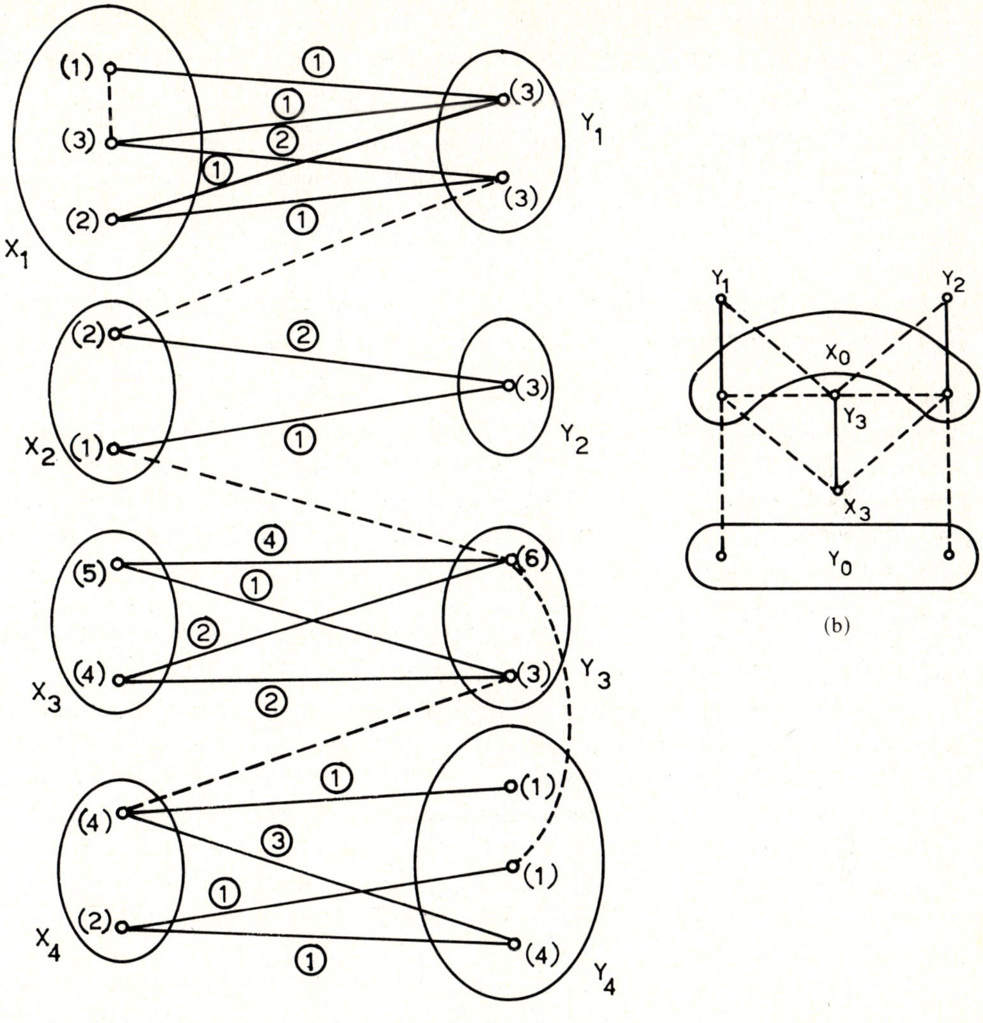

Fig. 4. The solid edges are active; the dotted ones are passive. The bracketed numbers are the node-weights. The circled numbers define a maximum b-matching.

Therefore,

$$S = Y_1 \cup Y_2 \cup X_3 \cup X_4 \quad \text{and} \quad T = X_1 \cup X_2 \cup Y_3 \cup Y_4;$$

or

$$S = Y_1 \cup Y_2 \cup X_3 \cup Y_4 \quad \text{and} \quad T = X_1 \cup X_2 \cup Y_3 \cup X_4.$$

Example 3.6 (Cf. Fig. 4(b)). The graph is unweighted; $X_1 = X_2 = \emptyset$.

$$f = x_0 \vee x_1 \vee x_2 \vee x_0x_3 \vee x_0\bar{x}_1 \vee x_0\bar{x}_2 \vee x_0\bar{x}_3 \vee \bar{x}_1\bar{x}_3 \vee \bar{x}_2\bar{x}_3$$

$$= x_0 \vee x_1 \vee x_2 \vee \bar{x}_3.$$

$$f=0 \Rightarrow x_0=x_1=x_2=0, \quad x_3=1,$$

$$S=Y_0\cup Y_1\cup Y_2\cup X_3 \quad \text{and} \quad T=X_0\cup Y_3.$$

It is worth pointing out that both Algorithms 1 and 2 have a linear programming stage (solving CWMP by a maximum flow algorithm) and a purely combinatorial stage (the labelling procedure in Algorithm 1 or the generation and the solution of the quadratic boolean equation in Algorithm 2). As mentioned above, both stages can be carried out in time polynomial in the number of nodes of G.

Theorem 3.7. *Given the weighted graph (G, b), one can build in $O(n^3)$ time a graph $\tilde{G}$ with fewer vertices than G and such that (G, b) is a b-KEG if and only if $\tilde{G}$ is a KEG.*

Proof. Let f be defined as in Algorithm 2. Now build $\tilde{G}$ as follows. For every variable x_i (resp. $\bar{x}_i$) of f create a node i (resp. $\bar{i}$). Every quadratic term of f gives rise to an edge linking the corresponding nodes of G. For every linear term x_i link node i to both j and $\bar{j}$, j being arbitrary. Finally, link every node j to $\bar{j}$. In [14] it is established that $f=0$ is consistent if and only if $\tilde{G}$ is a KEG. Hence, (G, b) is a b-KEG if and only if $\tilde{G}$ is a KEG. □

Example 3.8. (cf. Example 3.5). $f=\bar{x}_1x_2 \vee x_2\bar{x}_3 \vee \bar{x}_3x_4 \vee \bar{x}_3\bar{x}_4 \vee x_1$. The corresponding $\tilde{G}$ is drawn on Figure 5. One can choose $T=(1,2,\bar{3},4)$ or $T=(1,2,\bar{3},\bar{4})$.

$|T|=4=$ maximum cardinality of a matching.

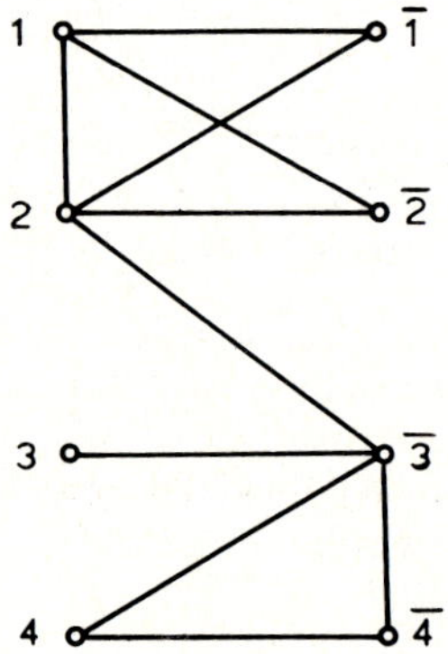

Fig. 5.

Acknowledgement

We are indebted to a referee for his valuable suggestions, which resulted in a simpler presentation of the results in Section 2.1. This work has been supported by NSERC.

References

[1] B. Aspvall, M.F. Plass and R.E. Tarjan, "A linear-time algorithm for testing the truth of certain quantified formulas", *Information Processing Letters* 8 (1979) 121–123.

[2] C. Berge, *Graphes et hypergraphes* (Dunod, Paris, 1970).

[3] R.W. Deming, "Independence numbers of graphs—An extension of the König-Egerváry theorem", *Discrete Mathematics* 27 (1979) 23-34.

[4] M.R. Garey and D.S. Johnson, *Computers and intractability* (Freeman, San Francisco, 1979).

[5] F. Gavril, "Testing for equality between maximum matching and minimum node covering", *Information Processing Letters* 6 (1977) 199–202.

[6] M. Gondran and M. Minoux, *Graphes et algorithmes* (Eyrolles, Paris, 1979).

[7] P.L. Hammer, P. Hansen and B. Simeone, "Vertices belonging to all or to no maximum stable sets of a graph", *SIAM Journal on Algebraic and Discrete Methods* 3 (1982) 511–522.

[8] F. Harary, *Graph theory* (Addison-Wesley, Reading, Mass., 1972).

[9] J.L. Kennington and R.V. Helgason, "*Algorithms for network programming*" (Wiley, New York, 1980).

[10] E.L. Lawler, *Combinatorial optimization: Networks and matroids* (Holt, Rinehart and Winston, New York, 1976).

[11] L. Lovász, "Minimax theorems for hypergraphs", in: C. Berge and D.K. Ray-Chaudhuri eds., *Hypergraph seminar 1972*, Lecture Notes 411 (Springer-Verlag, Berlin, 1972).

[12] W.M. Malhotra, M.P. Kumar and S.N. Maheshwari, "An $O(|V|^3)$ algorithm for finding maximum flows in networks", *Information Processing Letters* 6 (1978) 277–278.

[13] W.R. Pulleyblank, "Dual-integrality in *b*-matching problems" *Mathematical Programming Study* 12 (1980) 176–196.

[14] B. Simeone, *Quadratic 0-1 programming, boolean functions and graphs*, Doctoral Dissertation, University of Waterloo, 1979.

[15] F. Sterboul, "A characterization of the graphs in which the transversal number equals the matching number", *Journal of Combinatorial Theory, Series B* 27 (1979) 228–229.

[16] W.T. Tutte, "A short proof of the factor theorem for finite graphs", *Canadian Journal of Mathematics* 6 (1954) 347–353.

Mathematical Programming Study 22 (1984) 64–69
North-Holland

A POLYNOMIAL-TIME TEST FOR TOTAL DUAL INTEGRALITY IN FIXED DIMENSION

W. COOK

Department of Combinatorics and Optimization, University of Waterloo, Waterloo, Ontario, Canada N2L 3G1

L. LOVÁSZ,

Analizis I. tszk., Eötvös Lorand Tudományegyetem, Muzeum krt. 6–8, H-1088, Budapest, Hungary

A. SCHRIJVER,

Department of Econometrics, Tilburg University, P.O. Box 90153, Tilburg, Holland

Received 2 December 1982
Revised manuscript received 27 June 1983

In this note we show that, for any fixed number r, there exists a polynomial-time algorithm to test whether a given system of linear inequalities $Ax \leq b$ is totally dual integral, where A is an integer matrix of rank r.

Key words: Polynomial-Time, Algorithm, Total Dual Integrality, Hilbert Basis, Integer Linear Programming.

1. Introduction

Let A be a rational $m \times n$-matrix, and let b be a rational vector of length m. The system $Ax \leq b$ of linear inequalities is called *totally dual integral* if the minimum in the linear programming duality equation

$$\max\{cx \mid Ax \leq b\} = \min\{yb \mid y \geq 0, yA = c\} \tag{1}$$

has an integer optimal solution for each integer vector c for which the optima exist (cx and yb denoting inner products). Edmonds and Giles [3] motivated this concept by showing that if $Ax \leq b$ is totally dual integral and b is integer, then also the maximum in (1) has an integer optimal solution, for each vector c for which the optima exist. Equivalently, then the polyhedron

$$P = \{x \mid Ax \leq b\} \tag{2}$$

is the convex hull of the lattice points contained in it.

In this note we show that, for any fixed number r, there exists an algorithm which tests the total dual integrality of a given system $Ax \leq b$ of linear inequalities, where A is an integer matrix of rank r, in time bounded by a polynomial in the sizes of

A and b (in binary notation). The main tool here is Lenstra's polynomial-time algorithm for integer linear programming in fixed dimension [7].

Total dual integrality is closely connected to so-called 'Hilbert bases'. Following Giles and Pulleyblank [4], a set of integer vectors $a_1, \ldots, a_k$ is called a *Hilbert basis* if each integer vector in the convex cone spanned by $a_1, \ldots, a_k$ is a nonnegative integer combination of $a_1, \ldots, a_k$. Indeed, the system $Ax \leq b$ is totally dual integral if and only if for each minimal face F of P the rows of A which are active in F form a Hilbert basis (a row is *active in* F if the corresponding inequality in $Ax \leq b$ is satisfied with equality by each vector in F). In particular, all rows of A form a Hilbert basis if and only if the system $Ax \leq 0$ is totally dual integral.

It is easy to see that each pointed rational polyhedral cone C is spanned by a unique minimal Hilbert basis (minimal under inclusion)—take all nonzero integer vectors in C which cannot be expressed as a sum of other integer vectors in C (this is a finite set, as these vectors are all contained in the polytope $\{\lambda_1 b_1 + \cdots + \lambda_t b_t \mid 0 \leq \lambda_i \leq 1\ (1 \leq i \leq t)\}$, where $b_1, \ldots, b_t$ are integer vectors spanning C). This implies that each full-dimensional rational polyhedron is defined by a unique minimal totally dual integral system with integer left-hand sides (cf. [8]).

Since, if we fix the rank r of A, the number of minimal faces of P is polynomially bounded by the number of rows of A, and since they can be enumerated in polynomial time, testing total dual integrality for fixed rank in polynomial time can be reduced to the test of being a Hilbert basis for fixed rank in polynomial time.

Chandrasekharan [1] proved that there exists an algorithm which finds for any totally dual integral system $Ax \leq b$, with A integer, and for any integer vector c, an integer optimum solution y for the minimum in (1), in time polynomially bounded by the sizes of A, b and c (in binary notation).

The property of being a totally dual integral system, and that of being a Hilbert basis, is in the class co-NP, if the left-hand side coefficients (i.e. the entries of A) are integral. That is, there exists a polynomial-length proof for not being totally dual integral, or for not being a Hilbert basis. Indeed, if the integer vectors $a_1, \ldots, a_k$ do not form a Hilbert basis, there exists an integer vector c in the cone C spanned by $a_1, \ldots, a_k$ such that $c - a_i \notin C$ for all i for which a_i does not belong to the minimal face F of C, and such that c does not belong to F. Note that, given a vector c, these conditions can be checked with Khachiyan's method. Moreover, c can be chosen in the polytope $\{\lambda_1 a_1 + \cdots + \lambda_k a_k \mid 0 \leq \lambda_i \leq 1\ (1 \leq i \leq k)\}$, so that the size of c is polynomially bounded by the size of $a_1, \ldots, a_k$.

In fact it is the content of the theorem that if we fix the rank of $a_1, \ldots, a_k$, such a vector c can be found in polynomial time. (Parts of the proof of this theorem also occur in [2].)

Theorem. *For any fixed natural number r, there exists an algorithm which tests, for any integer matrix A of rank r and for any rational vector b, whether the system $Ax \leq b$ is totally dual integral, in time bounded by a polynomial in the sizes of A and b (in binary notation).*

Proof. I. Let A have rows $a_1, \ldots, a_m$. We first argue that we may assume that $b=0$, and that the convex cone C spanned by $a_1, \ldots, a_m$ is pointed and full-dimensional (equivalently, that the cone $\{x \mid Ax \leq 0\}$ is pointed and full-dimensional).

Indeed, we have to test whether in each minimal face F of $P=\{x \mid Ax \leq b\}$, the active rows of A form a Hilbert basis. Now we can enumerate all minimal faces of P as follows. Enumerate all collections of r linearly independent rows of A. If, say, $a_1, \ldots, a_r$ are linearly independent, we check whether $F=\{x \mid a_1x=b_1, \ldots, a_rx=b_r\}$ is contained in P, which can be done by checking whether $Ax_0 \leq b$ holds for some solution x_0 of $a_1x=b_1, \ldots, a_rx=b_r$. If $F \subseteq P$, then F is a minimal face of P. Let, say, $a_1x \leq b_1, \ldots, a_kx \leq b_k$ be those inequalities among $Ax \leq b$ which are active in F (i.e., which hold with equality for some vector, and hence for each vector, in F). We have to test whether $a_1, \ldots, a_k$ is a Hilbert basis. That is, if A' denotes the matrix with rows $a_1, \ldots, a_k$, we have to test whether $A'x \leq 0$ is totally dual integral. Therefore, since for fixed r the number of minimal faces is polynomially bounded by the size of A, we may assume that $b=0$.

Next we show that we may assume that the convex cone C spanned by $a_1, \ldots, a_m$ is pointed and full-dimensional. Let F be the unique minimal face of C, and let L be the linear hull of C. Let, say, $a_1, \ldots, a_t$ be the rows of A belonging to F (they can be determined in polynomial time as a row a_i of A belongs to F if and only if $-a_i$ belongs to C). Let $d := \dim F$ (and note that $r=\dim L$). Now there exists a unimodular matrix U such that $FU=\mathbb{R}^d \times 0^{n-d}$, and $LU=\mathbb{R}^r \times 0^{n-r}$. Such a matrix U can be found as follows. Choose d linearly independent vectors $v_1, \ldots, v_d$ from $a_1, \ldots, a_t$, and after that choose $r-d$ vectors $v_{d+1}, \ldots, v_r$ from $a_{t+1}, \ldots, a_m$ such that $v_1, \ldots, v_r$ are linearly independent. Let V be the matrix with rows $v_1, \ldots, v_r$, and determine, with the algorithm of Kannan and Bachem [5], a unimodular matrix U such that VU is in (lower-triangular) Hermite normal form. One easily checks that U has the required properties.

Now $Ax \leq 0$ is totally dual integral if and only if $AUx \leq 0$ is totally dual integral. So we may assume that $A=AU$, and that $F=\mathbb{R}^d \times 0^{n-d}$ and $L=\mathbb{R}^r \times 0^{n-r}$. Since now the last $n-r$ columns of A are zero, we may assume that $n=r$. It is easy to see that $Ax \leq 0$ is totally dual integral if and only if $a_1, \ldots, a_t$ form a Hilbert basis for F, and $A'x \leq 0$ is totally dual integral, where A' is the matrix consisting of the last $n-d$ columns of A.

Now the fact that F is a linear space implies that $a_1, \ldots, a_t$ is a Hilbert basis for F if and only if F is the convex cone generated by $a_1, \ldots, a_t$ (which is a given fact, by definition of $a_1, \ldots, a_t$), and $F \cap \mathbb{Z}^n$ is the lattice generated by $a_1, \ldots, a_t$ (which can be checked again with the algorithm of Kannan and Bachem). Indeed, necessity of these conditions is immediate. To see sufficiency, let z be an integer vector in F. Then $z=\nu_1a_1+\cdots+\nu_ta_t$ for certain integers $\nu_1, \ldots, \nu_t$. Moreover, for each $i=1, \ldots, t$, the vector $-a_i$ belongs to F, and can hence be written as a convex combination of $a_1, \ldots, a_t$. Therefore, $0=\mu_1a_1+\cdots+\mu_ta_t$ for positive rationals $\mu_1, \cdots, \mu_t$. By choosing M appropriately, $z=(\nu_1+M\mu_1)a_1+\cdots+(\nu_t+M\mu_t)a_t$ is a decomposition of z as a nonnegative integer combination of $a_1, \ldots, a_t$.

So it suffices to describe a test of total dual integrality for $A'x \leq 0$. That is, without loss of generality we may assume that the cone C generated by $a_1, \ldots, a_m$ is full-dimensional and pointed.

II. Here we describe an algorithm to test whether $a_1, \ldots, a_m$ is a Hilbert basis for the cone C generated by $a_1, \ldots, a_m$, where C is pointed and full-dimensional.

We first observe that $a_1, \ldots, a_m$ is a Hilbert basis for C if and only if the only integer vector in the set

$$C_0 = \{z \in C \mid z - a_i \notin C \text{ for } i = 1, \ldots, m\} \tag{3}$$

is the origin.

To prove necessity, let z be an integer vector in C_0. Then $z = \lambda_1 \alpha_1 + \cdots + \lambda_m a_m$ for nonnegative integers $\lambda_1, \ldots, \lambda_m$. As $z - a_i \notin C$ we know that $\lambda_1 = \cdots = \lambda_m = 0$, i.e., that $z = 0$.

To prove sufficiency, let z be an integer vector in C. Let λ_1 be the highest rational number such that $z - \lambda_1 a_1$ belongs to C. Next, let λ_2 be the highest rational number such that $z - \lfloor \lambda_1 \rfloor a_1 - \lambda_2 a_2$ belongs to C. If $\lambda_1, \ldots, \lambda_j$ have been found, let λ_{j+1} be the highest rational number such that $z - \lfloor \lambda_1 \rfloor a_1 - \lfloor \lambda_2 \rfloor a_2 - \cdots - \lfloor \lambda_j \rfloor a_j - \lambda_{j+1} a_{j+1}$ belongs to C. When $\lambda_1, \ldots, \lambda_m$ have been found, the vector $z - \lfloor \lambda_1 \rfloor a_1 - \cdots - \lfloor \lambda_m \rfloor a_m$ is an integer vector in C_0, and hence is the origin. This expresses z as a nonnegative integer combination of $a_1, \ldots, a_m$.

So it suffices to check whether the only integer vector in C_0 is the origin. To this end let $b_1, \ldots, b_t$ be vectors such that

$$C = \{z \mid b_j z \geq 0 \text{ for } j = 1, \ldots, t\}. \tag{4}$$

Since the rank $r = n$ of $a_1, \ldots, a_m$ is fixed, such $b_1, \ldots, b_t$ can be found in polynomial time (as each facet of C is determined by r linearly independent vectors from $a_1, \ldots, a_m$).

Now it follows trivially from (3) and (4) that

$$C_0 = \{z \in C \mid \text{for all } i = 1, \ldots, m \text{ there exists } j = 1, \ldots, t \text{ with } b_j z < b_j a_i\}. \tag{5}$$

So if Φ denotes the collection of all functions $\phi : \{1, \ldots, m\} \to \{1, \ldots, t\}$, then

$$C_0 = \bigcup_{\phi \in \Phi} \{z \mid b_j z \geq 0 \text{ for } j = 1, \ldots, t, \text{ and } b_{\phi(i)} z < b_{\phi(i)} a_i \text{ for } i = 1, \ldots, m\}. \tag{6}$$

This expresses C_0 as a union of convex sets, and we have to test whether the only integer vector in each of these convex sets is the origin. Below we shall see that this can be done in polynomial time (for fixed dimension) with Lenstra's algorithm for integer linear programming [7]. (Note that Φ generally has exponential size, even for fixed rank r of $a_1, \ldots, a_m$.)

Let Z be the collection of vectors z determined by n linearly independent equations from:

$$\begin{aligned} &b_j z = 0 \qquad (j = 1, \ldots, t), \\ &b_j z = b_j a_i \quad (j = 1, \ldots, t;\ i = 1, \ldots, m). \end{aligned} \tag{7}$$

Since n is fixed, we can enumerate and store Z in polynomial time. Next let Σ be the collection of all subsets $\{z_1, \ldots, z_n\}$ of Z such that:

$$\begin{aligned} &\text{(i) } z_1, \ldots, z_n \text{ are linearly independent,} \\ &\text{(ii) } z_1, \ldots, z_n \text{ belong to } C, \\ &\text{(iii) } \forall i = 1, \ldots, m\ \exists j = 1, \ldots, t\ \forall k = 1, \ldots, n\colon b_j z_k \leqslant b_j a_i. \end{aligned} \tag{8}$$

Again, Σ can be enumerated and stored in polynomial time. Define

$$\sigma(z_1, \ldots, z_n) = \text{convex hull}\{0, z_1, \ldots, z_n\} \backslash \text{convex hull}\{z_1, \ldots, z_n\}, \tag{9}$$

for $\{z_1, \ldots, z_n\}$ in Σ. We finally show:

$$C_0 = \bigcup_{\{z_1, \ldots, z_n\} \in \Sigma} \sigma(z_1, \ldots, z_n). \tag{10}$$

we are finished as soon as we have proved (10): with Lenstra's algorithm we can test, for each $\{z_1, \ldots, z_n\}$ in Σ, whether $\sigma(z_1, \ldots, z_n) \backslash \{0\}$ contains integer vectors. Hence we can test whether C_0 contains integer vectors other than the origin.

To prove (10), first observe that C_0 is bounded. Indeed, $C_0 \subseteq \{\lambda_1 a_1 + \cdots + \lambda_m a_m \mid 0 \leqslant \lambda_i < 1 \text{ for } i = 1, \ldots, m\}$. Now let $w \in C_0$. Then by (6) there exists a function ϕ in Φ such that w belongs to the convex set

$$P = \{z \mid b_j z \geqslant 0 \text{ for } j = 1, \ldots, t, \text{ and } b_{\phi(i)} z < b_{\phi(i)} a_i \text{ for } i = 1, \ldots, m\}. \tag{11}$$

Since P is bounded and nonempty, it remains bounded if we replace in (11) the sign $<$ by $\leqslant$, thus obtaining the closure $\bar{P}$ of P. Since $w \in P$ there exists an $\varepsilon > 0$ such that $(1+\varepsilon)w$ belongs to P. As $0 \in \bar{P}$, $(1+\varepsilon)w$ is a convex combination of some linearly independent vertices $z_1, \ldots, z_n$ of $\bar{P}$. Then $\{z_1, \ldots, z_n\} \in \Sigma$ and $w \in \sigma(z_1, \ldots, z_n)$.

To prove the reverse inclusion for (10), let $\{z_1, \ldots, z_n\} \in \Sigma$ and $w \in \sigma(z_1, \ldots, z_n)$. There exists $\varepsilon > 0$ such that $(1+\varepsilon)w \in \sigma(z_1, \ldots, z_n)$. By (8) (ii) w belongs to C. Moreover, by (8) (iii), there exists a function ϕ in Φ such that $b_{\phi(i)} z_k \leqslant b_{\phi(i)} a_i$ for $i = 1, \ldots, m$ and $k = 1, \ldots, n$. Since $b_j z_k \geqslant 0$ for all $j = 1, \ldots, t$ and $k = 1, \ldots, n$, and since $z_1, \ldots, z_n$ are linearly independent, we know that $b_{\phi(i)} a_i > 0$ for all $i = 1, \ldots, m$. Therefore, $(1+\varepsilon) b_{\phi(i)} w \leqslant b_{\phi(i)} a_i$ for $i = 1, \ldots, m$, implies that $b_{\phi(i)} w < b_{\phi(i)} a_i$ for $i = 1, \ldots, m$, and hence that w belongs to C_0. □

The problem is still open whether testing total dual integrality is co-NP complete or polynomially solvable (or both).

References

[1] R. Chandrasekharan, "Polynomial algorithms for totally dual integral systems and extensions", *Annals of Discrete Mathematics* 11 (1981) 39–51.

[2] R. Chandrasekharan and S. Shirali, "Total weak unimodularity: Testing and applications", Report University of Texas at Dallas, 1981.

[3] J. Edmonds and R. Giles, "A min–max relation for submodular functions on graphs", *Annals of Discrete Mathematics* 1 (1977) 185–204.

[4] F.R. Giles and W.R. Pulleyblank, "Total dual integrality and integer polyhedra", *Linear Algebra and Its Applications* **25** (1979) 191–196.

[5] R. Kannan and A. Bachem, "Polynomial algorithms for computing the Smith and Hermite normal forms of an integer matrix", *SIAM Journal on Computing* 8 (1979) 499–507.

[6] L.G. Khachiyan, "A polynomial algorithm in linear programming", *Doklady Akademii Nauk SSSR* 244 (1979) 1093–1096 (English translation: *Soviet Mathematics Doklady* 20 (1979) 191–194).

[7] H.W. Lenstra, Jr., "Integer programming with a fixed number of variables", *Mathematics of Operations Research* 8 (1983) 538-548.

[8] A. Schrijver, "On total dual integrality", *Linear Algebra and Its Applications* 38 (1981) 27–32.

Mathematical Programming Study 22 (1984) 70–85
North-Holland

INACCURACY IN QUASI-NEWTON METHODS: LOCAL IMPROVEMENT THEOREMS

J.E. DENNIS, Jr.*
Mathematical Sciences Department, Rice University, Houston, TX 77251, USA

Homer F. WALKER**
Department of Mathematics, University of Houston, Houston, TX 77004, USA

Received 3 March 1983
Revised manuscript received 4 April 1984

In this paper, we consider the use of bounded-deterioration quasi-Newton methods implemented in floating-point arithmetic to find solutions to $F(x)=0$ where only inaccurate F-values are available. Our analysis is for the case where the relative error in F is less than one. We obtain theorems specifying local rates of improvement and limiting accuracies depending on the nearness to Newton's method of the basic algorithm, the accuracy of its implementation, the relative errors in the function values, the accuracy of the solutions of the linear system for the Newton steps, and the unit-rounding errors in the addition of the Newton steps.

Key words: Quasi-Newton Methods, Local Convergence, Floating-Point Errors, Limiting Accuracy.

1. Introduction

This paper is the first of several anticipated reports on the effects of inaccuracy, computational or otherwise, on quasi-Newton methods for solving nonlinear algebraic equations. In this report, we consider bounded-deterioration quasi-Newton methods and discuss the effects on their performance of inaccuracy from all sources, including not only computer arithmetic, but also the differences among the ideal problem, its mathematical model, and the computer implementation of the model. In later work, we will explore in greater depth the effects of specific sources of inaccuracy on the performance of particular least-change secant methods; in fact, we will suggest modifications of the standard versions of these methods which enable them to deal more effectively with certain types of inaccuracy. Here, however, our concern is with the general case, and our objective is to investigate rates of improvement and limiting accuracies that can be obtained near solutions. We would like eventually to provide heuristics for deciding how accurately the nonlinear residuals must be evaluated at the current point in order to improve the solution estimate.

* Research sponsored by DOE DE-AS05-82ER13016, ARO DAAG-79-C-0124, NSF MCS81-16779. This work was supported in part by the International Business Machine Corporation, Palo Alto Scientific Center, Palo Alto, CA.
** Research sponsored by DOE DE-AS05-82ER13016.

This would be very useful for certain parameter identification problems in which a system of differential equations must be integrated numerically to compute the residuals at the current parameter estimates, and the cost of the integration escalates with the required accuracy.

Urabe (1956), Collatz (1966), Lancaster (1966), and Ypma (1983) also have considered the same kinds of effects on general iterative methods. Glad and Goldstein (1977) have considered the effects of objective function inaccuracy on the unconstrained minimization problem. They suggested an algorithm to achieve a limiting accuracy consistent with ours and with the limiting expression obtained by Ypma. Although the work was done independently, our results are especially close to Ypma's, but he follows the tradition of the earlier papers on general methods in relating each term in the iteration sequence being generated to the iterate that would be generated were there no errors from any source. Our results, based directly on the computed sequence, are more easily applicable to methods with memory such as Broyden's method. In fact, applicability to the sparse Broyden method was our guiding light, and it will be a particular focus of subsequent work.

The problem of interest here is the following:

Problem. Given $F: \Omega \subseteq \mathbb{R}^n \to \mathbb{R}^n$, find $x_* \in \Omega$ satisfying

$$F(x_*) = 0. \tag{1.1}$$

The function F can be thought of as the idealized problem being modeled on the computer, or as the problem programmed into the computer, but whose values are being approximated to be used by implementations of methods which are modeled after Newton's methods. Accordingly, the following is assumed for the idealized problem throughout the sequel:

The standard hypothesis. Let F be differentiable in an open convex neighborhood Ω of a point $x_* \in \mathbb{R}^n$ for which $F(x_*) = 0$ and $F'(x_*)$ is nonsingular. Let $\gamma \geq 0$ and $p \in (0, 1]$ be such that for $x \in \Omega$,

$$|F'(x) - F'(x_*)| \leq \gamma |x - x_*|^p,$$

where $|\cdot|$ denotes a norm on $\mathbb{R}^n$ and its subordinate operator norm on $\mathbb{R}^{n\times n}$.

This standard hypothesis is sufficient to insure that sequences of iterates produced by Newton's method converge locally to x_* with q-order $(1+p)$. (See, for example, Dennis and Schnabel (1983) or Ortega and Rheinboldt (1970).)

The methods of interest for solving (1.1) numerically are iterative methods of the general quasi-Newton form

$$x_{k+1} = x_k - B_k^{-1} F(x_k), \tag{1.2}$$

in which B_k is regarded as an approximation of the Jacobian matrix $F'(x_k)$. (Since our concern here is with local behavior, we consider only situations in which the full step $-B_k^{-1}F(x_k)$ would be taken for each k, if it could be computed.) For generality, it is desirable to specify as little as possible about the manner in which the approximate Jacobians B_k in (1.2) are chosen; but for applicability of the results, it is useful to give some carefully chosen structure to the approximation rule. Otherwise, each application ends up requiring a contextual reproof of the general result. Therefore, we assume that there is some nonsingular matrix B_* and an *update function* U, defined in a neighborhood $N \subseteq \mathbb{R}^n \times \mathbb{R}^{n\times n}$ of (x_*, B_*) and having values which are subsets of $\mathbb{R}^{n\times n}$, such that $B_{k+1} \in U(x_k, B_k)$ for each k if $(x_k, B_k) \in N$.

The notion of an update function has proved very useful in the analysis of quasi-Newton methods (see Broyden, Dennis and Moré (1973) or Dennis and Walker (1981)). It provides an essentially completely general framework within which all of the popular and successful quasi-Newton methods can be described. To understand the role of B_*, one should think of it as being a distinguished approximation of $F'(x_*)$ among some set of allowable approximants. For example, in the case of the nonlinear Jacobi iteration (see Ortega and Rheinboldt (1970)), in which each B_k is a diagonal matrix having the same diagonal as $F'(x_k)$, one might take B_* to be the diagonal matrix having the same diagonal as $F'(x_*)$. In many applications, one naturally takes $B_* = F'(x_*)$.

Our concern here is with the effects of the inaccuracies which inevitably appear in any practical implementation of an iteration (1.2) to solve (1.1). These inaccuracies can arise from the following sources: the evaluation of an approximation to $F(x_k)$, the solution of a linear system to obtain the step $-B_k^{-1}F(x_k)$, the addition of the step to x_k to obtain x_{k+1}, and the determination of B_{k+1} from $U(x_k, B_k)$. While inaccuracy can potentially cause problems at any stage of an iteration (1.2), it will almost certainly degrade the performace of the iteration near x_*. In particular, if the iterates approach, but never equal x_*, then we will see below that inaccuracy in (1.2) causes the speed of their approach to decrease until further iterations may no longer produce additional accuracy in the iterates. Thus, it is more appropriate to call our results *local improvement theorems* rather than *local convergence theorems.*

Our specific purpose in this paper is to investigate the local improvement properties of an iteration (1.2) in the presence of inaccuracy. In the following, we first offer a general model of a practical implementation of (1.2) which takes into account inaccuracies arising from the sources listed above. Then we analyze the performance of the model near x_* to determine rates of improvement, or convergence, and limiting accuracies that can ultimately be obtained. As they should, the results of our analysis reduce to the standard results (see Theorem A2.1 of Dennis and Walker (1981)) for an iteration (1.2) when no inaccuracy is present. Finally, to illustrate the range of our results and to place them in some perspective, we consider the special case of stationary iteration for linear systems; for example, the analysis given here is essentially a generalization of that of Moler (1967) for LU iterative improvement.

2. Local improvement analysis

Our model of a practical implementation of (1.2) which takes inaccuracy into account is the following:

Iteration 2.1. Given x_0 and B_0, do the following until x_k is acceptable, no further progress can be made, or the iteration is no longer feasible:

(2.1i) Call for $F(x_k)$ and receive $F_k = F(x_k) + N(x_k)$.

(2.1ii) If $k > 0$, then compute B_k which differs from an element of $U(x_{k-1}, B_{k-1})$ by M_k.

(2.1iii) Ask for the solution of $B_k s_k = -F_k$ and get s_k, the solution of $(B_k + \Delta_k)s_k = -F_k$.

(2.1iv) Try to add s_k to x_k and get $x_{k+1} = (I + \Gamma_k)(x_k + s_k)$ for a diagonal matrix Γ_k.

In (2.1i), $N(x_k)$ is to be thought of as 'noise' or inaccuracy associated with the computation and approximation of $F(x_k)$. Our assumptions on $N(x_k)$ are flexible enough for it to be regarded as the combination of noise from all the relevant sources that account for the difference between F_k and $F(x_k)$. For example, if (1.1) is of the form $G(x) - b = 0$, where G models some phenomenon and b is data, then $N(x)$ can be regarded at the choice of the user as a combination of any or all of the following: the modeling error in G, the errors in the computer evaluation of G, the errors in b, and the cancellation error in forming $G(x) - b$.

Our results are for the setting where F can be evaluated to at least one significant figure for any iterate from which improvement is expected at the next iteration. There is a considerable body of literature on *stochastic approximation* algorithms, which are indended for use in solving (1.1) when $N(x_k)$ is random noise, i.e., an observation on a random variable (see Young (1976)). The case in which $N(x_k)$ is random noise is certainly not excluded here, but in practice, stochastic approximation algorithms are perhaps of greatest interest when $N(x_k)$ is likely to be fairly large relative to $F(x_k)$. Here we wish to study the effects of relatively small amounts of noise on quasi-Newton methods, since these methods are in common use for problems in which such noise occurs.

Similarly, M_k in (2.1ii) represents inaccuracy in the determination of B_k. For the purposes of this paper, one can regard it as the least troubling of the inaccuracy terms in Iteration 2.1; however, we feel that it should not be omitted altogether. There are two cases in which it seems fair to take $M_k = 0$. The first is the obvious one in which $B_k = B_{k-1}$, an option allowed in many implementations of methods of the form (1.2). The second is that in which B_k is determined in a manner independent of B_{k-1}, e.g., as an analytic or finite-difference evaluation of $F'(x_k)$. In this case, one can choose to regard B_k as being exact with any inaccuracy incurred in its evaluation absorbed in Δ_k in (2.1iii). Also, it is sometimes possible by increasing the constants to absorb M_k into the bounded deterioration assumptions made below on U. In the following, we suggest thinking of Δ_k and M_k as separate.

The effect of Δ_k in (2.1iii) on the iteration depends on the size of $B_k^{-1}\Delta_k$, which in practice depends on the stability of the linear solution algorithm, the arithmetic in which it is implemented, and the conditioning of B_k with respect to inversion. We are not concerned here with the particular solution algorithm used, so long as it is stable, and so we hypothesize a uniform bound $|B_k^{-1}\Delta_k| \le \mu < 1$. In practice μ is usually much smaller than 1, and one can properly regard it as such in the following. The assumption that μ is independent of k is reasonable, because for many algorithms implemented in arithmetic with unit rounding error ε_A, one has

$$|B_k^{-1}\Delta_k| \le \varphi(n)\kappa(B_k)\varepsilon_A, \tag{2.2}$$

independent of the right-hand side of the linear equation being solved. Here $\kappa(B_k) = |B_k| \cdot |B_k^{-1}|$ embodies dependence on k, but we observe in Theorem 2.5 and Remark 1 below that $\{|B_k|\}$ and $\{|B_k^{-1}|\}$ stay uniformly bounded locally. If one wishes (2.2) to hold for *every* nonsingular B_k, then the forms of $\varphi(n)$ that apply to various algorithms can be found in Stewart (1973), Wilkinson (1963), or Wilkinson (1965). These forms provide worst-case bounds; in particular, there are pathological cases for Gaussian elimination with partial pivoting that require $\varphi(n)$ to grow exponentially with n. However, as Dongarra et al. (1979) point out for these algorithms, $|B_k^{-1}\Delta_k|$ grows much more slowly with n in practice. Consequently, our hypothesis on $|B_k^{-1}\Delta_k|$ seems sound. In fact, it is reasonable to think of $\mu \approx \kappa(B_*)\varepsilon_A$ in the context of this paper.

The term Γ_k in (2.1iv) is indended to account for error incurred in adding x_k to s_k in floating-point arithmetic with unit rounding error ε_A, and so we assume that $|\Gamma_k| \le \varepsilon_A$.

It is convenient to establish two technical lemmas before introducing our main result.

Lemma 2.2. *Under the standard hypothesis, for any* $x \in \Omega$,

$$|F(x) - F'(x_*)(x - x_*)| \le \frac{\gamma|x - x_*|^{p+1}}{p+1}.$$

Proof. Let $e = x - x_*$. Then

$$\begin{aligned}|F(x) - F'(x_*)(x - x_*)| &= |F(x) - F(x_*) - F'(x_*)e| \\ &= \left|\left\{\int_0^1 [F'(x - te) - F'(x_*)]\,\mathrm{d}t\right\} e\right| \\ &\le \gamma\left[\int_0^1 (1-t)^p\,\mathrm{d}t\right] \cdot |e|^{p+1}\end{aligned}$$

and the result follows from elementary calculus.

Our second lemma is intended to shed light on the change in the error in successive approximate solutions at a single step of an iteration of the form (2.1). For convenience and clarity, we introduce a 'single-step iteration' and direct the lemma toward it. Note that the constants r and r_* need not be less than 1. We will follow the lemma with a simple illustrative example.

Iteration 2.3. Suppose that x and B are given.

(2.3i) Call for $F(x)$ and receive $F = F(x) + N(x)$.

(2.3ii) Ask for the solution of $Bs = -F$ and get s, the solution of $(B+\Delta)s = -F$.

(2.3iii) Try to add s to x and get $x_+ = (I+\Gamma)(x+s)$ for a diagonal matrix Γ.

Lemma 2.4. *Let F satisfy the standard hypothesis and let $B_* \in \mathbb{R}^{n\times n}$ have the property that B_*^{-1} exists and $|I - B_*^{-1}F'(x_*)| \le r_*$. Let ε_A, ε_F and r_A be positive constants such that $|\Gamma| \le \varepsilon_A$, $|N(x)| \le \varepsilon_F|F(x)|$ and $\varepsilon_A|x_*| \le r_A|x - x_*|$. Suppose also that B satisfies $|B - B_*| \le \delta < 1/|B_*^{-1}|$, which implies that B is invertible, and that $|B^{-1}\Delta| \le \mu < 1$. Then Iteration 2.3 is well defined and $|x_+ - x_*| \le r|x - x_*|$, where*

$$r = \frac{1+\varepsilon_A}{(1-\mu)(1-|B_*^{-1}|\delta)}\Big\{[(1-\mu)|B_*^{-1}|\delta + \mu]|B_*^{-1}F'(x_*)|$$

$$+|B_*^{-1}|\left[\frac{\gamma(1+\varepsilon_F)|x-x_*|^p}{p+1} + \varepsilon_F|F'(x_*)|\right]\Big\} + (1+\varepsilon_A)r_* + r_A.$$

Proof. For convenience, denote $e = x - x_*$, $e_+ = x_+ - x_*$, $\beta = |B_*^{-1}|$ and $F'_* = F'(x_*)$, Note that since $|B_*^{-1}(B - B_*)| \le \beta|B - B_*| \le \beta\delta < 1$, B^{-1} exists and is given by

$$B^{-1} = [I + B_*^{-1}(B - B_*)]^{-1}B_*^{-1}. \tag{2.4}$$

It follows that

$$B^{-1} = B_*^{-1} - [I + B_*^{-1}(B - B_*)]^{-1}B_*^{-1}(B - B_*)B_*^{-1}. \tag{2.5}$$

Similarly, since $|B^{-1}\Delta| \le \mu < 1$, $(B+\Delta)^{-1}$ exists and is given by

$$(B+\Delta)^{-1} = (I + B^{-1}\Delta)^{-1}B^{-1}, \tag{2.6}$$

one can write

$$(B+\Delta)^{-1} = B^{-1} - (I + B^{-1}\Delta)^{-1}B^{-1}\Delta B^{-1}. \tag{2.7}$$

With (2.4) and (2.5), one obtains from (2.7) that

$$(B+\Delta)^{-1} = B_*^{-1} - [I + B_*^{-1}(B - B_*)]^{-1}B_*^{-1}(B - B_*)B_*^{-1}$$
$$- (I + B^{-1}\Delta)^{-1}B^{-1}\Delta[I + B_*^{-1}(B - B_*)]^{-1}B_*^{-1}. \tag{2.8}$$

Also, it follows from (2.4) that

$$|B^{-1}| \le \frac{\beta}{1-\beta\delta}$$

and from this inequality and (2.6) that

$$|(B+\Delta)^{-1}| \leq \frac{\beta}{(1-\mu)(1-\beta\delta)}. \tag{2.9}$$

Since $(B+\Delta)^{-1}$ exists, Iteration 2.3 is well-defined and yields

$$x_+ = (I+\Gamma)\{x-(B+\Delta)^{-1}[F(x)+N(x)]\},$$

from which it follows that

$$\begin{aligned} e_+ &= (I+\Gamma)\{e-(B+\Delta)^{-1}[F(x)+N(x)]\}+\Gamma x_* \\ &= (I+\Gamma)\{[I-(B+\Delta)^{-1}F'_*]e \\ &\quad -(B+\Delta)^{-1}[F(x)-F'_*e]-(B+\Delta)^{-1}N(x)\}+\Gamma x_* \end{aligned}$$

Substituting (2.8) in this expression gives

$$\begin{aligned} e_+ = (I+\Gamma)\{&(I-B_*^{-1}F'_*)+\{[I+B_*^{-1}(B-B_*)]^{-1}B_*^{-1}(B-B_*) \\ &+(I+B^{-1}\Delta)^{-1}B^{-1}\Delta[I+B_*^{-1}(B-B_*)]^{-1}\}B_*^{-1}F'_*\}e \\ &-(I+\Gamma)(B+\Delta)^{-1}\{[F(x)-F'_*e]+N(x)\}+\Gamma x_*; \end{aligned}$$

and using (2.9) and Lemma (2.2), one obtains

$$\begin{aligned} |e_+| &\leq (1+\varepsilon_A)\left\{r_* + \left[\frac{\beta\delta}{1-\beta\delta}+\frac{\mu}{(1-\mu)(1-\beta\delta)}\right]|B_*^{-1}F'_*|\right\}|e| \\ &\quad +(1+\varepsilon_A)\frac{\beta}{(1-\mu)(1-\beta\delta)}\left\{\frac{\gamma|e|^p}{p+1}+\varepsilon_F\left(\frac{\gamma|e|^p}{p+1}+|F'_*|\right)\right\}|e|+r_A|e| \\ &= \left\{\frac{(1+\varepsilon_A)}{(1-\mu)(1-\beta\delta)}\left[\{(1-\mu)\beta\delta+\mu\}|B_*^{-1}F'_*|\right.\right. \\ &\quad \left.\left. +\beta\left\{\frac{\gamma(1+\varepsilon_F)|e|^p}{p+1}+\varepsilon_F|F'_*|\right\}\right]+(1+\varepsilon_A)r_*+r_A\right\}|e|. \end{aligned}$$

We are used to thinking of relative errors, so ε_F is easily interpreted, and r_A is just the ratio of the best possible relative error ε_A in representing x_* to the relative error in the current approximator x.

When (2.3) is applied to a function whose values are more inaccurate than one supposes, or when the naive user asks for excessive accuracy in x, typical behavior is for the iteration sequence to thrash around and neither converge nor diverge. Often this behavior is misdiagnosed as failure to provide a sufficiently good initial guess. The preceding lemma suggests an alternate explanation which we will illustrate with a simple example.

Consider $F(x)=x^2-2$, and suppose that the subroutine to evaluate F were implemented in 3-digitit arithmetic, by which we mean that x was rounded to 3 significant decimal digits, x^2 was likewise rounded, and 2 was subtracted. Note that we should not expect an absolute error in F of less than 10^{-2} because of the final subtraction. In order to isolate the influence of function inaccuracy on local behavior, we used Newton's method and arithmetic of sufficient accuracy in computing the Newton step so that x_+, defined with the given inaccurate F, was correctly computed to 6 significant decimal digits. Thus, we took $\varepsilon_A = 5\times 10^{-6}$ at every step and we felt justified in always taking $\mu = 0$. This caused r, the rate of improvement in the above lemma, to be simplified in each step of the example to $(r)_k \approx (\varepsilon_F)_k + (r_A)_k$. We took $(r_A)_k = \varepsilon_A/|e_k|$; here and in the example, $|e_k|$ will denote the relative error in x_k with respect to the HP15C value of $\sqrt{2}$.

At $x_0 = 1$, we computed $F_0 = -1.0$, $\varepsilon_F = 0$, $r_A \approx 2\times 10^{-5}$, which gives no indication of problems. Indeed, $x_1 = 1.5$, $|e_1| \approx 0.061$, was a good step. We found $r_A \approx 8.2\times 10^{-5}$, $F_1 = 0.25$, $\varepsilon_F = 0$, again no indication of problems. And we made a good step, to $x_2 = 1.41667$, so $|e_2| \approx 0.0017$. This time $r_A \approx 2.9\times 10^{-3}$, $F_2 = 0.02$, $\varepsilon_F \approx 1.9 \approx r$, which predicts $|e_3| \approx 0.0033$. If we only knew it, we should stop here since $r > 1$. As predicted, the error did increase with $x_3 = 1.40961$, $|e_3| \approx 0.0033$. But this time, $r_A \approx 1.5\times 10^{-3}$, $F_3 = -0.01$, $\varepsilon_F \approx 0.23$. Notice that we have now been thrown out far enough from x_* so that the cancellation error in $F(x)$ is not as bad, and $r \approx \varepsilon_F \approx 0.23$. This predicts $|e_4| \approx 0.23|e_3| \approx 0.00075$, and in fact, $|e_4| \approx 0.00075$. Now we are too close again, and $\varepsilon_F \approx 2.4 \approx r$. This behavior continued through the eight iterations we made with x_6, moving away far enough to get F_6 a relative error $\varepsilon_F \approx 0.23$, which was sufficient to make x_7 move back toward x_*, but then F_7 had a relative error $\varepsilon_F \approx 2.5$ and x_8 moved off again.

Theorem 2.5 below is our main result. In it, we assume that there is a norm $\|\cdot\|$ on $\mathbb{R}^{n\times n}$ in which a *bounded deterioration* inequality (2.10) holds for the update function U. This assumption is not a serious restriction, because the update functions appropriate for essentially all of the popular and successful quasi-Newton methods exhibit such bounded deterioration (see Broyden, Dennis and Moré (1973)). The theorem asserts that the iterates x_k will improve as approximations to x_* at a q-linear rate until one of three conditions is violated. These conditions can be categorized as representing respectively: the inaccuracies of the problem (condition (ii)), the inaccuracies in the basic algorithm (condition (iii)), and the ratio of the accuracy of the current iterate to the unit rounding error of the arithmetic of the implementation (condition (i)). One interpretation of the theorem is that suggested by Lemma 2.4 and the example above: If the relative errors in x_k and F_k are not too great, and if B_k is a sufficiently good approximate Jacobian, then one can expect the step to x_{k+1} to be an improvement. An alternate to this one-step interpretation is to think of choosing a good enough basic algorithm, implementing it with sufficient accuracy, and supplying it with accurate enough function values, so that improvement is obtained at a fast enough rate r until condition (i) with $r_A = \varepsilon_A/\text{relx}$ is the first to fail, where relx is the user supplied convergence condition on $|x_k - x_*|/|x_*|$.

In Theorem 2.5, we let η denote a consant for which $|A| \leq \eta\|A\|$ for every $A \in \mathbb{R}^{n\times n}$.

Theorem 2.5. *Let F satisfy the standard hypothesis and let $B_* \in \mathbb{R}^{n\times n}$ have the property* *that B_*^{-1} exists and $|I - B_*^{-1}F'(x_*)| \le r_* < 1$. Let U be defined in a neighborhood $N = N_1 \times N_2$ of (x_*, B_*) where $N_1 \subset \Omega$ and N_2 contains only nonsingular matrices. Assume that there are nonnegative constants α_1, α_2 such that for each $(x, B) \in N$ and x_+ given by (2.3), $B_+ \in U(x, B)$ satisfies*

$$\|B_+ - B_*\| \le [1 + \alpha_1 \sigma(x, x_+)^p]\|B - B_*\| + \alpha_2 \sigma(x, x_+)^p \tag{2.10}$$

for $\sigma(x, x_+) = \max\{|x - x_*|, |x_+ - x_*|\}$. Let ε_A, μ, ε_F and r_A be nonnegative constants such that $\mu < 1$ and

$$1 > r_0 \equiv \frac{(1+\varepsilon_A)}{(1-\mu)}[\varepsilon_F|B_*^{-1}| \cdot |F'(x_*)| + \mu|B_*^{-1}F'(x_*)|] + (1+\varepsilon_A)r_* + r_A.$$

Then for any r such that $r_0 < r < 1$ and any ε_r, δ_r such that

(a) *$(x, B) \in N$ whenever $|x - x_*| \le \varepsilon_r$ and $|B - B_*| \le \delta_r$,*

(b) *$\delta_r < 1/|B_*^{-1}|$,*

(c) $$r \ge r_0 + \frac{(1+\varepsilon_A)|B_*^{-1}|}{(1-\mu)(1-|B_*^{-1}|\delta_r)} \times \left[\frac{(1+\varepsilon_F)}{p+1}\gamma\varepsilon_r^p + (|B_*^{-1}F'(x_*)| + \varepsilon_F|B_*^{-1}| \cdot |F'(x_*)|)\delta_r\right]$$

the following is true:

If $|x_0 - x_*| \le \varepsilon_r$ and $|B_0 - B_*| \le \delta_r$, then any iteration of the form (2.1) with $|B_k^{-1}\Delta_k| \le \mu$ and $|\Gamma_k| \le \varepsilon_A$ is well defined and satisfies

$$|x_{k+1} - x_*| \le r|x_k - x_*| \quad \text{and} \quad |B_k - B_*| \le \delta_r \tag{2.11}$$

until the first value of $k \ge 0$ for which one of the following fails to hold:

(i) $\varepsilon_A|x_*| \le r_A|x_k - x_*|$;

(ii) $|N(x_k)| \le \varepsilon_F|F(x_k)|$;

(iii) *if $k > 0$, then*

$$\|B_0 - B_*\| + \sum_{i=1}^{k}\|M_i\| + \frac{\alpha_2}{1-r^p}|x_0 - x_*|^p \le \frac{\delta_r}{\tau\eta},$$

where

$$\tau = \prod_{i=0}^{\infty}[1 + \alpha_1|x_0 - x_*|^p r^{pi}].$$

Remark 1. In light of (2.11) and the hypotheses of the theorem, one sees that not only $|B_k|$ but also $|B_k^{-1}|$, $|B_k + \Delta_k|$, and $|(B_k + \Delta_k)^{-1}|$ are bounded independent of k as long as (i), (ii), and (iii) hold.

Remark 2. Suppose that for each , $F(x_k)$ and B_k are determined exactly (so that $N(x_k) = 0$ and $M_k = 0$) and the step s_k, which may be determined inexactly, is added exactly to x_k (so that $\Gamma_k = 0$). Then one can take $\varepsilon_F = \varepsilon_A = 0$ and obtain from the conclusion of the theorem that $\{x_k\}$ converges to x_* q-linearly with rate constant at most r, provided

$$\|B_0 - B_*\| + \frac{\alpha_2}{1-r^p}|x_0 - x_*|^p \leq \frac{\delta_r}{\tau\eta}.$$

Proof. Suppose that the hypotheses of the theorem hold and that $|x_0 - x_*| \leq \varepsilon_r$ and $|B_0 - B_*| \leq \delta_r$ for some r such that $r_0 < r < 1$ and some positve ε_r, δ_r satisfying (a), (b) and (c). We prove the theorem by induction on k.

If (i) and (ii) hold for $k=0$ (and if $|B_0^{-1}\Delta_0| \leq \mu$ and $|\Gamma_0| \leq \varepsilon_A$), then an iteration of the form (2.1) is well-defined for $k=0$ and $|x_1 - x_*| \leq r|x_0 - x_*|$ by Lemma 2.4.

As an inductive hypothesis, suppose that for some $j > 0$, an iteration of the form (2.1) is well-defined and satisfies (2.11) for $k = 0, \ldots, j-1$. One has that

$$\sigma(x_{k+1}, x_k) = |x_k - x_*| \leq r^k|x_0 - x_*| \tag{2.12}$$

for $k = 0, \ldots, j-1$. If B_j is determined as in (2.1ii), then (2.12) and repeated applications of (2.10) yield

$$\|B_j - B_*\| \leq \tau_{jj}\|B_0 - B_*\| + \sum_{k=0}^{j-1} \tau_{jk}[\|M_{j-k}\| + \alpha_2 r^{p(j-k-1)}|x_0 - x_*|^p], \tag{2.13}$$

where

$$\tau_{jk} = \begin{cases} 1 & \text{if } k=0. \\ \prod_{i=1}^{k} [1 + \alpha_1 r^{p(j-i)}|x_0 - x_*|^p] & \text{if } 1 \leq k \leq j. \end{cases}$$

It follows from (2.13) that

$$|B_j - B_*| \leq \tau\eta\left\{\|B_0 - B_*\| + \sum_{k=1}^{j} \|M_k\| + \frac{\alpha_2}{1-r^p}|x_0 - x_*|^p\right\}.$$

Consequently, if (iii) is satisfied for $k=j$, then $|B_j - B_*| \leqq \delta_r$ and B_j^{-1} exists. If (i) and (ii) are also satisfied for $k=j$ (and if $|B_j^{-1}\Delta_j| \leqq \mu$ and $|\Gamma_j| \leq \varepsilon_A$), then Lemma 2.4 implies that x_{j+1} is well-defined by (2.1iii) and (2.1iv) and satisfies $|x_{j+1} - x_*| \leq r|x_j - x_*|$. This completes the induction, and the theorem is proved.

3. Stationary iteration for linear systems

We now consider the special case of (1.1) in which the system to be solved is linear, i.e., $F(x) = Ax - b$ for $A \in \mathbb{R}^{n\times n}$ and $b \in \mathbb{R}^n$. The iteration of interest is a stationary iteration

$$x_{k+1} = x_k - B^{-1}F(x_k),$$

in which $B \in \mathbb{R}^{n\times n}$ is a convenient invertible approximation of $A = F'(x_k)$. Of course, this iteration is just (1.2) with $B_k = B$ for each k. Classical examples of such an iteration are Jacobi iteration, Gauss–Seidel iteration, and iterative refinement. In theory, the convergence properties of this iteration are determined by how well B approximates A. For example, iterative refinement in its ideal form uses $B = A$ and therefore produces the exact solution in one step if no inaccuracy enters the iteration. In practice, of course, inaccuracy not only affects the rate of improvement of the iterates but also the limiting accuracy that can be obtained.

We discuss below the implications of the results of the preceding section for an iteration of this type. Our practical model is that obtained from Iteration 2.1 in the case of interest, i.e., the following:

Iteration 3.1. Given x_0, do the following until x_k is acceptable or no further progress can be made:

(3.1i) Call for $F(x_k)$ and receive $F_k = F(x_k) + N(x_k)$.

(3.1ii) Ask for the solution of $Bs_k = -F_k$ and get s_k, the solution of $(B+\Delta_k)s_k = -F_k$.

(3.1iii) Try to add s_k to x_k and get $x_{k+1} = (I+\Gamma_k)(x_k+s_k)$ for a diagonal matrix Γ_k.

The development here parallels and extends to a large extent the treatment of iterative refinement by Moler (1967). In particular, Iteration 3.1 reduces to his model when $B = A$; and our results share many of the same essential features in this case.

Our objective is to determine rates of improvement and limits of accuracy that can be obtained by Iteration 3.1 near $x_* = A^{-1}b$. Our plan is first to apply Theorem 2.5 in a straightforward way and then to obtain more refined conclusions by making more specific assumptions about the context in which the iteration is carried out and looking more closely at the bounds μ and ε_F. One has quite a bit of flexibility in applying Theorem 2.5, most of it centering around the choice of B_*. For example, B_* can be taken to be $F'(x_*) = A$ (even if this matrix cannot be observed), or the representation of A in the computer, if this differs from A, or the matrix B used to define the iteration matrix. In fact, it might be of interest to take B_* to be the matrix that would define an ideal iteration matrix and let B be the product of its computed factors. We leave the full exploration of these possibilities to the reader and take $B_* = B$ below, a choice which gives the sharpest results in the cases which we believe will most often be of interest.

We apply Theorem 2.5 with $F'(x_*) = A$, $B_* = B$ and $\gamma = \delta_r = 0$. If $|I - B^{-1}A| = r_* < 1$ and if ε_A, μ, ε_F and r_A are nonnegative constants such that $\mu < 1$ and

$$1 > r_0 \equiv \frac{(1+\varepsilon_A)}{(1-\mu)}[\varepsilon_F|B^{-1}||A| + \mu|B^{-1}A|] + (1+\varepsilon_A)r_* + r_A, \tag{3.2}$$

then Theorem 2.5 immediately implies that for any x_0, Iteration 3.1, with $|B^{-1}\Delta_k| \le \mu$ and $|\Gamma_k| \le \varepsilon_A$, is well-defined and the iterates $\{x_k\}$ satisfy $|x_{k+1} - x_*| \le r_0|x_k - x_*|$ as long as both $|N(x_k)| \le \varepsilon_F|F(x_k)|$ and $\varepsilon_A|x_*| \le r_A|x_k - x_*|$.

We note in passing that if one makes the alternative choices $F'(x_*) = B_* = A$, $\gamma = 0 = r_*$ and $\delta_r = |B - A|$, then for appropriate ε_A, μ, ε_F, r_A and δ_r, Theorem 2.5 yields the linear convergence rate constant

$$\frac{(1+\varepsilon_A)}{(1-\mu)(1-|A^{-1}|\delta_r)}[(1-\mu)|A^{-1}|\delta_r + \mu + \varepsilon_F \kappa(A)] + r_A.$$

But this rate constant is at least as great as r_0 given by (3.2), since

$$|B^{-1}A| \leq \frac{1}{1-|A^{-1}|\delta_r}, \qquad |B^{-1}||A| \leq \frac{\kappa(A)}{1-|A^{-1}|\delta_r}$$

$$\text{and} \quad |I - B^{-1}A| \leq \frac{|A^{-1}|\delta_r}{1-|A^{-1}|\delta_r}.$$

To draw more refined conclusions, let us consider more closely the two terms in brackets in (3.2), which are associated with the inaccuracy in evaluating F and the inaccuracy in solving linear systems with coefficient matrix A. Below, we obtain more detailed results from a useful expression for μ and a reasonable value of ε_F such that $|N(x_k)| \leq \varepsilon_F |F(x_k)|$ whenever $\varepsilon_A |x_*| \leq r_A |x_k - x_*|$. We will then be able to determine how much improvement, if any, can be guaranteed by an iteration of (3.1) when there is a relative error of at least ε_A / r_A in x_k. As is customary in discussions of iterative refinement, we assume that all computation except that involved in the computation of F is done in some 'working precision' arithmetic, while an approximate value of F is determined in some other (perhaps higher) precision arithmetic, and then rounded to working precision. In the following, we take $\varepsilon_1 = \varepsilon_A$ to be the unit rounding error in the working precision arithmetic and denote by ε_2 the unit rounding error in the arithmetic used in the evaluation of F. For convenience, we also assume that $|\cdot|$ is the l_∞ vector norm, i.e., that $|v| = \max_i |v_i|$ for $v = (v_1, \ldots, v_n)^{\mathrm{T}} \in \mathbb{R}^n$, although the results below hold without substantive change for any vector norm.

To determine a useful expression for μ, we suppose either that one of the standard factorization techniques is applied to B in arithmetic with unit rounding error ε_1, or that B is already given in factored form. In the first case, (2.2) holds with $B_k = B$ and $\varepsilon_A = \varepsilon_1$. Based on the discussion surrounding (2.2), one should be able to take μ far less than one unless A is very ill-conditioned. For convenience, we follow Moler (1967) and simple write

$$\mu = \rho \kappa(B) \varepsilon_1 \tag{3.3}$$

for an appropriate value of ρ, e.g., for the smallest ρ such that $|B^{-1}\Delta_k| \leq \rho\kappa(B)\varepsilon_1$ for all k. In the second case, one should have $|B^{-1}\Delta_k| \leq \rho\kappa(B)\varepsilon_1$ with a relatively small value of ρ, since Δ_k only accounts for the small errors in solving the factor systems. Thus we express μ by (3.3) in either case.

To determine ε_F, we take the novel point of view that F_k is determined in (3.1i) by performing an inaccurate evaluation of a function $\tilde{F}(x) = \tilde{A}x - \tilde{b}$ defined by

$\tilde{A} \in \mathbb{R}^{n \times n}$ and $\tilde{b} \in \mathbb{R}^n$ which are regarded as approximations of A and b, respectively. In practice, $\tilde{A}$ and $\tilde{b}$ arise as inaccurate representations within a computer of some ideal A and b associated with a problem of interest. For example, $\tilde{A}$ might be the floating-point representations of measured values of some independent variables A that produced observations of some dependent variable b whose floating-point representations are $\tilde{b}$. This point of view is taken not only for the sake of generality but also to illustrate another point which we feel is very important: the treatment of inaccuracy given here allows consideration of the inaccuracy which arises from the formulation of a problem itself, i.e., from the difference between the problem which one actually tries to solve and the problem which one ideally would like to solve. Of course, it is always possible to take $\tilde{A} = A$ and $\tilde{b} = b$, which is the traditional way. Proceeding from this point of view, we denote the inaccuracy in the evaluation of $\tilde{F}(x_k)$ by $\tilde{N}(x_k)$ and have $F_k = \tilde{F}(x_k) + \tilde{N}(x_k) = F(x_k) + N(x_k)$, where

$$N(x_k) = \tilde{F}(x_k) - F(x_k) + \tilde{N}(x_k). \tag{3.4}$$

Under the supposition that $\varepsilon_1 |x_*| \leq r_A |x_k - x_*|$, we derive reasonable relative bounds on $|\tilde{F}(x_k) - F(x_k)|$ and $|\tilde{N}(x_k)|$. In order to arrive at relatively simple inequalities which embody essential qualitative features, a few somewhat generous substitutions are made below. Also, it is assumed from here on that either $b \neq 0$ or $\tilde{b} = b$ in order that the error in $\tilde{b}$ as an approximation of b be expressible in relative terms. First, one has

$$|\tilde{F}(x_k) - F(x_k)| \leq |\tilde{A} - A||x_k| + |\tilde{b} - b|; \tag{3.5}$$

and, since $|x_k - x_*| \leq |A^{-1}||F(x_k)|$ and

$$|b| \leq |A||x_*| \leq \frac{r_A}{\varepsilon_1} \kappa(A)|F(x_k)|, \tag{3.6}$$

$$|\tilde{A} - A||x_k| + |\tilde{b} - b| \leq \left(1 + \frac{r_A}{\varepsilon_1}\right)[|\tilde{A} - A||A^{-1}| + \kappa(A)|\tilde{b} - b||b|^+]|F(x_k)|, \tag{3.7}$$

where the superscript '+' denotes pseudo-inverse, i.e., $|b|^+ = 0$ if $b = 0$ and $|b|^+ = |b|^{-1}$ otherwise. It follows from (3.7) that

$$|\tilde{F}(x_k) - F(x_k)| \leq \left(1 + \frac{r_A}{\varepsilon_1}\right)[|\tilde{A} - A||A^{-1}| + \kappa(A)|\tilde{b} - b||b|^+]|F(x_k)|. \tag{3.8}$$

Now, to bound $\tilde{N}(x_k)$, we make from here on the assumptions of Moler (1967) that $n\varepsilon_2 \leq 0.1$ and $1.06(1 + \varepsilon_1) \leq 1.2$ and obtain from the inequality *preceding* his (7), p. 318, that

$$|\tilde{N}(x_k)| \leq \varepsilon_1 |\tilde{F}(x_k)| + 1.2\varepsilon_2 |\tilde{b}| + 1.2(n+2)\varepsilon_2 |\tilde{A}||x_k|. \tag{3.9}$$

Using the triangle inequality, (3.5), (3.6) and (3.7), one sees from (3.9) that

$$\begin{aligned}|\tilde{N}(x_k)| &\le \varepsilon_1|F(x_k)| + 1.2\varepsilon_2|b| + 1.2(n+2)\varepsilon_2|A||x_k| + [\varepsilon_1 + 1.2(n+2)\varepsilon_2]\\ &\quad \times(|\tilde{A}-A||x_k| + |\tilde{b}-b|)\\ &\le \{[\varepsilon_1 + 1.2(n+3)(1+r_A/\varepsilon_1)\varepsilon_2\kappa(A)]\\ &\quad + [\varepsilon_1 + 1.2(n+2)\varepsilon_2](1+r_A/\varepsilon_1)\\ &\quad \times[|\tilde{A}-A||A^{-1}| + \kappa(A)|\tilde{b}-b||b|^+]\}|F(x_k)|. \end{aligned} \tag{3.10}$$

From (3.4), (3.8) and (3.10), one obtains the following:
If $\varepsilon_1|x_*| \le r_A|x_k - x_*|$, then $|N(x_k)| \le \varepsilon_F|F(x_k)|$, where

$$\begin{aligned}\varepsilon_F &= [\varepsilon_1 + 1.2(n+3)(1+r_A/\varepsilon_1)\varepsilon_2\kappa(A)]\\ &\quad + [1+\varepsilon_1 + 1.2(n+2)\varepsilon_2](1+r_A/\varepsilon_1)[|\tilde{A}-A||A^{-1}| + \kappa(A)|\tilde{b}-b||b|^+]. \end{aligned} \tag{3.11}$$

To proceed toward a detailed statement of the convergence properties of Iteration 3.1, we define for convenience

$$\sigma = \frac{1+\varepsilon_1}{1-\rho\varepsilon_1\kappa(B)}; \tag{3.12}$$

$$\tau = [1+\varepsilon_1 + 1.2(n+2)\varepsilon_2][|\tilde{A}-A||A^{-1}| + \kappa(A)|\tilde{b}-b||b|^+]; \tag{3.13}$$

$$\begin{aligned}C_1 &= \sigma\{[\varepsilon_1 + 1.2(n+3)\varepsilon_2\kappa(A) + \tau]|B^{-1}||A| + \rho\kappa(B)\varepsilon_1|B^{-1}A|\}\\ &\quad + (1+\varepsilon_1)|I - B^{-1}A|; \end{aligned} \tag{3.14}$$

$$C_2 = 1 + \frac{\sigma}{\varepsilon_1}[1.2(n+3)\varepsilon_2\kappa(A) + \tau]|B^{-1}||A|. \tag{3.15}$$

With μ and ε_F given by (3.3) and (3.11), respectively, (3.2) becomes

$$1 > r_0 \equiv C_1 + C_2 r_A. \tag{3.16}$$

Note that in order for (3.16) to be satisfied, it is necessary that $C_1 < 1$ and $r_A < r_A^* = (1-C_1)/C_2$.

Our results for Iteration 3.1 in the above context are stated in the following theorem, the proof of which is immediate from the preceding discussion.

Theorem 3.2. *Let ρ be such that $\rho\varepsilon_1\kappa(B) < 1$ and let σ, τ, C_1 and C_2 be defined by (3.12)–(3.15). For any x_0, the iterates $\{x_k\}$ produced by Iteration 3.1, with $|B^{-1}\Delta_k| \le \rho\varepsilon_1\kappa(B)$ and $|\Gamma_k| \le \varepsilon_1$, are well-defined and satisfy the following: If $C_1 < 1$ and $r_A < r_A^* = (1-C_1)/C_2$ and if $\varepsilon_1|x_*| \le r_A|x_k - x_*|$ for some k, then $|x_{k+1} - x_*| \le (C_1 + C_2 r_A)|x_k - x_*|$.*

We conclude with some remarks about Theorem 3.2. Under the assumption of the theorem, if $C_1 < 1$ and $r_A < r_A^* = (1 - C_1)/C_2$, then eventually a relative bound

$$\frac{|x_k - x_*|}{|x_*|} < \frac{\varepsilon_1}{r_A}$$

will be achieved for some k (provided $x_* \neq 0$). Note in particular that the infimum of such relative bounds is ε_1/r_A^* and that the q-linear convergence rate constant $(C_1 + C_2 r_A)$ approaches 1 as r_A approaches r_A^*. Also, C_1 and C_2 both decrease as ε_2 is decreased; therefore, greater rates of improvement and greater limiting accuracies can be obtained with smaller values of ε_2. Of course, the practical worth of taking ε_2 to be anything other than either ε_1 or the double precision choice $\varepsilon_2 = \varepsilon_1^2$ is questionable in most situations because of the cost of doing so. However, it is not difficult to imagine circumstances (especially in the case of iterative refinement as considered by Moler, in which $A = B = \tilde{A}$ and $b = \tilde{b}$) such that a very small choice of ε_2 results in much faster convergence and much greater limiting accuracy being assured by Theorem 3.2 than the choices $\varepsilon_2 = \varepsilon_1$ and $\varepsilon_2 = \varepsilon_1^2$.

4. Acknowledgements

We wish to thank Richard Byrd for goading us into this research. Since we obtained these results in the summer of 1982, we have had the benefit of discussing them with too many colleagues to name. Such discussions, especially with Trond Steihaug and Richard Tapia, where a great help in formulating the results to better communicate their implications. We would like to thank the referees for carefully reading the manuscript.

References

C.G. Broyden, J.E. Dennis Jr and J.J. Moré, "On the local and superlinear convergence of quasi-Newton methods", *Journal of the Institute of Mathematics and its Applications* 34 (1973) 223–245.

L. Collatz, *Functional analysis and numerical analysis* (Academic Press, New York, 1966).

J.E. Dennis Jr. and R.B. Schnabel, *Numerical methods for unconstrained optimization and nonlinear equations* (Prentice-Hall, Englewood Cliffs, 1983).

J.E. Dennis Jr. and H.F. Walker, "Convergence theorems for least-change secant update methods", *SIAM Journal on Numerical Analysis* 18 (1981) 949–987.

J.J. Dongarra, J.R. Bunch, C.B. Moler and G.W. Stewart, *LINPACK users' guide* (SIAM Publications, Philadelphia, 1979).

T. Glad and A. Goldstein, "Optimization of functions whose values are subject to small errors", *BIT; Nordisk Tidskrift for Informations-behandling* 17 (1977) 160–169.

P. Lancaster, "Error analysis for the Newton–Raphson method", *Numerische Mathematik* 9 (1966) 55–68.

C.B. Moler, "Iterative refinement in floating-point", *Journal of the Association for Computing Machinery* 14 (1967) 316–321.

J.M. Ortega and W.C. Rheinboldt, *Iterative solution of nonlinear equations in several variables* (Academic Press, New York, 1970).

G.W. Stewart, *Introduction to matrix computations* (Academic Press, New York, 1973).

M. Urabe, "Error estimation in numerical solution of equations by iterations process', *Journal of Sceince of Hiroshima University* 19 (1956) 479–489

J.H. Wilkinson, *Rounding errors in algebraic processes* (Prentice-Hall, Englewood Cliffs, 1963).

J.H. Wilkinson, *The algebraic eigenvalue problem* (Oxford University Press, Oxford, 1965).

P. Young, "Optimization in the presence of noise—a guided tour", in: L.C.W. Dixon, ed., *Optimization in action* (Academic Press, London, 1976) pp. 517–573.

T.J. Ypma, "The effect of rounding errors on Newton-like methods", *Institute of Mathematics and its Applications Journal for Numerical Analysis* 3 (1983) 109–118.

Mathematical Programming Study 22 (1984) 86–98
North-Holland

ORIENTATION OF MATRICES

J. FONLUPT and M. RACO
Laboratoire Artemis, U.S.M.G., B.P. 68, 38402 St. Martin D'Heres Cedex, France

Received 24 August 1983
Revised manuscript received 30 May 1984

Camion proved that every real-valued matrix A can be transformed by pivoting operations and nonzero multiplications of columns into a nonnegative matrix. In this paper we describe a finite algorithm to make this transformation, based on the results of Camion. Our main result is that when A is a totally unimodular matrix this transformation can be made by a polynomial algorithm.

Key words: Linear Algebra, Matroid Theory, Total Unimodularity.

1. Introduction

Let $\mathcal{M} = (E, \mathcal{F})$ be a matroid with $|E| = m$ elements, of rank n, representable over the real field. An interesting question concerning the representability of $\mathcal{M}$ is the following one: does there exist a representation of $\mathcal{M}$ such that all the vectors of this representation are nonnegative? A stronger formulation of this problem is: given a real $n \times m$ matrix A that we shall assume to be in standard form (there exists an $n \times n$ submatrix of A which is the identity matrix), can we transform this matrix into a nonnegative matrix by application of the two following operations:

- Pivoting operations similar to pivoting operation in linear programming. (α)
- Replacing a column vector by its negative. (β)

This operation will be called orientation of a matrix.

The answer to these two problems is affirmative. This result, due to Camion [1], was recently extended to some classes of oriented matroids, see [3] or [5]. The first part of this paper is the description of a finite algorithm to solve this problem. This algorithm, whose main ideas are in [1], is followed by some comments and a geometrical interpretation.

Contrasting with the general situation is the special case of regular matroids (regular matroids are those which can be represented by totally unimodular matrices). Here it is possible to use deeper results about the structure of matrices to provide a polynomial algorithm to transform a totally unimodular (t.u.) matrix A into a t.u. matrix $\bar{A}$ by successive applications of operations (α) and (β). This is developed in Section 3 of this paper.

The algorithm is based on the idea of connectivity for matroids or matrices. This important notion was developed by Tutte as a main tool for the study of matroids [8].

Our algorithm appears also as a consequence of a result concerning binary matroids without Fano minors. In this sense, this study is strongly related to some recent works concerning separating cocircuits in binary matroids [2, 8].

Note that the main interest of our algorithm is that the study of t.u. matrices $A = \|A_i^j\|$ with $A_i^j \in \{-1, 0, 1\}$ can be reduced to the study of (0, 1) t.u. matrices. This may be of great interest for new algorithms recognizing graphic or cographic matroids or even for deciding whether or not a given binary matroid is a regular matroid.

For instance it can be checked by inspection that the matroid R_{10} which plays a crucial role in the study of the structure of regular matroids, has a unique [7], standard representation as a (0, 1) t.u. standard matrix.

2. Orientation of real matrices

Definition

– Given a real $n \times m$ matrix A, a *basis* is an index subset of columns $J = \{j_1, j_2, \ldots, j_n\}$ such that the submatrix $A^J = [A^{j_1}, A^{j_2}, \ldots, A^{j_n}]$ is nonsingular.

– A is said to be in *standard form* with respect to a basis J if A^J is the $n \times n$ identity matrix

– A real matrix $\bar{A}$ is said to be *equivalent* to a matrix A, if we can transform A into $\bar{A}$ by pivoting operations and column multiplications by -1. Let A be an $n \times m$ matrix in standard form with respect to a basis. The set of all equivalent matrices to A will be denoted by $\varepsilon(A)$.

Remark. If $\bar{A} \in \varepsilon(A)$ and $\bar{A}'$ is obtained from $\bar{A}$ by multiplying the ith row by -1, then $\bar{A}' \in \varepsilon(A)$.

Proof. If e_i denotes the unit vector of $\mathbb{R}^n$ with 1 in the ith position and 0 elsewhere, then one of the columns of $\bar{A}$ is either e_i or $-e_i$. If this column is e_i, then multiply this column by -1, pivot on the only nonzero element of this column, and multiply the pivot column by -1.

The resulting matrix is $\bar{A}'$. If this column is $-e_i$ then $\bar{A}'$ is obtained by pivoting on the only nonzero element of this column. Thus, $\bar{A}'$ is obtained from $\bar{A}$ by pivoting operations and column multiplications by -1.

Definition

– A real matrix is said to be *well-oriented* if all its elements are nonnegative.

– *To orient* a real matrix A means to transform it into an equivalent well-oriented matrix $\bar{A}$.

The main theorem [Camion 1]. *Each real-valued matrix is equivalent to a well-oriented matrix.*

In his publication [1] Camion does not describe a complete algorithm to prove his result. However, in some respects his proof is algorithmic and in particular the lexicographic order used here was introduced by him.

Assume that A is in standard form with respect to a basis, and that each column of A contains at least one nonzero element. (Starting from an $n \times m$ matrix of rank n, it is always possible by pivoting operations to get from this matrix a matrix in standard form.)

Remark. $\varepsilon(A)$ is a finite set. This can be shown as follows: If A^+ denotes the set of column vectors of A and $A^- = -A^+$, a matrix equivalent to A is formed by m vectors of the set $A^+ \cup A^-$ with respect to a basis of n vectors in $A^+ \cup A^-$.

We can assume, without loss of generality, that the matrix A is in standard form with respect to the basis $J_0 = \{1, 2, \ldots, n\}$.

Let us define an ordering of the indices of the elements of A, before presenting the orientation algorithm.

(i, j) is said to be *less than* (i', j') if and only if $(i < i')$ or $(i = i'$ and $j < j')$.

The orientation algorithm

Input: an $n \times m$-matrix A in standard form with respect to a basis $J_0 = \{1, 2, \ldots, n\}$.

Output: a well-oriented matrix equivalent to A, in standard form with respect to a basis.

Step 1. Make positive the first nonzero element of each row and each column of A by the following operations: For each row i, $i := 1$ to n, do:

Let S_i^+ (resp. S_i^-) be the set of the column indices whose first non-zero elements are positive (resp. negative) and lie on the row i.

Let $S_i = S_i^+ \cup S_i^-$. We have $S_1 \cup S_2 \cup \cdots \cup S_n = \{1, 2, \ldots m\}$ and $S_i \cap S_j = \emptyset$, $i \neq j$.

(1.1) If the first nonzero element of row i is negative, then multiply this row by -1, and then multiply the column whose indices are in S_i^+ by -1.

If the first nonzero element of the row i is positive, then multiply the column whose indices are in S_i^- by -1. Let $\bar{A}$ be the resulting matrix. Go to Step 2.

Step 2. If $\bar{A}$ is well-oriented then stop. Otherwise, let (i, j) be the smallest pair of indices such that $\bar{A}_i^j$ is negative. Pivot at the (i, j)th position. Go to Step 1.

To prove that this algorithm is finite, we define a pre-order relation on all $n \times m$-matrices and we show that, after each iteration, the resulting matrix is strictly greater than the preceding ones, according to this order relation. $\varepsilon(A)$ being finite, our algorithm is finite.

If B is an $n \times n$-matrix, B is said to be *l-positive* (denoted by $B > 0$) if its least-indexed nonzero element is positive. Given two $n \times n$ matrices B and B', we say that B is *l-greater* than B' if $B - B' > 0$.

Let $J_0=\{1,2,\ldots,n\}$. The *l*-greater relation defined on all $n\times n$-matrices induces a pre-order relation on all $n\times m$-matrices as follows:

Given two $n\times m$-matrices A and $\bar{A}$, $\bar{A}$ is said to be *l-greater* than A if and only if $\bar{A}^{J_0}>A^{J_0}$.

Let us show that if A is transformed into $\bar{\bar{A}}$ by an iteration (steps 1 and 2) of the algorithm, then $\bar{\bar{A}}$ is *l*-greater than A. Let $\bar{A}$ be the result obtained by Step 1 on A. We have: $\bar{A}\geqslant A$ because a row (or a column) of A is multiplied by -1 only if its first nonzero element is negative. Thus, the least indexed negative element of this row (or column) becomes positive.

We execute Step 2 if and only if we detect a negative element, say at the (i,j)th position of $\bar{A}$.

Let $\bar{A}_r^j$ (resp. $\bar{A}_i^s$) be the first nonzero element of the jth column (resp. ith row) of A. Note that $s\in J_0$ since $\bar{A}_i^{J_0}$ is a row of the nonsingular matrix $\bar{A}^{J_0}$. The execution of Step 1 insures that $\bar{A}_r^j$ and $\bar{A}_i^s$ are strictly positive.

The least-indexed element of $\bar{A}$ modified by the pivoting operation of Step 2 is $\bar{A}_r^s$. It is transformed into:

$$\bar{\bar{A}}_r^s=\bar{A}_r^s-\bar{A}_r^j\cdot\frac{\bar{A}_i^s}{\bar{A}_i^j}>\bar{A}_r^s$$

therefore $\bar{\bar{A}}>\bar{A}$. We have $\bar{A}\geqslant A$ and $\bar{\bar{A}}>\bar{A}$. Thus $\bar{\bar{A}}>A$.

Geometrical interpretation

Recall that a subset K of the vector space $\mathbb{R}^n$ is said to be a pointed *cone* if:

$$x\in K \text{ and } \lambda\in\mathbb{R}^+ \text{ (positive real)} \;\Rightarrow\; \lambda x\in K.$$

If S is a family of vectors of $\mathbb{R}^n$, a positive combination of elements of S is a vector of the form:

$$\lambda_1x_1+\lambda_2x_2+\cdots+\lambda_kx_k$$

where

$$\lambda_i\geqslant 0 \text{ and } x_i\in S,\quad i=1,2,\ldots,k.$$

The *cone generated by S*, is the set of all positive combinations of elements of S. If $K(S)$ denotes the cone generated by S, the set

$$K(S)\cup -K(S)$$

is called the *bicone generated by S*.

Let an $n\times m$-matrix A be of rank n. Consider the column A^j of A as the representation of a vector $v_j\in\mathbb{R}^n$ by its coordinates in the canonical basis, for $1\leqslant j\leqslant m$.

Let $A^+=\{v_1,v_2,\ldots,v_n\}$ and $A^-=-A^+$. Note that:

(a) To write A in standard form with respect to a basis J, means to choose for basis the vectors $\{v_j\}_{j\in J}$ and express the other columns of A with respect to this basis.

(b) Given A in standard form with respect to J, to pivot on $A_i^j \neq 0$ ($j \notin \mathcal{J}$) means to eliminate from basis J the vector containing the 1 in its ith position and to include in the basis the vector v_j. This pivoting operation transforms A to standard form with respect to the basis:

$$J' = J - \{i\} + \{j\}.$$

The theorem states that among the vectors of $A^+ \cup A^-$, there exist n linearly independent vectors such that the bicone generated by these n vectors contains all the vectors of the set $A^+ \cup A^-$.

Final remarks

(1) The lexicographic order used here is similar to many lexicographic orders used in linear programming. Note that any rule which would improve the submatrix A^{J_0} according to the lexicographic order from an iteration to the next one would work as well as the given rule.

(2) This algorithm is finite although we do not know if it is polynomial or not. In fact, we do not know if there is a polynomial time algorithm for orienting a rational matrix, or that this problem is NP-hard.

We shall set the following conjecture:

Conjecture. Let A be an $n \times m$ standard matrix with integer coefficients. There exists a polynomial algorithm for transforming A into an oriented standard matrix $\bar{A}$.

Remark. As in linear programming, it may be possible that the easiest way to prove this conjecture is to use some special methods as the ellipsoid method [4], which is the only known way to solve in polynomial time some combinatorial problems.

3. Orientation of totally unimodular matrices

Preliminaries

All the definitions and notations not given here can be found in Welsh [9].

Let $\mathcal{M} = (E, \mathcal{F})$ be a representable matroid over a field K of rank n and let $A \in K^{n \times m}$ be a representable matrix of $\mathcal{M}$. The columns of A are indexed by $E = \{1, 2, \ldots, m\}$ and a subset $J \subset E$ is independent if an only if the column vectors of the submatrix A^J are linearly independent.

If $K = \mathrm{GF}(2)$ (the binary field), $\mathcal{M}$ is a binary matroid and A is a binary matrix. If A has an identity submatrix of size $n \times n$, A is a standard representative matrix (SRM) for $\mathcal{M}$. If A^{J_0} is the $n \times n$ identity matrix, J_0 is a basis of $\mathcal{M}$ and conversely: if J_0 is a basis we can find a (SRM) for $\mathcal{M}$ denoted A such that A^{J_0} is the identity submatrix of $\mathcal{M}$.

If $\mathcal{M} = (E, \mathcal{F})$ is represented by a matrix A, the linear subspace of K^m generated by the rows of A is the cocircuit space. Cocircuits of $\mathcal{M}$ are the minimal subsets of E which are supports of nonzero elements of the cocircuit space. For binary matroids we shall identify the cocircuits of $\mathcal{M}$ with the elements of the cocircuits space (except the vector $\vec{O}$) of minimal support. In particular if A is a SRM over the binary field, the rows of A are cocircuits of A.

If $X \subset E$, $\mathcal{M}\backslash X$ (resp. $\mathcal{M}/X$) denotes the matroid obtained from $\mathcal{M}$ by deleting (contracting) X. For $X \subset E$, the closure of X denoted $\mathrm{cl}(X)$ is the largest subset Y of E containing X such that $r(Y) = r(X)$ where r denotes the rank function of M.

A matrix A whose coefficients are 0, +1 or −1 is said to be *totally unimodular* if every square submatrix of A has a determinant equal to +1, 0 or −1.

It is well known that pivoting operations on a matrix preserve total unimodularity (see [1]).

A matroid $\mathcal{M} = (E, \mathcal{F})$ is *regular* if and only if it can be represented over the rational field by a totally unimodular matrix [8].

Connectedness

A matroid $\mathcal{M} = (E, \mathcal{F})$ is *separable* if there exists a proper subset S of E ($S \neq \emptyset$ and $S \neq E$) such that $r(S) + r(E\backslash S) = r(E)$ where r denotes the rank function of $\mathcal{M}$; otherwise $\mathcal{M}$ is *nonseparable* or *connected.*

A matroid can always be written in a unique way as a sum of nonseparable matroids: $\mathcal{M} = \mathcal{M}_1 \oplus \cdots \oplus \mathcal{M}_s$ where

$$\mathcal{M}_i = (E_i, \mathcal{F}_i), \quad i = 1, \ldots, s, \quad \bigcup_{i=1}^{s} E_i = E; \quad E_i \cap E_j = \emptyset \ \forall 1 \leq i, j \leq s.$$

The matroids, $\mathcal{M}_i$, $1 \leq i \leq s$ are called *components* of $\mathcal{M}$.

Using the relations between representable matroids and matrices we can define in a similar way a connected matrix and the components of a matrix (sometimes called blocks). A cocircuit Y is *separating* if the matroid $\mathcal{M}\backslash Y$ is separable and $\mathcal{M}$ is not.

Given a SRM a for $\mathcal{M}$ the components of A can be easily calculated. To each row i of A we associate a vertex v_i of a graph $G = (V, E)$. Two rows i_1 and i_2 are adjacent if there exists a column j such that $A^j_{i_1} \neq 0$ and $A^j_{i_2} \neq 0$.

Two vertices of G are linked by an edge if the corresponding rows are adjacent. Finding the components of $\mathcal{M}$ is equivalent to finding the connected components of G.

So we can define a chain between two row $\bar{\imath}$ and $\bar{\imath}'$ as a sequence $i_1, j_1, \ldots, i_p, j_p, \ldots, i_k$, where the j_p are columns indices, the i_p are row-indices, $A^{j_p}_{i_p} \neq 0$, $A^{j_p}_{i_{p+1}} \neq 0$, $1 \leq p \leq k-1$, and $i_1 = \bar{\imath}$ and $i_k = \bar{\imath}'$. k is the length of the chain.

A chain between two rows i and $\bar{\imath}'$ is minimal if no proper subsequence of it is a chain between i and $\bar{\imath}'$. If $L = i_1, j_1, \ldots, i_k$, is a minimal chain, the submatrix induced by L has exactly two nonzero elements in each column and each row except for the rows i_1 and i_k which have only one nonzero element.

An interesting lemma

The binary matrix

$$F=\begin{bmatrix}1&0&0&1&1&0&1\\0&1&0&1&0&1&1\\0&0&1&0&1&1&1\end{bmatrix}$$

is called the *Fano matrix* and $M(F)$ the *Fano matroid.* If $M(A)$ has a Fano minor then A can be transformed by pivoting operations into a (SRM) containing a standard submatrix which is Fano.

Lemma (dominant cocircuit lemma). *Let $\mathcal{M}$ be a binary matroid without Fano minor, and let C_1 be a nonseparating cocircuit of $\mathcal{M}$. Then for each $e\in C_1$, there exists a cocircuit C_2 of M such that*

$$[C_1\backslash \mathrm{cl}(e)]\subset C_2 \quad \textit{and} \quad C_2\cap \mathrm{cl}(e)=\emptyset.$$

Proof. There exists a SRM $A=[I, N]$ such that: e is the first column index of the identity matrix I and C_1 is represented by the first row of A.

By a suitable permutation of columns of A, N has the form shown in Fig. 1, where P (resp. Q) is the submatrix of N formed by the rows $2, 3, \ldots, n$ and by the columns of N containing a 1 (resp. a 0) in the 1st row.

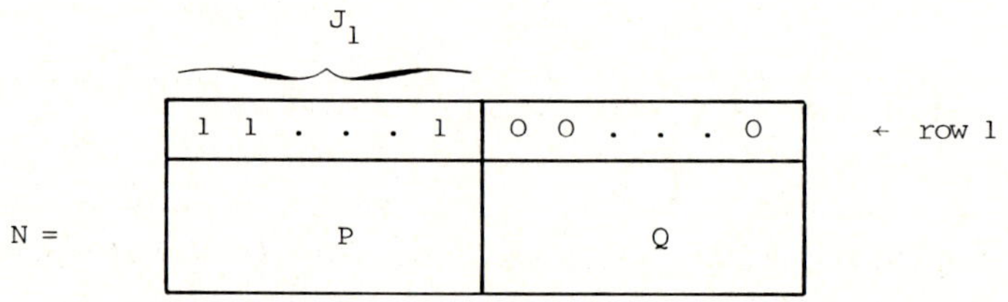

Fig. 1.

Since C_1 is nonseparating, the matrix Q is connected. Let J_1 (resp. J_0) be the set of column indices of P (resp. Q). Note that the pivoting operations applied on the submatrix Q do not affect the first row of N.

Without loss of generality, we can assume by a permutation of columns that the first n columns of A constitute the identity submatrix I.

After each pivoting operation, the new matrix will be denoted A. Given two vectors $x=(x_1, x_2, \ldots, x_k)$, $y=(y_1, y_2, \ldots, y_k)$ and a subset $J\subset\{1, 2, \ldots, k\}$, we say that x *dominates* y *in* J if and only if $s(y)\cap J\subset s(x)\cap J$ and $s(y)\cap J\neq s(x)\cap J$ where $s(\cdot)$ denotes the support of the vector.

We present a *pivoting algorithm* which transforms A into a matrix where the first row is C_1 and the second row dominates all the other rows in J_1. We present here an *iteration of the algorithm*:

Choose a row $i_1 \in \{2, 3, \ldots, n\}$ such that

$$|s(i_1) \cap J_1| = \max\{|s(i) \cap J_1| : j \in 2, 3, \ldots, n\}.$$

we can assume that $i_1 = 2$. Let us set $\gamma = |s(2) \cap J_1|$.

If $s(2) \cap J_1 \supset s(i) \cap J_1$ $\forall i \in \{3, \ldots, r\}$ then row 2 satisfies the lemma; if not, let K be the set of row indices of P not dominated by row 2 in J_1. Since Q is connected, for each $k \in K$ there exists a chain, and, therefore, a minimal chain linking row 2 to row k.

Let us choose a shortest of these chains (in terms of number or rows), and let us call it L. By a suitable reordering of the rows we can assume that:

$$L = (2, j_1, 3, j_2, \ldots, r-1, j_{r-2}, r)$$

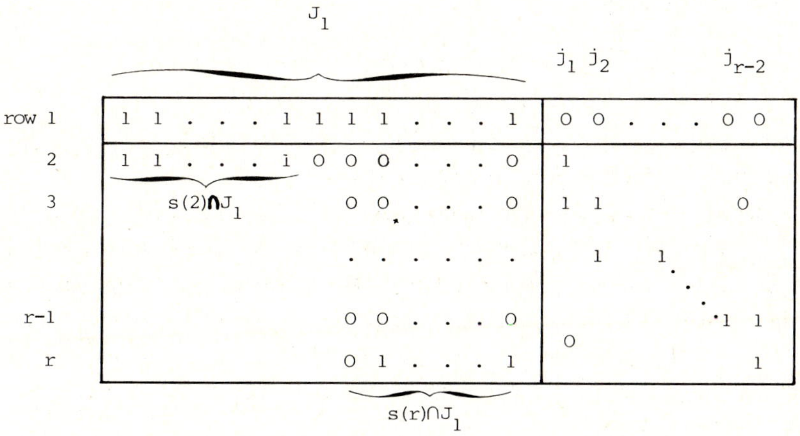

Fig. 2

The submatrix whose column index set is $[s(r) \backslash s(2)] \cap J_1$ and whose row index set is $\{3, \ldots, r-1\}$ is a zero-matrix by our choice of L (see Fig. 2).

If $r > 3$, when we pivot on (r, j_{r-2}) the 1's found in the intersection of the rth row and the columns whose indices are in $[s(r) \backslash s(2)] \cap J_1$ appear in the $(r-1)$th row.

Thus, the length of L decreases by one. So reiterating this process of pivoting we are led to the situation where the length of the chain L is 2. Then, the submatrix whose column index set is

$$\{(s(2) \cup s(r)) \cap J_1\} \cup \{j_1\}$$

and whose row index set is $\{1, 2, 3\}$ has the form shown in Fig. 3. If all the elements represented by an asterisk are equal to 0 then we can increase the cardinality of the support of the second row in J_1: $s(2) \cap J_1$ by pivoting on the 1 at the $(3, j_1)$ position. If all the elements represented by an asterisk are equal to 1 the cardinality of the support of the third row is larger than the cardinality of the support of the second row.

	J_1									j_1
row 1	1	.	.	.	.	.	1	1	1	0
2	1	1	1	1	.	.	0	0	0	1
3	*	*	*	*	.	.	1	1	1	1

Fig. 3

If some elements represented by an asterisk are equal to 1 and other elements are equal to 0, then a Fano matrix is detected and situation never occurs if M has no Fano minor.

$$\begin{vmatrix} \boxed{1} & \boxed{1} \cdots \boxed{1} & 1 \cdots 1 & 0 \\ \boxed{1} & \boxed{1} \cdots 0 & 0 \quad\; 0 & \boxed{1} \\ \boxed{1} & 0 \quad\;\; \boxed{1} & 1 \quad\; 1 & \boxed{1} \end{vmatrix}$$

End of an iteration

Therefore, with the indicated pivoting operations, if the Fano matrix does not appear, we construct a row d such that $|s(d) \cap J_1| > \gamma$.

This iteration is repeated until a dominant row in J_1 is found. The first row (representing the cocircuit C_1) and the first column of the identity matrix (representing the element e) remain unchanged during the application of the algorithm.

Let C_2 be the support of the dominant row d obtained at the end of the algorithm. C_2 is a cocircuit which satisfies the theorem. Therefore, the dominant cocircuit lemma enables us to deduce the following theorem (Tutte):

Corollary. *A nonseparable binary matroid $\mathcal{M}$, of rank at least* 3 *and without Fano minor, contains a separating cocircuit.*

Proof. Given a cocircuit C_1 of the matroid M, if C_1 is separating, then it satisfies the theorem; if not, let $e \in C_1$. Consider the (SRM) A obtained at the end of the previous algorithm of construction of the dominant cocircuit, and let B be the basis represented by the $n \times n$ identity matrix.

$e \in B$ and $\text{cl}(e) \cap C_2 = \emptyset$. Thus, $\text{cl}(e)$ is a component of $M \backslash C_2$. Since M is of rank at least 3, then B contains at least 3 elements. By the construction of C_2, this cocircuit intersects B in a single element. This is also true for C_1.

Let g be an element of B, neither in C_1 nor in C_2. Thus g belongs to a component of $M \backslash C_2$ distinct from $\text{cl}(e)$. So $M \backslash C_2$ has at least two components. Therefore C_2 is a separating cocircuit of M.

Remark. Another algorithmic proof of this theorem is given by Cunningham (see [2]). This proof is very similar to our proof: the main difference is that we try to construct a cocircuit C_2 such that $|C_2 \cap C_1|$ is as large as possible. On the contrary Cunningham constructs a cocircuit C_2 such that $|C_2 \cap C_1|$ is as small as possible. But if at the end of our algorithm we pivot on an element on the first row corresponding to C_1 and on a column corresponding to an element $e \in C_1 \cap C_2$, we replace the cocircuit C_2 by a cocircuit C_2' such that $|C_1 \cap C_2'|$ is as small as possible (and the converse is also true. In fact the two algorithms given in [2] and here have the same complexity $O(m^2 \times n^2)$.

Orientation of totally unimodular matrices

We prove in this final subsection that there exists a polynomial algorithm for orienting a totally unimodular matrix.

Definition. A block is defined by:
- an $n_1 \times m_1$ totally unimodular matrix. $A = [I, N]$.
- A row i_1 of A called reference row, whose coefficients are 0 or 1.
- If J_1 is the support of row i_1, J_1 is a nonseparating cocircuit.
- If we call J the set of columns of A, we shall set:

$$P = A^{J_1} \quad \text{and} \quad Q = A^{J \setminus J_1}$$

A block will be defined by the following notation: (i_1, P, Q).

Remarks. (a) Note that the regular matroid $\mathcal{M}$ associated to A is also a binary matroid without Fano minor [8]. It can easily be proved that, if we consider the binary matrix $\text{Im}(A)$ obtained by replacing the nonzero elements (resp. the zero element) of the (SRM) A by the nonzero element (resp. the zero element) of GF(2), $\text{Im}(A)$ is a (SRM) for the binary matroid $\mathcal{M}$. Thus, if A' is the matrix obtained from A after pivoting over Q on a nonzero element and $\bar{A}'$ is the matrix obtained from $\text{Im}(A)$ after pivoting over GF(2) on the element in the same position, $\bar{A}' = \text{Im}(A')$. (For a presentation of these results, see [1].)

A consequence is that, in order to find a dominant cocircuit when we work with matrix A, we apply the same algorithm as in section *b*, but the pivot operations are done with reference to the rational field. (Nonzeroes elements will remain in the same positions.) So, at the end of the algorithm, A is transformed into a matrix $\bar{A}$ where a row i_2 is dominant in J_1 (see Fig. 4).

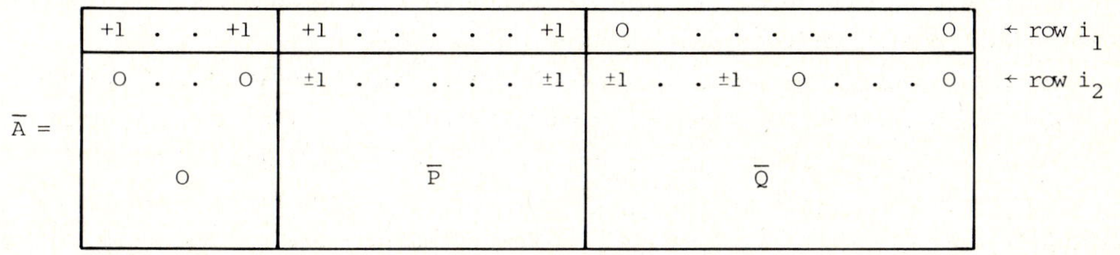

Fig. 4

(b) All the pivoting operations concern columns of $J \backslash J_1$ and the row i is not modified during this algorithm. We can now define a procedure for a block (i_1, P, Q).

Procedure 'Block'

(1) By application of the algorithm on page 93 with the modification of remark (a), find a dominant row i_2 associated to a dominant cocircuit. Let J_2 be the support of row i_2.

(2) As A is a t.u. matrix, if $A_{i_1}^{j_1} \neq 0$ and $A_{i_2}^{j_2} \neq 0$ for $j_1 \in J_1$ and $j_2 \in J$ $A_{i_2}^{j_1} \cdot A_{i_2}^{j_2} = +1$, otherwise the submatrix of A defined on rows i_1, i_2 and columns j_1, j_2 would have a determinant equal to ± 2. Therefore by an eventual multiplication of row i_2 by -1, transform row i_2 into a row where $A_{i_2}^{j} = 0$ or $+1$ for $j \in J_1$. By an eventual multiplication of columns of $J_2 \backslash J_1$ transform row i_2 into a nonnegative row.

(3) If $J \backslash (J_1 \cup J_2) = \emptyset$. STOP. End of the procedure.

(4) If $J \backslash (J_1 \cup J_2) \neq \emptyset$, let $Q_1, Q_2, \ldots, Q_s$ be the components of the submatrix $A^{J \backslash (J_1 \cup J_2)}$. If I_l is the set of rows of the submatrix Q_1, let P_1 be the submatrix $A_{I_l}^{J_2}$. Define the blocks (i_2, P_l, Q_l), $1 \leq l \leq s$.

END of procedure 'Block'

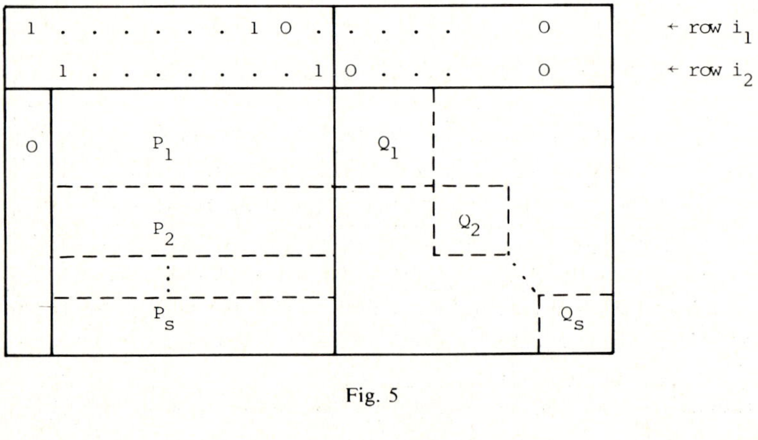

Fig. 5

We can now describe the orientation algorithm:

Input $A = [I, N]$.

Initialization: Take the first row of A. Let J_1 be the support of this row.

(1) By multiplying some columns $j \in J_1$ of A by -1 transform row (1) into a nonnegative row.

(2) If J_1 is not a separating cocircuit, define a unique block: $(1, A^{J_1}, A^{J \backslash J_1})$.

End of the initialization step:

(3) If J_1 is a separating cocircuit, let $Q_1, \ldots, Q_s$ be the components of $A^{J \backslash J_1}$. If I_l is the set of rows of submatrix Q_l, let us set $P_l = A_{I_l}^{J_1}$. Define the blocks $(1, P_l, Q_l)$, $1 \leq l \leq s$.

End of the initialization step.

General iteration: As long as there exists a block not yet studied, apply to it the procedure 'Block'.
End of the algorithm.

Justification
Let us assume that at step t of the general iteration, the following assertions are true.

(α) Some rows of A have been classified at a previous step to step t as reference rows and they have remained unchanged from these steps to step t. Therefore we shall continue to call them reference rows. In particular these rows are nonnegative rows (see procedure 'block'). Moreover, without loss of generality we can assume that these rows are rows $1, \ldots, k$.

Let us call J_i the support of row i, $1 \leqslant i \leqslant k$, we shall set $\bar{J}_k = \bigcup_{i=1}^{k} J_i$ and $\bar{I}_k = \{1, \ldots, k\}$.

(β) The number of existing blocks not yet studied is equal to the number of connected components of $A^{J \setminus \bar{J}_k}$.

Each row i which is not a reference row belongs to exactly one block (i_l, P_l, Q_l) and if i_l is the reference row of this block $s(i) \cap \bar{J}_k \subset s(i_l)$. (See Fig. 6.)

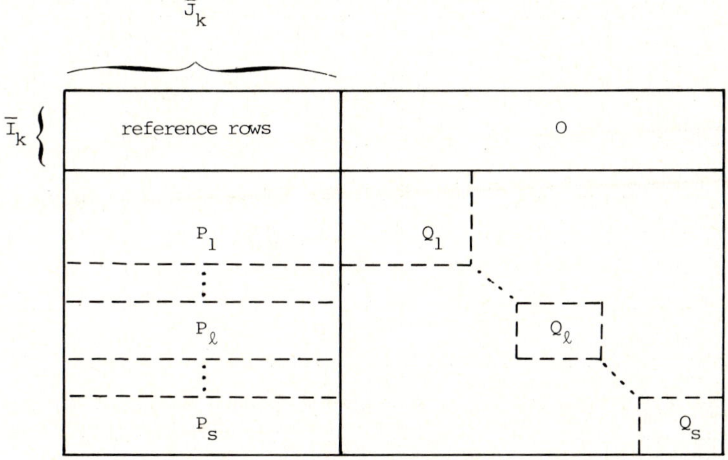

Fig. 6

Let us prove that these properties hold at step $t+1$ of the general iteration. At Step $t+1$ we can assume that we apply procedure 'block' to the block (i_l, P_l, Q_l).

If J_l is the set of columns of Q_l and I_l the set of rows of Q_l, the procedure 'block' will transform only the submatrices P_l and Q_l. In particular all the other existing blocks remain unchanged as well as the reference rows. The new reference row that we can assume to be the row $k+1$ is nonnegative. Therefore assertion (α) remains true.

On the other hand, the new blocks obtained at the end of the procedure 'block' are produced by some subdivision of (P_l, Q_l) and the dominating cocircuit lemma insures that (β) will remain true.

This induction proves also that the algorithm is finite and bounded by the number n of rows of A since at each call of the procedure 'block' the number of reference rows increases by one.

Note also that procedure 'block' is called as long as the union of the support of all the references rows obtained so far is not J. This implies that there are exactly n calls to the procedure 'block', and at the end of the algorithm all the rows of the matrix are reference rows and are nonnegative rows. Thus, at the end of this algorithm, the matrix A is well oriented.

Efficiency of the algorithm

A pivoting operation in A requires a number of operations bounded by $O(n \times m)$. Finding a minimal chain requires a number of operations bounded by $O(n \times m)$ (see [2]). The complexity of the algorithm is given by the construction of the dominant row. For this we need at most $m \times n$ pivots (n pivots to increase by one the cardinality of $s(1) \cap s(i_2)$ and mn pivots to find a dominant row). So the number of operations in the procedure 'block' is bounded by $O(n^2 \times m^2)$. As there are exactly n calls to this procedure the number of elementary operations to orient an $n \times m$ totally unimodular matrix is $O(n^3 \times m^2)$. Certain material presented in this paper can be found in reference [6].

Acknowledgement

We are grateful to Professor William Cunningham for his suggestions and comments, and for providing his article [2] as reference.

References

[1] P. Camion, "Modules unimodulaires", *Journal of Combinatorial Theory* 4 (1968) 301–362.
[2] W.H. Cunningham, "Separating cocircuits in binary matroids", *Linear Algebra and Its Applications* 43 (1982) 69–86.
[3] J. Edmonds, *Seminaire à l'Université de Grenoble* (Grenoble, France, June 1981).
[4] M. Grötschel, L. Lovasz and A. Schrijver, "The ellipsoid method and its consequences in combinatorial optimisation", *Combinatorica* 1 (1981) 169–197.
[5] A. Mandel, "The topology of oriented matroids", Ph.D. Thesis, University of Waterloo (Waterloo, 1982).
[6] M. Raco, Thèse 3ème Cycle, l'Université de Grenoble (Grenoble, France, June 1983).
[7] P.D. Seymour, "Decomposition of regular matroids", *Journal of Combinatorial Theory* B28 (1980) 305–359.
[8] W.T. Tutte, "Lecutres on matroids", *Journal of Research, National Bureau of Standards* B69 (1965) 1–47.
[9] D.J.A. Welsh, *Matroid theory* (Academic Press, New York, 1976).

Mathematical Programming Study 22 (1984) 99–112
North-Holland

COVERING DIRECTED AND ODD CUTS

András FRANK and Éva TARDOS
Research Institute for Telecommunication, Gábor Á.u.65, Budapest II, Hungary H-1026

András SEBŐ
Computer and Automation Institute, Hungarian Academy of Sciences, Kende u.13–17, Budapest XI, Hungary H-1502

Received 24 May 1983
Revised manuscript received 27 October 1983

Analogous pairs of theorems are investigated concerning coverings of directed and odd cuts. One such pair of results is the Lucchesi–Younger theorem on directed cuts and Seymour's theorem on odd cuts. Here we strengthen these results (incidently providing a simple proof of Seymour's theorem). For example, the minimum cardinality of a T-join in a graph $G=(V, E)$ is proved to equal the maximum of $\sum q_T(V_i)/2$ over all partitions of V where $q_T(X)$ is the number of T-odd components of $V-X$. Moreover, if G is bipartite, there is an optimal partition arising from a partition of the two parts. Secondly some orientation problems of undirected graphs are discussed. The results also emphasize the analogy between strong connectivity and parity conditions.

Key words: Odd Cut, Directed Cut, Matching, Chinese Postman, Orientation.

1. Introduction

In this paper we investigate analogous pairs of theorems concerning directed cuts of digraphs and odd cuts (or T-cuts) of undirected graphs. For example, the Lucchesi–Younger theorem asserts that the minimum cardinality of a covering of directed cuts (i.e. a set of edges meeting all directed cuts) is equal to the maximum number of pairwise disjoint directed cuts. Seymour's theorem states the same for odd cuts of bipartite graphs. Here we shall strengthen these results.

For a fixed directed cut $F=\delta(X)$, a minimax formula is proved for $c'(X)$, the minimum of $|C \cap F|$ over all coverings C of directed cuts. Making use of the supermodularity of c' we prove a variant of the Lucchesi–Younger theorem in which the optimal packing of directed cuts forms a chain.

An analogous minimax formula is proved for the minimum of $|C \cap F|$ over all T-joins C (from which, incidently, a simple proof for Seymour's theorem follows). Here the main result is: the minimum cardinality of a T-join in a bipartite graph $G=(V_1, V_2; E)$ is equal to the maximum of $\sum q_T(X_i)$ over all partitions of V_1 where $q_T(X)$ denotes the number of T-odd components of $G-X$.

This theorem easily implies Seymour's theorem as well as the Berge–Tutte formula. It also provides a formula for the minimum number of edges in a graph G such that G becomes Eulerian by doubling these edges (i.e. for the minimum cardinality of a Chinese postman tour). This minimum is equal to the maximum of $\sum q_0(V_i)/2$ over all partitions of V where $q_0(X)$ is the number of components in $G-X$ of odd degree.

We also deal with several problems concerning orientations of undirected graphs. Frank and Gyárfás [4] gave good characterizations for the existence of strongly connected orientations with certain side-conditions. Here analogous results are proved when strong connectivity is replaced by certain parity restrictions.

We shall use the terms graph and digraph for undirected and directed graph, respectively. By an edge we mean an undirected edge while directed edges are called arrows. Throughout we work with connected graphs and weakly connected digraphs. A *mixed* graph may contain both directed and undirected edges.

For a mixed graph $G=(V, E)$ the *coboundary* $\delta(X)$ of $X \subseteq V$ is the set of edges and arrows entering X, $S(X)$ is the set of edges and arrows incident with at least one node of X, $E(X)$ is the set of edges and arrows induced by X. We denote by $d(X, Y)$ the number of edges and arrows with one end node in $X-Y$ and the other one in $Y-X$ and we set $d(X)=d(X, V-X)$. The number of (weak) components of $V-X$ is denoted by $c(X)$.

Let $G=(V, E)$ be a digraph. The *in-degree* $\rho(X)$ is the number of arrows entering X. A *kernel* X is a subset of nodes with no arrows leaving X. The coboundary $\delta(X)\,(\neq\emptyset)$ of a kernel X is called a *directed cut* (or dicut) determined by X. A *covering* C (of directed cuts) is a subset of arrows meeting all the directed cuts.

Let $G=(V, E)$ be a graph and $T \subseteq V$ a subset of even cardinality. A set $X \subseteq V$ is *T-odd* if $|X \cap T|$ is odd. $q_T(X)$ denotes the number of T-odd components in $G-X$. In particular $q_V(X)$ is the number of components in $G-X$ of odd cardinality. For a T-odd set X the coboundary $\delta(X)$ is called an *odd cut* or *T-cut*. (We shall prefer the latter.) A subset C of edges is called a *T-join* if the number of edges in C incident with v is odd exactly when $v \in T$. Obviously a T-join and a T-cut have an odd number of edges in common.

An *orientation* of undirected graph G is a digraph arising from G by orienting its edges.

A family $\mathcal{F}$ of subsets is called *half-disjoint* if every element of the ground set is in at most two members of $\mathcal{F}$. For a vector $u \in \mathbb{R}^V$ set $u(X)=\sum(u(x)\colon x \in X)$.

2. Orientations I

Our starting point is a result of Frank and Gyárfás [4] concerning orientations of graphs which satisfy lower and upper bound restrictions on the in-degree. Given a graph $G=(V, E)$, let l and u be two integral functions on V such that $0 \leq l \leq u$.

Theorem 1. a. *There is an orientation of G with $\rho(v) \leq u(v)$ for $v \in V$ iff $|E(X)| \leq u(X)$ for $X \subseteq V$.*

b. *There is an orientation of G with* $\rho(v) \geq l(v)$ *for* $v \in V$ *iff* $|S(X)| \geq l(X)$ *for* $X \subseteq V$.

c. *There is an orientation of G with* $l(v) \leq \rho(v) \leq u(v)$ *for* $v \in V$ *iff there is one with* $\rho(v) \geq l(v)$ *for* $v \in V$ *and there is one with* $\rho(v) \leq u(v)$ *for* $v \in V$.

An earlier orientation result is due to H. Robbins [10].

Theorem 2. *A graph has a strongly connected orientation iff there is no isthmus.*

A slight strengthening of this theorem is

Theorem 2a [2]. *In a mixed graph G there is an orientation of the undirected edges which results in a strongly connected digraph iff there is no isthmus and G contains no directed cuts.*

Theorem 2a can be interpreted so that any partial orientation of an isthmus-free graph can be continued in order to get a strongly connected orientation provided that there is either an undirected edge or an already oriented edge leaving X for every subset X ($\neq \emptyset, V$).

The following characterization for the existence of a strongly connected orientation satisfying lower and upper bound conditions on the in-degrees was presented in [4].

Theorem 3. *Suppose G is an undirected graph with no isthmus.* a. *There is a strongly connected orientation of G with* $\rho(v) \leq u(v)$ *for* $v \in V$ *iff* $|E(X)| + c(X) \leq u(X)$ *for* $X \subseteq V$.

b. *There is a strongly connected orientation of G with* $\rho(v) \geq l(v)$ *for* $v \in V$ *iff* $|S(X)| - c(X) \geq l(X)$ *for* $X \subseteq V$.

c. *There is a strongly connected orientation of G with* $l(v) \leq \rho(v) \leq u(v)$ *for* $v \in V$ *iff there is one with* $\rho(v) \leq u(v)$ *for* $v \in V$ *and there is one with* $l(v) \leq \rho(v)$ *for* $v \in V$.

In this paper we shall prove some analogous theorems in which the property of strong connectivity is replaced by a certain parity condition—namely, given a subset T of nodes, find an orientation for which $\rho(v)$ is odd exactly if $v \in T$ and ρ satisfies lower and upper bound conditions. Without loss of generality one can suppose that $l(v)$ and $u(v)$ are odd exactly when $v \in T$. Thus we assume that $l(v) \equiv u(v) \pmod 2$ for $v \in V$. Call an orientation *u-congruent* if $\rho(v) \equiv u(v) \pmod 2$ for $v \in V$. Obviously u-congruent and l-congruent mean the same. The following simple observation is analogous to Robbins' theorem.

Theorem 4. *A connected graph* $G = (V, E)$ *has a u-congruent orientation iff*

$$|E(V)| \equiv u(V) \pmod 2. \tag{1}$$

Proof. Orient the edges so that the number of nodes v with $\rho(v) \not\equiv u(v) \pmod 2$ is as small as possible. If there are no such nodes, we are done. Otherwise, there are at least two, say u and w. Choose a path from u to w (in the undirected graph), and reverse the orientation on the edges of this path. This increases the number of nodes v with $\rho(v) \equiv u(v) \pmod 2$. □

Next let us deal with u-congruent orientations satisfying an upper bound restriction for the indegrees. Before the reader thinks this problem too artificial we remark that the 1-factor problem in a graph can be formulated in this way, as follows. Given a graph $H=(V, E)$, place a new node on every edge. Let $u(v)=1$ for original nodes and $u(v)=2$ for new nodes. There is a 1-to-1 correspondence between the 1-factors of H and the u-congruent orientations of the enlarged graph G satisfying $\rho \leq u$. Namely, an edge ab of H is in the 1-factor if and only if the in-degree of the middle node of ab is 0. Actually, Tutte's 1-factor theorem easily follows from the following theorem which can be considered as a counterpart to Theorem 3a. On the other hand, in the proof we rely on Tutte's theorem.

Theorem 5a. *There exists a u-congruent orientation of G for which $\rho(v) \leq u(v)$ if and only if*

$$|E(X)| + o(X) \leq u(X) \quad \text{for } X \subseteq V \tag{2}$$

where $o(X)$ denotes the number of components C (called u-odd) of $G-X$ for which $u(C) \not\equiv |S(C)| \pmod 2$.

Proof. (Necessity.) In any u-congruent orientation, any u-odd set is left by an odd number of arrows. Thus at least $o(X)$ arrows enter X and we have

$$|E(X)| + o(X) \leq \sum (\rho(v) \colon v \in X) \leq u(X).$$

(Sufficiency.) We deduce sufficiency from Tutte's 1-factor theorem by means of an elementary construction. First place a new node v^e on every edge e, then blow up every original node v to a complete graph of $u(v)$ nodes. More exactly, define $G'=(V', E')$ as follows.

$$V' = \{v_i \colon v \in V \text{ and } 1 \leq i \leq u(v)\} \cup \{v^e \colon e \in E\},$$

$$E' = \{v_i v_j \colon \text{for } v \in V \text{ and } 1 \leq i < j \leq u(v)\}$$

$$\cup \{(v^e v_i \colon e \in E, v \in V, e \text{ incident with } v\}.$$

Claim. *G has a u-congruent orientation for which $\rho(v) \leq u(v)$ $(v \in V)$ iff G' has a 1-factor.*

Proof. If G' has a 1-factor F and an edge of F incident to v^e is $v^e v_i$, orient e in G toward v. The orientation of G defined this way clearly satisfies the upper bound requirement. Furthermore, an even number of nodes is paired by F in the set of nodes in G' corresponding to a node $v \in V$ therefore the in-degree of v in the orientation is congruent to $u(v) \pmod 2$.

Conversely, the same correspondence shows that a u-congruent orientation with $\rho(v) \leq u(v)$ $(v \in V)$ determines a 1-factor of G'. ◇

By Tutte's theorem, a graph has a 1-factor iff for every subset $S \subseteq V'$, the number $q_{V'}(S)$ of components of $G' - S$ with an odd number of nodes is at most $|S|$. Let S be a minimal subset of V' violating the Tutte condition.

Claim. *For every $v \in V$, S either contains all the copies in G' corresponding to v or none of them.*

Proof. To the contrary, let $v \in V$ and $1 \leq i, j \leq u(v)$ be such that $v_i \in S$, $v_j \notin S$. Set $S_1 = S - v_i$. Since v_i and v_j are adjacent and the neighbors of v_i and v_j are the same, the components of $G' - S_1$ are the components of $G' - S$ with the only exception that the component containing v_j will also contain v_i. Thus $q_{V'}(S_1) \geq q_{V'}(S) - 1 > |S| - 1 = |S_1|$ contradicting the minimal choice of S. ◇

Claim. *S does not contain any new node v^e.*

Proof. For otherwise, set $S_1 = S - v^e$. Since every node of G was replaced by a complete graph, v^e can be connected with at most two components, say C_1, C_2, of $G' - S$. If both of them are odd, $C_1 \cup C_2 \cup \{v^e\}$ is odd, as well, whence we have $q_{V'}(S_1) \geq q_{V'}(S) - 1 > |S| - 1 = |S_1|$ contradicting the minimal choice of S. ◇

Using these claims one can easily check that $X = \{v \colon v_i \in S\}$ violates (2). □

An analogous result for l-congruent orientations with lower bounds simply follows from the preceding theorem.

Theorem 5b. *A graph $G = (V, E)$ has an l-congruent orientation for which $\rho(v) \geq l(v)$ $(v \in V)$ iff $|S(X)| - o(X) \geq l(X)$ for $X \subseteq V$ where $o(X)$ denotes the number of l-odd components of $G - X$.*

Somewhat surprisingly, the counterpart of Theorem 3c is not true. The graph following in Fig. 1 has a u-congruent orientation with upper bound u and an l-congruent orientation with lower bound l but no u-congruent orientation exists satisfying the upper bound and the lower bound condition at the same time. (At the nodes the first number denotes the lower bound l and the second number denotes the upper bound u.) However we have the following more complicated characterization. Suppose again that $u(v) \equiv l(v)$ (mod 2) $(v \in V)$.

Theorem 6. *A graph $G = (V, E)$ has a u-congruent orientation for which $l(v) \leq \rho(v) \leq u(v)$ for $v \in V$ iff*

$$u(A) - l(B) \geq p(A, B) + |E(A)| - |S(B)| \tag{3}$$

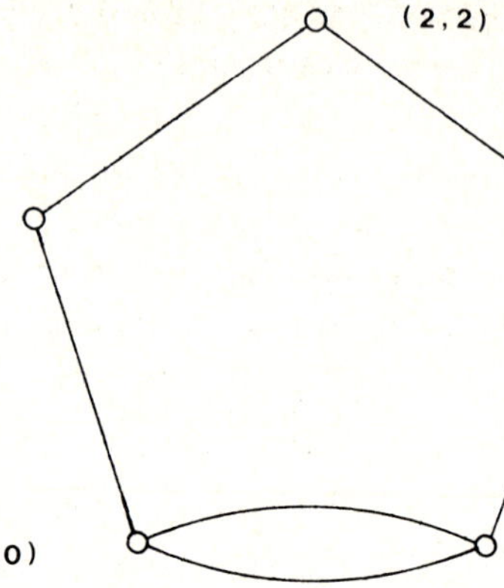

Fig. 1.

for disjoint subsets A, $B \subseteq V$, *where* $p(A, B)$ *denotes the number of those components* C *of* $V-(A\cup B)$ *for which* $|E(C)|+d(C, A)\not\equiv l(C)$ (mod 2).

The graph in the figure does not satisfy (3): let A consist of the two lower nodes and B of the upper most node.

Proof. The proof is similar to that of Theorem 5a so we outline only the sufficiency. Place a new node v^e on every edge e and blow up every original node v to a graph consisting of $l(v)$ isolated nodes and a complete graph of $u(v)-l(v)$ nodes. One can see that G has a u-congruent orientation ρ for which $l(v)\leqslant\rho(v)\leqslant u(v)$ ($v\in V$) iff G' has a 1-factor.

If there is no 1-factor of G', there exists a subset $S\subset V'$ violating the Tutte condition. Let S be minimal. A simple argument shows that, for every $v\in V$, S either contains all the nodes in G' corresponding to v or none of them. Set $A=\{v\in V\colon v_i\in S\}$ and $B=\{v\in V$: there is no edge $e\in E$ incident with v such that $v^e\not\in S\}$. The proof is completed by showing that $A\cap B=\emptyset$ and A, B violate condition (3). □

3. Orientations II

Let $G=(V, E)$ be an isthmus-free graph and X ($\neq\emptyset$, V) a specified subset of nodes. In this section we describe a min–max formula for the minimum number $c'(X)$ of arrows entering X in strongly connected orientations of G.

First of all observe that the components of X and $V-X$ can be shrunken into singletons without changing $c'(X)$. This is so because shrinking does not destroy the strong connectivity, on the other hand Theoreom 2.2a shows that a strongly connected orientation of the shrunken graph can be extended to a strongly connected orientation of G. Thus we can suppose that G is a bipartite graph with parts V_1,

V_2 where V_1 denotes the set formed by shrinking the components of X. Recall the definition of $c(X)$. The next lemma occurs in [3, 7].

Lemma 1. $c(A)+c(B) \leq c(A \cup B)+c(A \cap B)+d(A, B)$.

Theorem 2. *If $G=(V_1, V_2; E)$ is a bipartite graph, the minimum number of arrows entering V_1 among all strongly connected orientations of G is equal to*

$$\max(\textstyle\sum c(X_i)\text{: } \{X_i\} \text{ partitions } V_1,\ X_i \neq \emptyset).$$

Proof. (max ≤ min.) For a strongly connected orientation ρ, $\rho(X) \geq c(X)$ $(X \neq \emptyset)$ therefore $\rho(V_1)=\sum \rho(X_i) \geq \sum c(X_i)$.

(max = min.) We describe an algorithm that, given a strongly connected orientation with in-degree function ρ, either finds a better orientation ρ' (i.e. $\rho'(V_1)<\rho(V_1)$) or else finds a partition $\{X_i\}$ of V_1 for which $\rho(V_1)=\sum c(X_i)$.

Call a set *tight* (with respect to ρ) if $\rho(X)=c(X)$. (Note that V is always tight.)

Lemma 3. *If A, B are intersecting tight sets, $A \cap B$, $A \cup B$ are also tight.*

Proof. Using Lemma 1 we get

$$\rho(A)+\rho(B)=c(A)+c(B) \leq c(A \cap B)+c(A \cup B)+d(A, B) \leq \rho(A \cap B) + \rho(A \cup B)+d(A, B)=\rho(A)+\rho(B)$$

from which we have $c(A \cap B)=\rho(A \cap B)$ and $c(A \cup B)=\rho(A \cup B)$. ◇

An easy consequence is

Lemma 4. *If a family of tight sets forms a connected hypergraph the union is tight. The intersection $P(v)$ of all tight sets containing a node v is tight.* ◇

Let ρ be a strongly connected orientation of G. There may be two cases.

Case 1: For every $x \in V_1$, $P(x) \subseteq V_1$. Let $X_1, X_2, \ldots, X_n$ denote the components of the hypergraph $\{P(x)\text{: } x \in V_1\}$. Then $\{X_i\}$ forms a partition of V_1 and, by Lemma 4, each X_i is tight. Thus $\rho(V_1)=\sum \rho(X_i)=c(X_i)$, i.e. for the given orientation we have the required equality.

Case 2: There are nodes $x \in V_1$ and $y \in V_2$ such that $y \in P(x)$. Let P be a directed path from y to x. Reverse the orientation of the arrows of P. We get another orientation for which $\rho'(V_1)=\rho(V_1)-1$. Furthermore the new orientation is still strongly connected, for otherwise there would have been a set A with $\rho(A)=1$, $x \in A$, $y \notin A$ contradicting the assumption that $y \in P(x)$. □

Corollary 5. *The minimum over all strongly connected orientations of a graph $G=$*

(V, E) of the number of arrows entering a specified subset X is equal to $\max(\sum c(X_i)$: *$\{X_i\}$ partitions X and no edges lead between distinct X_i's, $X_i \neq \emptyset$).*

The next corollary concerns only c and not the orientation but we do not see any direct proof.

Corollary 6. *In a bipartite graph $G=(V_1, V_2; E)$,* $\max(\sum c(X_i)$: $X_i \neq \emptyset$, $\{X_i\}$ *partitions* $V_1) = \max(\sum c(Y_j)$: $Y_j \neq \emptyset$, $\{Y_j\}$ *partitions* V_2).

We conclude this section by showing that Lemma 1 is also true for c'.

Theorem 7. $c'(A)+c'(B) \leq c'(A \cap B)+c'(A \cup B)+d(A, B)$.

Proof. It suffices to prove the theorem when $d(A, B)=0$. If $d(A, B)>0$, place a new node on the middle of every edge between $A-B$ and $B-A$. The new nodes do not belong to A and B. Obviously this transformation increases $c'(A \cup B)$ by $d(A, B)$ but it does not affect $c'(A)$, $c'(B)$, and $c'(A \cap B)$. Furthermore there is no edge between $A-B$ and $B-A$ any more. Thus the theorem follows from the special case when $d(A, B)=0$ and henceforth we assume this.

The next lemma is due to Dunstan and Lovász [2, 7].

Lemma 8. *Let $\mathcal{F}$ be a family of subsets with $\emptyset \notin \mathcal{F}$ and let f be a function on $\mathcal{F}$ such that $X, Y \in \mathcal{F}$, $X \cap Y \neq \emptyset$ imply that $X \cap Y$, $X \cup Y \in \mathcal{F}$ and $f(X)+f(Y) \leq f(X \cap Y)+f(X \cup Y)$. Then the family*

$$\mathcal{F}' = \{\textstyle\bigcup F_i\text{: and } F_i \in \mathcal{F},\ F_i\text{'s are disjoint}\} \cup \{\emptyset\}$$

is closed under taking union and intersection. Moreover, let f' be the function on $\mathcal{F}'$ defined by

$$f'(F) = \max(\textstyle\sum f(F_i)\text{: } \{F_i\} \text{ partitions } F,\ F_i \in \mathcal{F}) \quad \text{if } F \in \mathcal{F}' - \{\emptyset\}, \qquad f'(\emptyset)=0.$$

Then, for $X, Y \in \mathcal{F}'$,

$$f'(X)+f'(Y) \leq f'(X \cup Y)+f'(X \cap Y). \qquad \diamond$$

Let us orient the edges of G so that no arrow enters A and B and set $\mathcal{F}=\{X\colon X \neq \emptyset, \rho(X)=0\}$. By Corollary 5, it is obvious for $X \in \mathcal{F}$ that $c'(X)=\max(\sum c(X_i)\colon \{X_i\}$ partitions X, $X_i \in \mathcal{F})$. By Lemma 1, Lemma 8 applies to $\mathcal{F}$ with $f=c$ and we are done. $\square$

4. Directed cuts, T-cuts

A fundamental theorem on directed cuts is due to Lucchesi and Younger [8].

Theorem 1. *In a digraph the minimum cardinality of a covering of directed cuts is equal to the maximum number of pairwise disjoint directed cuts.*

For a covering C and a dicut F by definition we have $|C \cap F| \geq 1$. We are going to determine the minimum of $|C \cap F|$ over all coverings C.

An answer can be read out from Theorem 1, namely, the minimum in question is equal to the maximum number of pairwise disjoint directed cuts in F. The main content of Theorem 2 is that the maximum is determined by certain disjoint kernels.

Theorem 2. *In a digraph, given a specified directed cut* $F = \delta(X)$, $\min(|C \cap F|$: C *covering*$) = \max(\sum c(X_i)$: $\bigcup X_i = X$, X_i*'s are disjoint kernels*$)$.

Proof. We can suppose that G does not contain isthmuses.

Lemma. *If C is a covering minimal with respect to inclusion, reorienting the elements of C results in a strongly connected digraph.*

Proof. Delete the orientation of the elements of C. We get a mixed graph to which Theorem 2.2a applies. In the resulting strongly connected reorientation of G each element of C must have direction reversed. For C is minimal and so for each $e \in C$ there is a directed cut T of G such that $T \cap C = \{e\}$ for each $e \in C$ and in the strongly connected reorientation of G, T is no longer a directed cut. ◇

This lemma and Corollary 3.5 proves the theorem. □

Notice that in Theorem 2, if G does not contain isthmuses, the minimum is exactly $c'(X)$. Thus we can use the notation $c'(X)$ for the minimum in Theorem 2 when G may contain isthmuses. With the help of c', the Lucchesi–Younger theorem can be formulated in a more compact form.

Theorem 3. *In a digraph the minimum cardinality of a covering of directed cuts is equal to* $\max(\sum c'(X_i)$: X_i*'s are kernels,* $X_1 \subset X_2 \subset \cdots \subset X_k$ *and no arrow enters more than one* $X_i)$.

Proof. Theorem 1 provides a covering C and a family $\mathcal{F}$ of disjoint directed cuts $\delta(X_i)$ such that $|C| = |\mathcal{F}| \leq \sum c'(X_i) \leq |C|$, i.e. $|C| = \sum c'(X_i)$. Making use of Theorem 3.7 we can apply a well-known trick [11] to 'uncross' $\{X_i\}$. Namely, let us choose $\{X_i\}$ in such a way that $|C| = \sum c'(X_i)$, $\delta(X_i)$ are pairwise disjoint, and $\sum |X_i|^2$ is maximal. We claim for each i, j that $X_i \subset X_j$ or $X_j \subset X_i$. If this were not true for X_1 and X_2, say, in the directed cut family replace X_1 and X_2 by $X_1 \cap X_2$ and $X_1 \cup X_2$. Applying Theorem 3.7 we get $|C| \geq \sum c'(X'_i) \geq \sum c'(X_i) = |C|$, i.e. $\sum c'(X'_i) = |C|$ and $\sum |X'_i|^2 > \sum |X_i|^2$, a contradiction. □

Next, let us consider a T-cut theorem of Lovász analogous to Theorem 1.

Theorem 4 [5]. *In a graph the minimum cardinality of a T-join is equal to the one-half of the maximum number of half-disjoint T-cuts.*

Unfortunately the analogy is not quite perfect. K_4 shows that the minimum cardinality of a T-join may be strictly bigger than the maximum number of pairwise disjoint T-cuts. As we shall see, however, this nicer min-max form is true for bipartite graphs.

As in the directed case, set up the following problem. Given a fixed coboundary $F = \delta(X)$, determine $\min|C \cap F|$ over all T-joins C. Parallel to Theorem 2 we have

Theorem 5. $\min|C \cap F|$ *over T-joins C is equal to* $\max \sum q_T(X_i)$ *over all partitions* $\{X_i\}$ *of X such that no edge connects distinct* X_i*'s.*

Like in the digraph case, first we reduce the problem to bipartite graphs, as follows.

Form a bipartite graph $G' = (V_1, V_2; E)$ by shrinking the components of X and $V - X$ into singletons. Let T' consist of those new nodes which came from a T-odd component. After the shrinking, a T-join of G transforms to a T'-join of G' and conversely, since a graph has a T-join iff its components are T-even, a T-join of G' can be extended in V_1 and V_2 to a T-join of G. Hence the minimum value for G' is the same as the minimum value for G, and therefore what we need is to find in G' a T'-join of minimum cardinality.

Theorem 4 of course gives an answer, the following result of Seymour, however, provides a better one.

Theorem 6 [13]. *In a bipartite graph* $G = (V_1, V_2; E)$ *the minimum cardinality of a T-join is equal to the maximum number of pairwise disjoint T-cuts.*

From this one can easily deduce Theorem 4 by placing a new node on the middle of each edge and keeping T unchanged.

Here we shall prove another formula, similar to that in Theorem 2, for the minimum cardinality of a T-join in a bipartite graph. Incidently, we get a proof for Theorem 6 which looks much simpler than the earlier proofs.

Theorem 7. *In a bipartite graph* $G = (V_1, V_2; E)$ *the minimum cardinality of a T-join is equal to*

$$\max \sum q_T(X_i)$$

taken over all partitions $\{X_i\}$ *of* V_1.

This theorem is considered as the main result of this section. It immediately implies Theorem 5. The nontrivial max $\geq$ min part of Theorem 6 also follows easily since the T-odd components of $G - X_i$ define disjoint T-cuts. Two further consequences will be mentioned at the end of this section.

At this point we can observe a difference between T-cuts and directed cuts. For T-cuts the bipartite case (Theorem 6) is stronger than the non-bipartite one (Theorem 2) while Theorem 3 does not seem to imply the Lucchesi–Younger theorem.

Before proving Theorem 7 we point out that the minimal T-join problem is equivalent to finding a negative circuit in a ± 1-weighted graph. To be more precise suppose the edges of G are assigned weights $+1$ or -1.

By the *length* $w(P)$ of a path or circuit P we mean the sum of its edge weights. A circuit is called *negative* if its length is negative. The *distance* $\lambda(u, v)$ of two nodes u, v is the minimum length of a (simple) path between u and v.

Theorem 6′. *In a ± 1-weighted bipartite graph there is no negative circuit if and only if there is a set of disjoint cuts such that every negative edge occurs in one of them and each cut contains only one negative edge.*

Theorem 7′. *In a ± 1-weighted bipartite graph $G = (V_1, V_2; E)$ there is no negative circuit if and only if there is a partition $X_1, X_2, \ldots, X_k$ of V_1 such that no component of $G - X_i$ is entered by more than one negative edge.*

Observe that a T-join F is minimal iff there is no negative circuit with respect to w where $w(e) = -1$ if $e \in F$ and $w(e) = +1$ if $e \in E - F$. Now Theorem 7′ implies Theorem 7 since a set X is T-odd if X is entered by exactly one edge of F. Conversely, for a given ± 1-weighting w set $F = \{e\colon w(e) = -1\}$ and define

$$T = \{v\colon \text{an odd number of negative edges is incident with } v\}.$$

Since no negative circuit exists, F is a minimal T-join; therefore by Theorem 7 there is a partition $\{X_1, \ldots, X_k\}$ of V_1 for which $|F| = \sum qT(X_i)$. This partition will do for Theorem 7′ as well.

The equivalence between Theorems 6 and 6′ is seen similarly.

Proof of Theorem 7′. The if part is straightforward. To see the other direction, the next lemma is crucial.

Lemma. Let $G = (V_1, V_2; E)$ ($|V_1 \cup V_2| \geq 3$) be a simple bipartite graph and w a ± 1-weighting on E such that no negative circuit exists and there is a negative path between every two nodes in the same class V_i. Then G is a tree and w is -1 everywhere.

Proof. Let x_0 be an arbitrary node. Let $m = \min \lambda(x_0, x)$ and choose a path $P_0 = (x_0, x_1, \ldots, x_n)$ from x_0 to x_n such that $w(P_0) = m$ and $\lambda(x_0, x_i) > m$ for $0 \leq i < n$. Obviously $m < 0$ and $w(x_{n-1}x_n) = -1$. ◇

By induction the next claim implies the lemma.

Claim. *$x_{n-1}x_n$ is the only edge incident with x_n.*

Proof. Suppose x_ny is another edge ($y \neq x_{n-1}$). Now $w(x_ny)$ cannot be -1 since then, if y is on P_0, $P_0[yx_n] \cup x_ny$ is a negative circuit, if y is not on P_0, $P_0 + x_ny$ is a path of length $m-1$.

By hypothesis there is a negative path P between x_{n-1} and y. By parity, $w(P) \leq -2$. P goes through x_n, for otherwise P and $x_{n-1}x_n$, x_ny would form a negative circuit. Moreover P traverses the edge $x_{n-1}x_n$. For otherwise the length of segment $P[x_{n-1}x_n]$ is at least $+1$ therefore the length of $P[x_ny]$ is at most -3. Thus $P[x_ny]$ and x_ny would form a negative circuit.

We also see that $w(P[x_n, y]) \geq -1$ and $w(P[x_n, y]) \leq -1$, i.e. $C = P[x_n, y] + x_ny$ is a circuit of length 0 and x_{n-1} is not on C.

We claim that x_n is the only common node of C and P_0. If not, let x be the next one on P_0 starting at x_n. By the choice of P_0, $w(P[x_n, x]) < 0$. Hence the length of both arcs between x and x_n on C is positive contradicting that $w(C) = 0$.

Now P_0 and $P[x_ny]$ have one node in common, namely x_n, therefore $P_0 \cup P[x_n, y]$ is a simple path from x_0 to y the length of which is $m-1$, a contradiction. ◇

Turning back to the proof of Theorem 7′, we can suppose that G is simple and $|V_1 \cup V_2| \geq 3$. There are two cases.

Case 1: There are two nodes x, y in the same class V_i with no negative path between them.

Identify x and y into a single new node z. The resulting bipartite graph has no negative circuit. By induction there is a partition with the desired property. After splitting up z the same partition of V_1 satisfies the requirements.

Case 2: There is a negative path between any two nodes in the same class V_i.

Apply the lemma. The required partition consists of the singletons of V_1. □

Next we exhibit a consequence of Theorem 7.

Theorem 8. *In a graph $G = (V, E)$ the minimum cardinality of a T-join is equal to $\frac{1}{2} \max \sum q_T(V_i)$ over all partitions $\{V_i\}$ of V.*

Proof. Place a new node on every edge, keep T the same and apply Theorem 7 to the resulting bipartite graph. □

If T consists of the nodes of odd degree, Theorem 8 provides

Corollary. *The minimum number of edges in a graph G such that G becomes Eulerian by doubling these edges, is equal to $\frac{1}{2} \max \sum q_0(V_i)$ over all partitions of V where $q_0(X)$ denotes the number of components C in $V - X$ for which $d(C)$ is odd.*

It is interesting that the same function q_0 plays a central role in Mader's famous A-path theorem [9]: Given a subset $A = \{v_1, \ldots, v_k\}$ of nodes in $G = (V, E)$, the maximum number of pairwise edge-disjoint paths joining distinct nodes of S is equal to $\frac{1}{2}\min\sum(d(V_i)-q_0(\cup V_i))$ over all families of disjoint subsets V_i with $V_i \cap A=\{v_i\}$.

Formulating Theorem 8 in terms of negative circuits we get

Theorem 8′. *In a ±1-weighted graph $G=(V, E)$ there is no negative circuit iff there is a partition $\{V_i\}$ of V such that each component of $V-V_i$ is entered by at most one negative edge, and, for each i, no negative edge has both ends in V_i.*

This easily implies

Berge–Tutte formula. *In a graph $G=(V, E)$ the maximum cardinality of a matching is $\frac{1}{2}\min(-q_V(X)+|V-X|)$.*

Proof. To prove the nontrivial inequality max ≥ min, we have to show a matching F and a subset $X \subseteq V$ such that each component of $V-X$ contains at most one node not covered by F and for every node v in X there is an odd component C_v of $V-X$ (with $C_v \neq C_u$ if $u \neq v$) and an edge in F connecting v and C_v.

Let F be a maximal matching. Adjoin a new node r to the graph and lead new edges from r to the nodes exposed by F. Define $w(e)=-1$ if $e \in F$ or e is a new edge and $w(e)=+1$ if $e \in E-F$. Since F is maximal this graph does not contain a negative circuit.

Consider the partition provided by Theorem 8′. Suppose V_1 contains r. It is easy to see that F and $X=V_1-r$ satisfies the requirements. □

Remark. Originally this paper was written by A. Frank and É. Tardos. Their proof for Theorem 7′ made use of the deep result of Seymour (Theorem 6) and was rather complicated. Before the paper was finished, A. Sebő discovered that Theorem 7′ can be proved directly and this approach is much better since it provides a simple proof for Seymour's theorem as well. Therefore we have chosen to include his proof. He also established some fundamental structural properties of T-joins. See [12].

References

[1] F.D.J. Dunstan, "Matroids and submodular functions", *Quarterly Journal of Mathematics* 27 (1976) 339–347.

[2] A. Frank, "Combinatorial algorithms, algorithmic proofs" (in Hungarian), Ph.D. Thesis, Eötvös L. University (Budapest, 1975).

[3] A. Frank, "How to make a digraph strongly connected", *Combinatorica* 1 (1981) 145–153.

[4] A. Frank and A. Gyárfás, "How to orient the edges of a graph", in: A. Hajnal and V.T. Sós, eds., *Colloquia Mathematica Societatis János Bolyai 18, Combinatorics* (North-Holland, Amsterdam, 1978) pp. 353–364.

[5] L. Lovász, "2-matchings and 2-covers of hypergraphs", *Acta Mathematica Academiae Scientiarum Hungaricae* 26 (1975) 433–444.

[6] L. Lovász, "Flats in matroids and geometric graphs", in: P.J. Cameron, ed., *Combinatorial Surveys: Proceedings of the sixth British Combinatorial Conference* (Academic Press, New York, 1979) pp. 45–86.

[7] L. Lovász, *Combinatorial Problems and Exercises* (Akadémiai Kiadó, Budapest, 1979).

[8] C.L. Lucchesi and D.H. Younger, "A minimax relation for directed graphs", *Journal of the London Mathematical Society (2)* 17 (1978) 369–374.

[9] W. Mader, "Ueber die Maximalzahl kantendisjunkter A-wege", *Archiv der Mathemaik* 30 (1978) 325–336.

[10] H.E. Robbins, "A theorem on graphs with an application to a problem of traffic control", *American Mathematical Monthly* 46 (1939) 281–283.

[11] A. Schrijver, "Total dual integrality directed graphs, crossing families and sub- and supermodular functions", in: *Proceedings of the Silver Jubilee Conference on Combinatorics* (held in Waterloo, June 1982, Academic Press).

[12] A. Sebő, "On the structure of odd-joins", in preparation.

[13] P.D. Seymour, "On odd cuts and plane multicommodity flows", *Proceedings of the London Mathematical Society* 42 (1981) 178–192.

Mathematical Programming Study 22 (1984) 113–131
North-Holland

SUBMODULAR SYSTEMS AND RELATED TOPICS

Satoru FUJISHIGE
Institute of Socio-Economic Planning, University of Tsukuba, Sakura, Ibaraki 305, Japan

Received 8 March 1983
Revised manuscript received 8 September 1983

Let $\mathscr{D}$ be a distributive lattice formed by subsets of a finite set with set union and intersection as the lattice operations, and let f be a submodular function on $\mathscr{D}$. The pair $(\mathscr{D}, f)$ is called a submodular system and is a generalization of a (poly-)matroid. The present paper makes a survey of the author's earlier work on submodular systems and provides a unifying view and some useful observations on related topics such as geometries on posets, generalized polymatroids, boundary hypermatroids, submodular functions on crossing families, submodular flows, strongly connected orientations of graphs, Lovász's extension of set functions, minimization of submodular functions etc. We also show a new approach to the problem of minimizing submodular functions.

Key words: Submodular Functions, Supermodular Functions, Matroids, Polymatroids, Submodular Systems, Polyhedra, Distributive Lattices, Posets, Crossing Families.

1. Introduction

Matroids originated by H. Whitney [61] and others have extensively been studied since W.T. Tutte's papers [57, 58] appeared (also see [59]) and have widely been recognized as one of the most important fields in combinatorics (cf. [60]). Moreover, the results on matroids have successfully been applied to practical engineering problems and many practical problems motivated new theoretical researches on matroids (cf. [19, 32–35, 42, 44, 56]).

A submodular system is a pair $(\mathscr{D}, f)$ of a distributive lattice $\mathscr{D}$ and a submodular function f on $\mathscr{D}$ and is a generalization of a (poly-)matroid [5]. The present paper will make a survey of the author's earlier work on submodular systems and provide a unifying view on related topics such as geometries on posets [8, 9], generalized polymatroids [12], boundary hypermatroids [52, 55], submodular functions on crossing families [16, 21], submodular flows [4, 7, 13, 15, 63], strongly connected orientations of graphs [11, 26], Lovász's extension of set functions [41], minimization of submodular functions etc.

Sections 2–6 will be devoted basically to a survey of the author's earlier work together with its related topics on submodular systems and will not contain new results but some useful observations on the related topics. In Section 7 we shall show a new approach to the problem of minimizing submodular functions.

Written while the author was at Institut für Ökonometrie und Operations Research, Universität Bonn, West Germany.

2. Submodular/supermodular systems, polyhedra and duality

Let E be a finite set and denote by 2^E the set of all the subsets of E. Also let a collection $\mathscr{D}$ of subsets of E be a *distributive lattice* with set union and intersection as the lattice operations, i.e., for any $X, Y \in \mathscr{D}$ we have $X \cup Y, X \cap Y \in \mathscr{D}$. Such a collection $\mathscr{D}$ is sometimes called a ring family or a lattice family in the literature. A function f from $\mathscr{D}$ to the set $\mathbb{R}$ of reals is called a *submodular function* on $\mathscr{D}$ if for each pair of $X, Y \in \mathscr{D}$

$$f(X)+f(Y) \geq f(X \cup Y)+f(X \cap Y). \tag{2.1}$$

It may be noted here that $\mathbb{R}$ can be any totally ordered additive group such as the set of integers or rationals in the following arguments except for the one in Section 7. A pair $(\mathscr{D}, f)$ of a distributive lattice $\mathscr{D} \subseteq 2^E$ and a submodular function $f: \mathscr{D} \to \mathbb{R}$ is called a *submodular system* and f is called the *rank function* of $(\mathscr{D}, f)$. We assume throughout the present paper that $\mathscr{D} \subseteq 2^E$, $\emptyset \in \mathscr{D}$ and $f(\emptyset) = 0$.

Given a submodular system $(\mathscr{D}, f)$, define a polyhedron $P(f)$ by

$$P(f) = \{x \mid x \in \mathbb{R}^E, \forall X \in \mathscr{D}: x(X) \leq f(X)\}, \tag{2.2}$$

where $\mathbb{R}^E$ is the set of vectors $x = (x(e): e \in E)$ with coordinates indexed by E and $x(e) \in \mathbb{R}$ $(e \in E)$ and

$$x(X) = \sum_{e \in X} x(e). \tag{2.3}$$

We call $P(f)$ the *submodular polyhedron* associated with the submodular system $(\mathscr{D}, f)$ (cf. [18], [19]). When $E \in \mathscr{D}$, define

$$B(f) = \{x \mid x \in P(f), x(E) = f(E)\}, \tag{2.4}$$

which is called the *base polyhedron* associated with $(\mathscr{D}, f)$. The base polyhedron $B(f)$ is the set of all maximal vectors in $P(f)$, where the order among vectors is the one with respect to the ordinary vector lattice, i.e., $x \leq y$ if and only if $\forall e \in E$: $x(e) \leq y(e)$. A vector in $B(f)$ is called a *base* of $(\mathscr{D}, f)$. It should be noted that different submodular systems give different submodular polyhedra and different base polyhedra (if they are defined) and that each submodular polyhedron and each base polyhedron uniquely determine the submodular system which gives them.

If we consider the order which is dual to the ordinary order among reals, or if we reverse the inequality signs, in the above definitions of a submodular function, a submodular system and the submodular polyhedron and the base polyhedron associated with the submodular system, then by replacing f by g we have the definitions of a *supermodular function* $g: \mathscr{D} \to \mathbb{R}$, a *supermodular system* $(\mathscr{D}, g)$ and the *supermodular polyhedron* $P(g)$ and the *base polyhedron* $B(g)$ associated with the supermodular system $(\mathscr{D}, g)$. A function which is submodular and supermodular is called a *modular function*. Figure 2.1 shows two-dimensional examples of submodular and supermodular polyhedra and their base polyhedra.

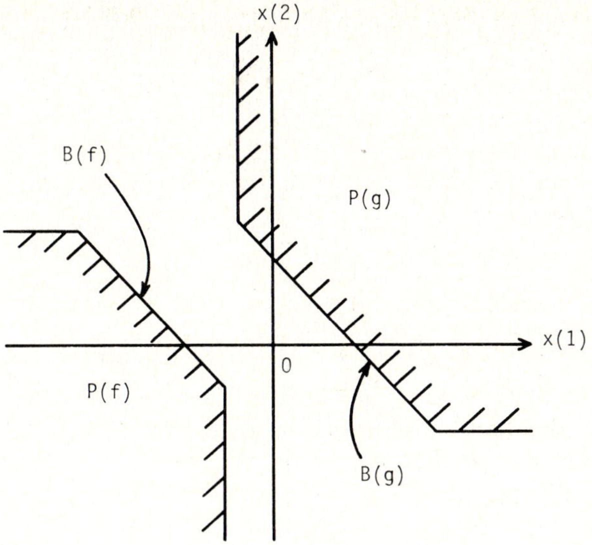

Fig. 2.1. A submodular polyhedron, a supermodular polyhedron and their base polyhedra.

Let $(\mathscr{D}, f)$ be a submodular system with $E \in \mathscr{D} \subseteq 2^E$ and define

$$\bar{\mathscr{D}} = \{E - X \mid X \in \mathscr{D}\}, \tag{2.5}$$

$$f^{\#}(E - X) = f(E) - f(X) \quad (X \in \mathscr{D}). \tag{2.6}$$

Here, $\bar{\mathscr{D}}$ is the dual distributive lattice of $\mathscr{D}$ and $f^{\#} : \bar{\mathscr{D}} \to \mathbb{R}$ is called the *dual supermodular function* of $f : \mathscr{D} \to \mathbb{R}$ and $(\bar{\mathscr{D}}, f^{\#})$ the *dual supermodular system* of $(\mathscr{D}, f)$ (cf. [18, 49]). Similarly, we define the *dual submodular function* $g^{\#} : \bar{\mathscr{D}} \to \mathbb{R}$ of a supermodular function $g : \mathscr{D} \to \mathbb{R}$ and call $(\bar{\mathscr{D}}, g^{\#})$ the *dual submodular system* of $(\mathscr{D}, g)$. A relationship between $(\mathscr{D}, f)$ (or $(\mathscr{D}, g)$) and its dual $(\bar{\mathscr{D}}, f^{\#})$ (or $(\bar{\mathscr{D}}, g^{\#})$) is given by the following

Proposition 2.1

$$(f^{\#})^{\#} = f, \qquad (g^{\#})^{\#} = g, \tag{2.7}$$

$$B(f) = B(f^{\#}), \qquad B(g) = B(g^{\#}). \tag{2.8}$$

(See Fig. 2.2.) The relation (2.8) corresponds to the polymatroid duality [43] (also cf. [18, 49]).

Submodular systems $(\mathscr{D}, f)$ with $\mathscr{D} = 2^E$ are called *hypermatroids* by N. Tomizawa who has been developing a theory of hypermatroids in a series of papers, among which are [49]–[55], where the hyperspace is a space of hypermatroids.

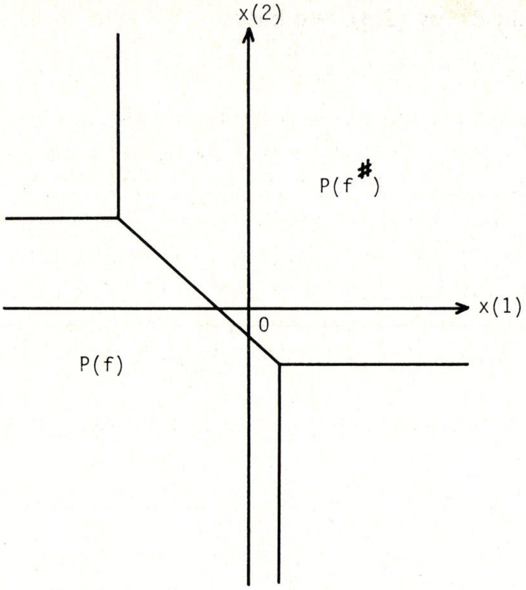

Fig. 2.2. An example of a submodular polyhedron and its dual supermodular polyhedron.

3. Fundamental operations on submodular systems

Let $(\mathscr{D}, f)$ be a submodular system. A *reduction* of $(\mathscr{D}, f)$ by a vector $x \in \mathbb{R}^E$ is a submodular system whose submodular polyhedron is given by

$$\{y \mid y \in P(f), \forall e \in E: y(e) \leqslant x(e)\}. \tag{3.1}$$

The rank function f^x of the reduction is expressed as

$$f^x(X) = \min\{f(Y) + x(X - Y) \mid X \supseteq Y \in \mathscr{D}\} \tag{3.2}$$

for each $X \subseteq E$. When $E \in \mathscr{D}$, a *contraction* of $(\mathscr{D}, f)$ by a vector $x \in P(f)$ is a submodular system whose base polyhedron is given by

$$\{y \mid y \in B(f), \forall e \in E: y(e) \geqslant x(e)\}. \tag{3.3}$$

The rank function f_x of the contraction is expressed as

$$f_x(X) = \min\{f(Y) - x(Y - X) \mid X \subseteq Y \in \mathscr{D}\} \tag{3.4}$$

for each $X \subseteq E$. It should be noted that by reduction and contraction the underlying distributive lattice becomes the Boolean lattice 2^E and that a contraction of $(\mathscr{D}, f)$ by $x \in P(f)$ corresponds to a reduction of the dual supermodular system $(\bar{\mathscr{D}}, f^{\#})$ by x.

A *translation* of $(\mathscr{D}, f)$ by a vector $x \in \mathbb{R}^E$ is a submodular system whose rank function is given by $f + x : \mathscr{D} \to \mathbb{R}$, where x should be considered as a set function (a modular function) on 2^E by (2.3). We have

$$P(f+x)=P(f)+\{x\}, \tag{3.5}$$

where the sum in the right-hand side denotes the vector sum. Note that a contraction of $(\mathscr{D}, f)$ by $x \in P(f)$ followed by a translation by $-x$ corresponds to an ordinary contraction of a (poly-)matroid (cf. [20]). The combinatorial structure of the submodular polyhedron and the base polyhedron is invariant with respect to translations and the rank function becomes monotone increasing by an appropriate translation. Therefore, any results obtained in (poly-)matroid theory which are invariant with respect to translations can easily be extended to submodular systems. For example, from the polymatroid intersection theorem [5] we have

Proposition 3.1 (The intersection theorem). *For any two submodular systems $(\mathscr{D}_i, f_i)$ $(i=1,2)$,*

$$\min\{f_1(X)+f_2(E-X) \mid X \in \mathscr{D}_1, E-X \in \mathscr{D}_2\} = \max\{x(E) \mid x \in P_1(f_1) \cap P(f_2)\}.$$

(Here, by convention, if there is no X such that $X \in \mathscr{D}_1$ and $E-X \in \mathscr{D}_2$, the minimum value of the left-hand side is regarded as $+\infty$ and in that case the maximum in the right-hand side is regarded as the supremum which attains $+\infty$.)

For two submodular systems $(\mathscr{D}_i, f_i)$ $(i=1,2)$ a *sum* of $(\mathscr{D}_i, f_i)$ $(i=1,2)$ is a submodular system whose rank function is given by $f_1+f_2 : \mathscr{D}_1 \cap \mathscr{D}_2 \to \mathbb{R}$ and we have

$$P(f_1+f_2)=P(f_1)+P(f_2). \tag{3.6}$$

A translation is a special case of a sum.

It is well known [1] that for a distributive lattice $\mathscr{D} \subseteq 2^E$ with $E \in \mathscr{D}$ there uniquely exist a partition $\Pi=\{A_1, A_2, \ldots, A_k\}$ of E and a partial order $\preccurlyeq$ on Π such that $X \in \mathscr{D}$ if and only if X is expressed as

$$X=\bigcup\{A_i \mid A_i \in \mathscr{I}\} \tag{3.7}$$

for an ideal $\mathscr{I}$ of the partially ordered set (poset) $(\Pi, \preccurlyeq)$. (Here, $\mathscr{I} \subseteq \Pi$ is an ideal of the poset $(\Pi, \preccurlyeq)$ if $A_i \preccurlyeq A_j \in \mathscr{I}$ implies $A_i \in \mathscr{I}$ for every A_i, $A_j \in \Pi$.) For a submodular system $(\mathscr{D}, f)$, by identifying each $A_i \in \Pi$ with an element a_i, we obtain a distributive lattice $\mathscr{D}' \subseteq 2^{E'}$ with $E'=\{a_1, a_2, \ldots, a_k\}$ and a submodular function $f' : \mathscr{D}' \to \mathbb{R}$. We call the pair $(\mathscr{D}', f')$ the *simplification* of $(\mathscr{D}, f)$. When the partition Π associated with $\mathscr{D}$ is the partition of E into singletons, we call $\mathscr{D}$ and $(\mathscr{D}, f)$ *simple*. When $\mathscr{D} \subseteq 2^E$ is simple, $\mathscr{D}$ is given by the set of ideals of a poset $\mathscr{P}=(E, \preccurlyeq)$ and is denoted by $\mathbf{2}^{\mathscr{P}}$.

For any partition $\hat{\Pi}=\{\hat{A}_1, \hat{A}_2, \ldots, \hat{A}_l\}$ of E a subset X of E is said to be *compatible with* $\hat{\Pi}$ if, for each $\hat{A}_i \in \hat{\Pi}$, $\hat{A}_i \cap X \neq \emptyset$ implies $\hat{A}_i \subseteq X$. Define

$$\mathscr{D}(\hat{\Pi})=\{X \mid X \in \mathscr{D}, X \text{ is compatible with } \hat{\Pi}\} \tag{3.8}$$

and denote the restriction of f to $\mathscr{D}(\hat{\Pi})$ by $f_{\hat{\Pi}}$. The pair $(\mathscr{D}(\hat{\Pi}), f_{\hat{\Pi}})$ is called the *aggregation* of $(\mathscr{D}, f)$ by $\hat{\Pi}$. Aggregations play a fundamental role in a decomposition theory for submodular systems [25].

4. The structure of base polyhedra

Suppose $(\mathscr{D}, f)$ is a submodular system with $E \in \mathscr{D}$. Then we have the following easy propositions.

Proposition 4.1. *The base polyhedron $B(f)$ is pointed (or has extreme points) if and only if $\mathscr{D} = \mathbf{2}^{\mathscr{P}}$ for some poset $\mathscr{P} = (E, \preccurlyeq)$ (i.e., $\mathscr{D}$ is simple).*

Proposition 4.2. *The base polyhedron $B(f)$ is bounded if and only if $\mathscr{D} = 2^E$ (i.e., $\mathscr{D}$ is simple and complemented).*

The next proposition follows from the so-called greedy algorithm [6] for base polyhedra (cf. [27, 41, 47]).

Proposition 4.3 (The extreme point theorem). *Suppose $\mathscr{D} = \mathbf{2}^{\mathscr{P}}$ for some poset $\mathscr{P} = (E, \preccurlyeq)$ (i.e., $\mathscr{D}$ is simple). Then a base $x \in B(f)$ is an extreme point of $B(f)$ if and only if for a maximal chain*

$$\mathscr{C}: \quad \emptyset = S_0 \subsetneq S_1 \subsetneq \cdots \subsetneq S_n = E \tag{4.1}$$

in $\mathscr{D}$ we have

$$x(S_i - S_{i-1}) = f(S_i) - f(S_{i-1}) \quad (i = 1, 2, \ldots, n). \tag{4.2}$$

Suppose $(\mathscr{D}, f)$ is a simple submodular system and define

$$D(e) = \bigcap\{X \mid e \in X \in \mathscr{D}\} \quad (e \in E), \tag{4.3}$$

$$\bar{\alpha}(e) = f(D(e)) - f(D(e) - \{e\}) \quad (e \in E). \tag{4.4}$$

Then, by the submodularity of f and Proposition 4.3, the vector $\bar{\alpha}$ is the least upper bound (or the join) of all the extreme points of $B(f)$ in the vector lattice $\mathbb{R}^E$ (cf. [27], and [49] when $\mathscr{D} = 2^E$). In particular, for each $X \in \mathscr{D}$,

$$f(X) \leqslant \bar{\alpha}(X), \tag{4.5}$$

since there is an extreme base $x \in B(f)$ such that $x(X) = f(X)$. Because of (4.5), $\bar{\alpha}$ can be used for estimating an upper bound of f. Moreover, denote by $\underline{\alpha}$ the greatest lower bound (or the meet) of all the extreme points of $B(f)$ in the vector lattice $\mathbb{R}^E$, which is given similarly as (4.3) and (4.4) in a dual form:

$$D^*(e) = \bigcup\{X \mid e \notin X \in \mathscr{D}\} \quad (e \in E), \tag{4.6}$$

$$\underline{\alpha}(e) = f(D^*(e) \cup \{e\}) - f(D^*(e)) \quad (e \in E). \tag{4.7}$$

It easily follows from Proposition 4.3 that f is monotone nondecreasing if and only if $\underline{\alpha} \geqslant \mathbf{0}$.

In general, $B(f)$ is expressed as the vector sum of the convex hull $Q(f)$ of extreme points of $B(f)$ and the recession cone $C(f)$ of $B(f)$:

$$B(f) = Q(f) + C(f), \tag{4.8}$$

where we assume $\mathcal{D}$ is simple, and

$$C(f) = \{x \mid x \in \mathbb{R}^E, \forall X \in \mathcal{D}: x(X) \leq 0, x(E) = 0\} \tag{4.9}$$

(cf. [45, 48]). Note that the recession cone (4.9) is determined only by the underlying distributive lattice $\mathcal{D}$.

It follows from Proposition 4.3 that if the rank function f is integer-valued and has a unit-increase property (i.e., for each X, $Y \in \mathcal{D}$ with $X \subseteq Y$ and $|Y - X| = 1$, $f(X) \leq f(Y) \leq f(X) + 1$), then all the extreme points of $B(f)$ are $\{0, 1\}$-vectors. Therefore, we can develop a matroid-like theory for a submodular system $(\mathcal{D}, f)$ with the rank function having the above-mentioned properties, which corresponds to *Faigle's geometry on a poset* $\mathcal{P} = (E, \preccurlyeq)$ with $\mathcal{D} = \mathbf{2}^{\mathcal{P}}$ [8, 9]. In fact, Proposition 4.3 gives a polyhedral characterization of the family of bases of Faigle's geometry.

A. Frank [12] introduced a concept of a *generalized polymatroid.* Suppose that a submodular system $(\mathcal{D}_1, f')$ and a supermodular system $(\mathcal{D}_2, g')$ with $\mathcal{D}_i \subseteq 2^{E'}$ $(i = 1, 2)$ satisfy

$$\forall X \in \mathcal{D}_1, \forall Y \in \mathcal{D}_2: \qquad X - Y \in \mathcal{D}_1, \quad Y - X \in \mathcal{D}_2,$$
$$f'(X) - g'(Y) \geq f'(X - Y) - g'(Y - X). \tag{4.10}$$

Then the polyhedron $P(f', g')$ defined by

$$P(f', g') = \{x \mid x \in \mathbb{R}^{E'}, \forall X \in \mathcal{D}_1: x(X) \leq f'(X), \forall Y \in \mathcal{D}_2: x(Y) \geq g'(Y)\} \tag{4.11}$$

is called a generalized polymatroid. (The same polyhedron is also considered by R. Hassin [30] for the case when $\mathcal{D}_1 = \mathcal{D}_2 = 2^{E'}$.) Generalized polymatroids are characterized by the following proposition [23] which is implicit in [12] (also see [46]).

Proposition 4.4. *The projection of the base polyhedron $B(f)$ of a submodular system $(\mathcal{D}, f)$ with $E \in \mathcal{D} \subseteq 2^E$ along an axis $e \in E$ on the hyperplane $x(e) = 0$ is a generalized polymatroid $P(f', g')$ in $\mathbb{R}^{E'}$ with $E' = E - \{e\}$, where*

$$\mathcal{D}_1 = \{X \mid e \notin X \in \mathcal{D}\} \subseteq 2^{E'}, \tag{4.12}$$

$$\mathcal{D}_2 = \{E - X \mid e \in X \in \mathcal{D}\} \subseteq 2^{E'}, \tag{4.13}$$

f' is a restriction of f to $\mathcal{D}_1$ and g' is a restriction of $f^{\#}$ to $\mathcal{D}_2$.

Conversely, every generalized polymatroid in $\mathbb{R}^{E'}$ is obtained in this way. For each generalized polymatroid $P(f', g')$ with $\mathcal{D}_i \subseteq 2^{E'}$ $(i = 1, 2)$ such a base polyhedron $B(f)$ with $\mathcal{D} \subseteq 2^E$ and $E = E' \cup \{e\}$ is unique up to translation along the new axis e, and the two polyhedra $P(f', g')$ and $B(f)$ are isomorphic with each other.

It follows from Proposition 4.4 that extreme points, extreme rays, faces etc. of $P(f', g')$ are characterized by the corresponding results for $B(f)$. Moreover, since the greedy algorithm works for $B(f)$, it also works for $P(f', g')$ (cf. [30]).

As was noted in Section 3, results, obtained for polymatroids, which are invariant with respect to translations can easily be extended to submodular systems. For example, let us consider elementary transformations of bases of a submodular system $(\mathcal{D}, f)$.

For any base $x \in B(f)$ and $e \in E$, define

$$\mathrm{dep}(x, e) = \bigcap \{X \mid e \in X \in \mathcal{D},\ x(X) = f(X)\} \quad (\neq \emptyset). \tag{4.14}$$

The function $\mathrm{dep}: B(f) \times E \to \mathcal{D}$ is called the *dependence function* associated with $(\mathcal{D}, f)$ (cf. [17]). Furthermore, for any $e' \in \mathrm{dep}(x, e) - \{e\}$, define

$$\tilde{c}(x, e, e') = \min\{f(X) - x(X) \mid e \in X \in \mathcal{D},\ e' \notin X\} \quad (>0) \tag{4.15}$$

which is called the *exchange capacity* associated with $x \in B(f)$, $e \in E$ and $e' \in \mathrm{dep}(x, e) - \{e\}$.

Proposition 4.5. *For any $x \in B(f)$, $e \in E$, $e' \in \mathrm{dep}(x, e) - \{e\}$, and $d \in R$ with $0 < d \leq \tilde{c}(x, e, e')$ we have*

$$x + d\chi_e - d\chi_{e'} \in B(f), \tag{4.16}$$

where χ_e is the unit vector in $\mathbb{R}^E$ with $\chi_e(e) = 1$.

The transformation from $x \in B(f)$ into $x + d\chi_e - d\chi_{e'} \in B(f)$ in (4.16) is called an *elementary transformation* of the base $x \in B(f)$.

The following proposition is important from an algorithmic point of view.

Proposition 4.6. *For any two bases $x, y \in B(f)$ x can be transformed into y by repeated elementary transformations such that each component $x(e)$ with $x(e) < y(e)$ monotonically increases and each component $x(e)$ with $x(e) > y(e)$ monotonically decreases.*

Consider a network $N = (G = (V, A), c)$ with an underlying graph $G = (V, A)$ with a vertex set V and an arc set A and with a nonnegative capacity function $c: A \to \mathbb{R}$. Define a submodular function $\hat{c}: 2^V \to \mathbb{R}$ by

$$\hat{c}(U) = \sum \{c(a) \mid a \in A,\ \partial^+ a \in U,\ \partial^- a \in V - U\} \tag{4.17}$$

for each $U \subseteq V$, where $\partial^+ a$ ($\partial^- a$) is the initial (terminal) vertex of an arc a. We call $\hat{c}$ the *cut function* of N. Also, a function $\phi: A \to R$ is called a *flow* in N if for each $a \in A$

$$0 \leq \phi(a) \leq c(a). \tag{4.18}$$

Moreover, define a function (or a vector) $\partial' \phi: V \to \mathbb{R}$ associated with a flow ϕ in N

by

$$\partial'\phi(v) = \sum_{a\in\delta^+ v} \phi(a) - \sum_{a\in\delta^- v} \phi(a) \tag{4.19}$$

for each $v \in V$, where $\delta^{\pm} v = \{a \mid a \in A, \partial^{\pm} a = v\}$. $\partial'\phi$ is called the *boundary* of ϕ. It can be shown that the set of boundaries of flows in N is a base polyhedron associated with a submodular system $(2^V, \hat{c})$([28, 52]) which is called a *boundary hypermatroid* [55]. A flow ϕ in N with $\partial'\phi = \mathbf{0}$ is called a *circulation* in N. There is a trivial circulation $\mathbf{0}$ in N.

Now, consider a network $N' = (G = (V, A), \underline{c}, \bar{c})$ with an underlying graph G and lower and upper capacity functions $\underline{c}, \bar{c}: A \to \mathbb{R}$ $(\underline{c} \leq \bar{c})$. A function $\phi: A \to \mathbb{R}$ is a flow in N' if $\underline{c}(a) \leq \phi(a) \leq \bar{c}(a)$ for each $a \in A$. Define

$$c = \bar{c} - \underline{c} \quad (\geq \mathbf{0}) \tag{4.20}$$

and also define $\hat{c}: 2^V \to \mathbb{R}$ by (4.17) using c in (4.20). Then the set of boundaries of flows in N' is the base polyhedron associated with a submodular system $(2^V, \hat{c} + \partial'\underline{c})$ which is a translation of $(2^V, \hat{c})$ by $\partial'\underline{c}$. Note that there exists a circulation in N' if and only if $\mathbf{0} \in B(\hat{c} + \partial'\underline{c})$. Since $\partial'\underline{c} \in B(\hat{c} + \partial'\underline{c})$, there exists a circulation in N' if and only if the base $\partial'\underline{c}$ is transformed into $\mathbf{0}$ by repeated elementary transformations as in Proposition 4.6, which gives a standard algorithm for finding a circulation in N' (cf. [31, 10]).

5. Submodular functions on crossing families

We say subsets X and Y of E *cross* if the four sets $X \cap Y$, $X \cap (E - Y)$, $(E - X) \cap Y$, $(E - X) \cap (E - Y)$ are nonempty. A family $\mathcal{F}$ of subsets of E is called a *crossing family* if for each crossing pair of $X, Y \in \mathcal{F}$ we have $X \cup Y, X \cap Y \in \mathcal{F}$. A function $f: \mathcal{F} \to \mathbb{R}$ is called a *submodular function on the crossing family* $\mathcal{F}$ if for each crossing pair of $X, Y \in \mathcal{F}$,

$$f(X) + f(Y) \geq f(X \cup Y) + f(X \cap Y). \tag{5.1}$$

Proposition 5.1 [21]. *Let f be a submodular function on a crossing family $\mathcal{F} \subseteq 2^E$ with $\emptyset$, $E \in \mathcal{F}$ and $f(\emptyset) = 0$. Define*

$$B(f) = \{x \mid x \in \mathbb{R}^E, \forall X \in \mathcal{F}\colon x(X) \leq f(X), x(E) = f(E)\} \tag{5.2}$$

and suppose $B(f) \neq \emptyset$. Then there exists a unique submodular system $(\mathcal{D}_1, f_1)$ such that

$$B(f) = B(f_1). \tag{5.3}$$

Moreover, each $f_1(X)$ $(X \in \mathcal{D}_1)$ is expressed as a linear combination of $f(Y)$'s $(Y \in \mathcal{F})$ with nonnegative integral coefficients for $Y \in \mathcal{F} - \{E\}$ and a (possibly negative) integral coefficient for $Y = E$. In particular, if f is integer-valued, then f_1 is also integer-valued.

From Proposition 5.1 we can easily see that if f is a nonnegative integer-valued submodular function on a crossing family $\mathcal{F}$ and the polyhedron given by

$$B(f_0^1) = \{x \mid x \in B(f), \forall e \in E: 0 \leq x(e) \leq 1\} \tag{5.4}$$

is nonempty, then $B(f_0^1)$ is the base polyhedron obtained by the reduction of $B(f)$ by the vector $\mathbf{1} = (\mathbf{1}(e) = 1: e \in E)$ and the contraction by the zero vector $\mathbf{0}$ and is a base polyhedron of a matroid, or from the polyhedral characterization of matroids by Edmonds [5]

$$\{X \mid X \subseteq E, \forall Y \in \mathcal{F}: |X \cap Y| \leq f(Y), |X| = f(E)\} \tag{5.5}$$

is a family of bases of a matroid. This is a result by A. Frank and É. Tardos [16].

Let $\hat{N} = (G = (V, A), \underline{c}, \bar{c}, f)$ be a network with an underlying graph G, lower and upper capacity functions $\underline{c}, \bar{c}: A \to \mathbb{R}$ and a submodular function f on a crossing family $\mathcal{F}$ formed by subsets of the vertex set V. Suppose $\emptyset, V \in \mathcal{F}$ and $f(\emptyset) = f(V) = 0$. A function $\phi: A \to \mathbb{R}$ is called a *feasible flow* in $\hat{N}$ if, for each, $a \in A$ $\underline{c}(a) \leq \phi(a) \leq \bar{c}(a)$ and $\partial'\phi \in B(f)$, where $\partial'\phi$ is the boundary of ϕ and $B(f)$ is given by (5.2) with E replaced by V. Given a cost function $\gamma: A \to R$ the *submodular flow problem* is to find a feasible flow ϕ in $\hat{N}$ which minimizes the cost $\sum\{\gamma(a)\phi(a) \mid a \in A\}$ among all feasible flows in $\hat{N}$, which was considered by J. Edmonds and R. Giles [7]. (The term 'submodular flow' was introduced by U. Zimmermann [63].) Because $B(f)$ is a base polyhedron, the submodular flow problem can be regarded as an independent flow problem considered in [17] (also see [21]) and this fact makes it possible to adopt the augmenting path method and the negative cycle method for the submodular flow problem (cf. [15, 63, 4]). Moreover, the problem of finding a feasible flow in $\hat{N}$ [13] can be reduced to that of finding a common base of two base polyhedra $B(f)$ and $B(\hat{c} + \partial'\underline{c})$, where $B(\hat{c} + \partial'\underline{c})$ is the set of boundaries of flows in $N' = (G = (V, A), \underline{c}, \bar{c})$ (see Section 4). Related arguments were also made in [38] and [64] (also see [39, 40]).

Now, let us consider a network $N = (G = (V, A), c)$ with a uniform capacity $c(a) = 1$ $(a \in A)$. Then any $\{0, 1\}$-flow ϕ in N uniquely determines the reorientation of $G = (V, A)$ under the correspondence that $\phi(a) = 1$ if and only if a is reoriented. $G = (V, A)$ is called *strongly k-connected* $(k \geq 1)$ if G remains strongly connected after deleting arbitrary $k-1$ arcs from G. Let $\hat{c}: 2^V \to \mathbb{R}$ be the cut function of N and define

$$\hat{c}_{(k)}(U) = \begin{cases} \hat{c}(U) - k & (U \in 2^V - \{\emptyset, V\}), \\ 0 & (U \in \{\emptyset, V\}) \end{cases} \tag{5.6}$$

for a positive integer k. Then $\hat{c}_{(k)}$ is a submodular function on a 'crossing family' 2^V and defines a base polyhedron $B(\hat{c}_{(k)})$ due to Proposition 5.1 [26]. $B(\hat{c}_{(k)})$ is called the *k-abridgment* of $B(\hat{c})$ in [51, 53]. It can easily be seen that

$$B(\hat{c}_{(k)}) = \{\partial'\phi \mid \phi \text{ defines a strongly } k\text{-connected reorientation of } G\}. \tag{5.7}$$

Here, we assume that the underlying totally ordered additive group R is the set of integers. Also note that the strong k-connectedness of a reorientation of G depends only on the boundary $\partial'\phi$ of ϕ which defines the reorientation. Because of this fact and Proposition 4.6 we obtain a theorem due to A. Frank [14]:

Let G' and G'' be strongly k-connected reorientations of G. Then, there exists a sequence of strongly k-connected reorientations $G' = G_0, G_1, \ldots, G_m = G''$ of G such that, for each $i = 1, \ldots, m$, G_i is obtained by reversing arcs in a directed path or a directed cycle in G_{i-1}.

Note that an elementary transformation of a base (or a boundary $\partial'\phi$) in $B(\hat{c}_{(k)})$ corresponds to a transformation of the reorientation of G (defined by ϕ) by reversing arcs in a directed path and possibly directed cycles and that reversing arcs in a directed cycle does not change the boundary $\partial'\phi$.

6. Submodular/supermodular functions viewed from convex analysis

Let $(\mathscr{D}, f)$ be a submodular system and $(\mathscr{D}, g)$ be a supermodular system. Define a function $f^*: \mathbb{R}^E \to \mathbb{R}$ by

$$f^*(x) = \max\{x(X) - f(X) \mid X \in \mathscr{D}\} \quad (x \in \mathbb{R}^E). \tag{6.1}$$

We call f^* the *convex conjugate function* of the submodular function f. Note that the term $x(X)$ can be regarded as an inner product of x and the characteristic vector $\chi(X)$ of X and that (6.1) is in a form similar to the ordinary definition of the convex conjugate function of a convex function (see [45, 48]). Similarly, define the *concave conjugate function* $g^*: \mathbb{R}^E \to \mathbb{R}$ of the supermodular function g by

$$g^*(x) = \min\{x(X) - g(X) \mid X \in \mathscr{D}\} \quad (x \in \mathbb{R}^E). \tag{6.2}$$

Proposition 6.1 (A Fenchel-type min-max theorem) [22].

$$\min\{f(X) - g(X) \mid X \in \mathscr{D}\} = \max\{g^*(x) - f^*(x) \mid x \in \mathbb{R}^E\}. \tag{6.3}$$

Here, the maximum value in the right-hand side is attained by an integral vector x if f and g are integer-valued.

Proposition 6.1 follows from Proposition 3.1 and is closely related to Frank's discrete separation theorem [15]:

Let $(\mathscr{D}, f)$ and $(\mathscr{D}, g)$ respectively, be a submodular system and a supermodular system. If, for each $X \in \mathscr{D}$, $g(X) \leq f(X)$, then there exists a vector $x \in \mathbb{R}^E$ such that for each $X \in \mathscr{D}$ $g(X) \leq x(X) \leq f(X)$. Furthermore, the vector $x \in \mathbb{R}^E$ can be integral if f and g are integer-valued.

Proposition 6.1 leads us to a definition of a subgradient of a submodular function as follows. A vector $x \in \mathbb{R}^E$ is a *subgradient* of a submodular function $f: \mathscr{D} \to \mathbb{R}$ at

$A \in \mathcal{D}$ if for each $X \in \mathcal{D}$

$$x(X) - x(A) \leq f(X) - f(A). \tag{6.4}$$

Denote by $\partial f(A)$ the set of all subgradients of f at $A \in \mathcal{D}$ and call it the *subdifferential* of f at $A \in \mathcal{D}$.

Figure 6.1 shows a two-dimensional example of subdifferentials of $f : \mathcal{D} \to \mathbb{R}$ with $\mathcal{D} = 2^{\{1,2\}}$. In general, distinct subdifferentials $\partial f(X)$ and $\partial f(Y)$ ($X, Y \in \mathcal{D}, X \neq Y$) do not have any common interior points and the collection of subdifferentials $\partial f(X)$ ($X \in \mathcal{D}$) forms a division of $\mathbb{R}^E$. Moreover, when $f(\emptyset) = 0$, $\partial f(\emptyset)$ is the submodular polyhedron $P(f)$ associated with $(\mathcal{D}, f)$ and when $E \in \mathcal{D}$, $\partial f(E)$ is the supermodular polyhedron $P(f^{\#})$ associated with the dual supermodular system $(\bar{\mathcal{D}}. f^{\#})$.

Note that, by definition (6.4), $\mathbf{0} \in \partial f(A)$ for $A \in \mathcal{D}$ if and only if A is a minimizer of f. In the example shown in Fig. 6.1 the zero vector $\mathbf{0}$ is in the interior of $\partial f(\emptyset)$, so that the empty set $\emptyset$ is the unique minimizer of f.

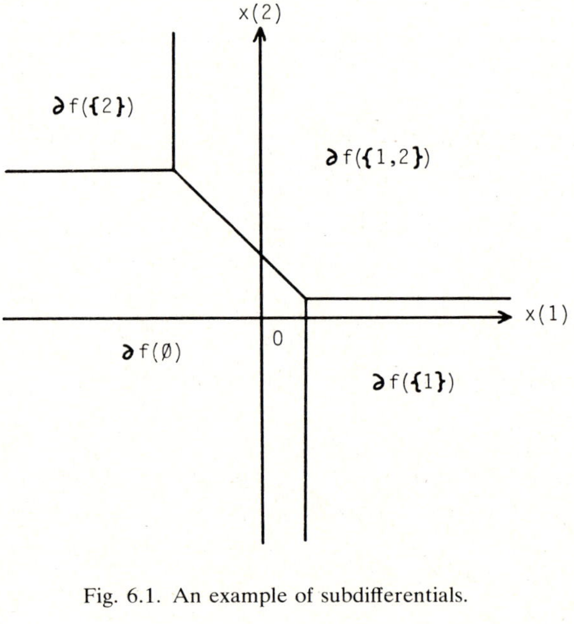

Fig. 6.1. An example of subdifferentials.

The above definitions of a convex conjugate function and a subdifferential of a submodular function can be interpreted as a convex conjugate function and a subdifferential, in an ordinary sense of convex analysis, of a convex function as follows.

Recently, L. Lovász [41] has defined an extension of a set function, which can easily be adapted to a function on a distributive lattice. Let $b : E \to \mathbb{R}$ be a nonnegative monotone nonincreasing function from a poset $\mathcal{P} = (E, \preccurlyeq)$ to $(R, \leq)$, where $\mathcal{D} = \mathbf{2}^{\mathcal{P}}$. Note that the set of such functions (or vectors) b is the polar cone, denoted by

$C^*(\mathscr{D})$, of a cone

$$C(\mathscr{D}) = \{x \mid x \in \mathbb{R}^E, \forall X \in \mathscr{D}: x(X) \leq 0\}. \tag{6.5}$$

Then there uniquely exist a chain

$$\emptyset \neq A_1 \subsetneq A_2 \subsetneq \cdots \subsetneq A_k \qquad (k \geq 0), \tag{6.6}$$

in $\mathscr{D} = \mathbf{2}^{\mathscr{P}}$ and $\lambda_i > 0$ $(i = 1, \ldots, k)$ such that

$$b = \sum_{i=1}^{k} \lambda_i \chi(A_i), \tag{6.7}$$

where $\chi(A_i)$ is the characteristic vector of $A_i \subseteq E$ and when $b = \mathbf{0}$, (6.6) is the empty chain and then the empty sum (6.7) is, by convention, equal to the zero vector. For any function $f: \mathscr{D} \to \mathbb{R}$ with $f(\emptyset) = 0$, define

$$\hat{f}(b) = \sum_{i=1}^{k} \lambda_i f(A_i) \tag{6.8}$$

according to the expression (6.7) for $b \in C^*(\mathscr{D})$. We call $\hat{f}$ *Lovász's extension of f.*

Denote by $P(\mathscr{D})$ the convex hull of the characteristic vectors $\chi(A)$ $(A \in \mathscr{D})$ in $\mathbb{R}^E$. Note that $P(\mathscr{D}) \subseteq C^*(\mathscr{D})$.

Proposition 6.2. *For a submodular function $f: \mathscr{D} \to \mathbb{R}$ with $f(\emptyset) = 0$, Lovász's extension $\hat{f}$ of f is convex on $C^*(\mathscr{D})$ and*

$$\min\{f(X) \mid X \in \mathscr{D}\} = \min\{\hat{f}(x) \mid x \in P(\mathscr{D})\}.$$

(Cf. [41] *when $\mathscr{D} = 2^E$.)*

Now, define a *truncated* Lovász's extension $\hat{\hat{f}}$ of f by

$$\hat{\hat{f}}(x) = \begin{cases} \hat{f}(x) & (x \in P(\mathscr{D})), \\ +\infty & (x \in \mathbb{R}^E - P(\mathscr{D})). \end{cases} \tag{6.9}$$

Then, the following propositions for a submodular function $f: \mathscr{D} \to \mathbb{R}$ with $f(\emptyset) = 0$ easily follow from Proposition 6.2 (see [24]).

Proposition 6.3. *For each $A \in \mathscr{D}$,*

$$\partial f(A) = \partial \hat{\hat{f}}(\chi(A)),$$

where $\partial \hat{\hat{f}}(\chi(A))$ is the subdifferential, in an ordinary sense of convex analysis, of the convex function $\hat{\hat{f}}$ at $\chi(A)$.

Proposition 6.4. *f^* defined by* (6.1) *is the convex conjugate function of $\hat{\hat{f}}$ in an ordinary sense of convex analysis.*

Proposition 6.1 except for the integrality follows from Proposition 6.4 and Fenchel's duality theorem for convex and concave functions [45, 48].

7. Minimization of submodular functions

It has been shown by M. Grötschel, L. Lovász and A. Schrijver [29] that the problem of minimizing an integer-valued submodular function $f : \mathcal{D} \to \mathbb{R}$ can be solved in polynomial time by the so-called ellipsoid method [36] using an oracle for function evaluation. Lovász [41] has also suggested minimizing a submodular function f by minimizing its extension $\hat{f}$. Moreover, R. E. Bixby, W. H. Cunningham and D. M. Topkis [2] have shown a possible approach to the problem by adapting a result by Cunningham [3] for minimizing translated matroid rank functions.

In this section we show another approach to the problem of minimizing a submodular function. Here, we assume that R is the set of reals or rationals (a totally ordered field) and $f : \mathcal{D} \to \mathbb{R}$ is an integer-valued submodular function.

Given a submodular system $(\mathcal{D}, f)$ with $\emptyset$, $E \in \mathcal{D}$ and $f(\emptyset) = 0$, consider a special quadratic programming problem

$$\text{(P):} \qquad \text{Minimize} \quad \sum_{e \in E} x(e)^2 \tag{7.1}$$

$$\text{subject to} \quad x \in B(f), \tag{7.2}$$

where $B(f)$ is the base polyhedron associated with $(\mathcal{D}, f)$. It is well known that the collection of all the minimizers of f is a (distributive) sublattice of $\mathcal{D}$. The optimal solution of problem (P) gives the unique minimal and maximal elements of the sublattice as follows.

Proposition 7.1. *Let $\hat{x}$ be the optimal solution of* (P) *and define*

$$A_- = \{e \mid e \in E, \hat{x}(e) < 0\}, \tag{7.3}$$

$$A_0 = \{e \mid e \in E, \hat{x}(e) \leq 0\}. \tag{7.4}$$

Then A_- is the unique minimal minimizer of f and A_0 is the unique maximal minimizer of f.

Proposition 7.1 easily follows from the result in [20].

It may be interesting to point out a relationship between the optimal solution $\hat{x}$ of (P) and subdifferentials of f. The following fact also easily follows from [20]. Let $c_1 < c_2 < \cdots < c_p$ be the distinct values of $\hat{x}(e)$ $(e \in E)$ and define

$$A_i = \{e \mid e \in E, \hat{x}(e) \leq c_i\} \quad (i = 1, 2, \ldots, p), \tag{7.5}$$

$$A_0 = \emptyset. \tag{7.6}$$

Then

$$\emptyset = A_0 \subsetneq A_1 \subsetneq \cdots \subsetneq A_p = E \tag{7.7}$$

is a chain in $\mathcal{D}$ and, for each $i = 1, 2, \ldots, p$,

$$c_i = \frac{f(A_i) - f(A_{i-1})}{|A_i - A_{i-1}|}. \tag{7.8}$$

On the other hand, let L be the line in $\mathbb{R}^E$ which goes through the origin $\mathbf{0}$ and the point $\mathbf{1}=(\mathbf{1}(e)=1\colon e\in E)$. Then a point $x_t=(x_t(e)=t\colon e\in E)$ on L belongs to $\partial f(\emptyset)$ for a sufficiently small t (possibly negative with a sufficiently large absolute value) and belongs to $\partial f(E)$ for a sufficiently large t. Subdifferentials

$$\partial f(A_0)=\partial f(\emptyset), \partial f(A_1), \ldots, \partial f(A_p)=\partial f(E) \tag{7.9}$$

given according to (7.7) are exactly those subdifferentials of f whose interiors have nonempty intersections with the line L. In other words, there is a division of the interval $(-\infty, +\infty)$ given by

$$I_0=(-\infty, t_1], I_1=[t_1, t_2], \ldots, I_{p-1}=[t_{p-1}, t_p], I_p=[t_p, +\infty) \tag{7.10}$$

such that $-\infty<t_1<t_2<\cdots<t_{p-1}<t_p<+\infty$ and, for each $t\in(t_k, t_{k+1})$ $(0\leqslant k\leqslant p)$, x_t belongs to the interior of $\partial f(A_k)$, where $t_0=-\infty$ and $t_{p+1}=+\infty$. Moreover,

$$t_i=c_i \quad (i=1,2,\ldots,p), \tag{7.11}$$

where c_i $(i=1,2,\ldots,p)$ are those appearing in (7.5). It can also be shown that the collection of A's in $\mathcal{D}$ for which $\partial f(A)\cap L\neq\emptyset$ forms a sublattice of $\mathcal{D}$. (This sublattice is the one in (7.12).) These results are closely related to the principal partition of polymatroids and its generalization (cf. [20, 32, 56]).

It should be noted that if $\hat{x}$ is the optimal solution of (P), then the optimal solution of (P) with $B(f)$ translated along the line L by $x_t=(x_t(e)=t\colon e\in E)$ for arbitrary $t\in R$ is given by $\hat{x}+x_t$. Therefore, finding the optimal solution $\hat{x}$ of (P) is much more than minimizing f. However, by adapting an algorithm in [20] we can find $\hat{x}$ in time polynomial in $|E|$ if we use an oracle for minimizing submodular functions. It is also known that problem (P), a (special) quadratic programming problem, can be solved in polynomial time by the ellipsoid method [37].

Problem (P) is the problem of finding a minimum-norm point in $B(f)$ and a solution algorithm for such a problem is proposed by P. Wolfe [62]. For simplicity, suppose $\mathcal{D}=2^E$, i.e., $B(f)$ is bounded. Then Wolfe's algorithm generates simplices $\hat{S}=\langle x_i\mid i\in I\rangle$ with distinct extreme points x_i $(i\in I)$ of $B(f)$ as the vertices of $\hat{S}$ such that

the minimal norm point x' in the affine subspace generated by x_i $(i\in I)$ belongs to the relative interior of $\hat{S}$. (*)

If $x'=\mathbf{0}$, then x' is the optimal solution $\hat{x}$. Otherwise, the dimension of a simplex $\hat{S}$ is increased by one by adding to $\hat{S}$ a new extreme point of $B(f)$ which minimizes a linear objective function with a coefficient vector x' (if the dimension does not increase, x' is the optimal solution $\hat{x}$), and then we find a subsimplex (call it $\hat{S}$ again) of the augmented simplex in a suitable way such that the above property (*) is satisfied. Therefore, Wolfe's algorithm requires that linear objective functions can easily be minimized over the polyhedron $B(f)$. This is the case for the base polyhedron, so that Wolfe's algorithm can be applied to problem (P) with some possible modification. (Wolfe's algorithm can be generalized so that it can work for unbounded base polyhedra when $\mathcal{D}\neq 2^E$.)

When Wolfe's algorithm terminates, it gives us the optimal solution $\hat{x}$ together with a set of extreme points x_i $(i \in I)$ of $B(f)$ such that $\hat{x}$ is in the relative interior of the simplex $\langle x_i \mid i \in I \rangle$. For each $i \in I$ define a poset $\mathscr{P}_i = (E, \preccurlyeq_i)$ as follows: $e_1 \preccurlyeq_i e_2$ if and only if $e_1 \in \mathrm{dep}(x_i, e_2)$. It is easy to see that this defines a poset, which is also treated in [2]. Also define a distributive lattice $\hat{\mathscr{D}}$ as

$$\hat{\mathscr{D}} = \bigcap_{i \in I} \mathbf{2}^{\mathscr{P}_i} \tag{7.12}$$

and choose an arbitrary maximal chain

$$\emptyset = \hat{A}_0 \subsetneq \hat{A}_1 \subsetneq \cdots \subsetneq \hat{A}_q = E \tag{7.13}$$

in $\hat{\mathscr{D}}$. Note that a maximal chain (7.13) is easily found from posets $\mathscr{P}_i$ $(i \in I)$ represented by Hasse diagrams. (For the construction of each Hasse diagram see [2].) Now, the optimal solution $\hat{x}$ of (P) is expressed as

$$\hat{x}(e) = \frac{f(\hat{A}_j) - f(\hat{A}_{j-1})}{|\hat{A}_j - \hat{A}_{j-1}|} \tag{7.14}$$

for each $e \in \hat{A}_j - \hat{A}_{j-1}$ $(j = 1, 2, \ldots, q)$. (The proof of (7.14) is not difficult if we see the relationship between the distributive lattice $\hat{\mathscr{D}}$ and $\mathrm{dep}(\hat{x}, e)$ $(e \in E)$, which is also employed in [3] (also see [17, 20]).) The chain in (7.7) is a subchain of any maximal chain in $\hat{\mathscr{D}}$. It should be noted that $\hat{A}_i$'s in (7.13) which minimize f on the chain (7.13) are minimizers of f on $\mathscr{D}$ and, in fact, that all the minimizers of f on $\mathscr{D}$ are given by the interval $[A_-, A_0]$ of $\hat{\mathscr{D}}$. Consequently, minimizers of f can be determined only by the associated discrete structure $\hat{\mathscr{D}}$ if we are given the set of extreme points x_i $(i \in I)$ of $B(f)$ such that the relative interior of the simplex $\langle x_i \mid i \in I \rangle$ contains the optimal $\hat{x}$.

The above-mentioned discrete characterization of the optimal solution may be useful in checking the computational error and possibly in reforming the final solution to get a correct one. Though based on very limited computational experience, the author believes that Wolfe's algorithm is a 'practical' method for minimizing a submodular function, where 'practical' may not mean 'good' in a strict sense of computational complexity theory.

Acknowledgments

The author is very grateful to Professor Masao Iri and Professor Nobuaki Tomizawa for many useful conversations and discussions on the subject of this paper. The present work is also benefited by conversations with the members and visiting members of Institut für Operations Research, Universität Bonn. Thanks are also due to the referees for their comments and careful reading of the paper. The present research is supported by the Alexander von Humboldt fellowship, West Germany (1982/83).

References

[1] G. Birkhoff, Rings of sets, *Duke Mathematical Journal* 3 (1937) 443–454.

[2] R.E. Bixby, W.H. Cunningham and D.M. Topkis, The partial order of a polymatroid extreme point, Report No. 82250-OR, Institut für Ökonometrie und Operations Research, Universität Bonn (December 1982).

[3] W.H. Cunningham, Testing membership in matroid polyhedra, Report No. 81207-OR, Institut für Ökonometrie und Operations Research, Universität Bonn (November 1981).

[4] W.H. Cunningham and A. Frank, A primal–dual algorithm for submodular flows, Report No. 82224-OR, Institut für Ökonometrie und Operations Research, Universität Bonn (July 1982).

[5] J. Edmonds, Submodular functions, matroids, and certain polyhedra, Proceedings of the Calgary International Conference on Combinatorial Structures and Their Applications (Gordon and Breach, New York, 1970), pp. 69–87.

[6] J. Edmonds, Matroids and the greedy algorithm, *Mathematical Programming* 1 (1971) 127–136.

[7] J. Edmonds and R. Giles, A min-max relation for submodular functions on graphs, *Annals of Discrete Mathematics* 1 (1977) 185–204.

[8] U. Faigle, The greedy algorithm for partially ordered sets, *Discrete Mathematics* 28 (1979) 153–159.

[9] U. Faigle, Geometries on partially ordered sets, *Journal of Combinatorial Theory Ser. B* 28 (1980) 26–51.

[10] L.R. Ford, Jr. and D.R. Fulkerson, Flows in networks (Princeton University Press, Princeton, N.J., 1962).

[11] A. Frank, How to make a digraph strongly connected, Combinatorica 1 (1981) 145–153.

[12] A. Frank, Generalized polymatroids, Proceedings of the Sixth Hungarian Combinatorial Colloquium (Eger, 1981), to appear.

[13] A. Frank, Finding feasible vectors of Edmonds–Giles polyhedra, Report No. 81205-OR, Institut für Ökonometrie und Operations Research, Universität Bonn (September 1981).

[14] A. Frank, A note on k-strongly connected orientations of an undirected graph, *Discrete Mathematics* 39 (1982) 103–104.

[15] A. Frank, An algorithm for submodular functions on graphs, *Annals of Discrete Mathematics* 16 (1982) 189–212.

[16] A. Frank and É. Tardos, Matroids from crossing families, Proceedings of the Sixth Hungarian Combinatorial Colloquium (Eger, 1981), to appear.

[17] S. Fujishige, Algorithms for solving the independent-flow problems, *Journal of the Operations Research Society of Japan* 21 (1978) 189–204.

[18] S. Fujishige, The independent-flow problems and submodular functions (in Japanese), *Journal of the Faculty of Engineering, University of Tokyo* A-16 (1978) 42–43.

[19] S. Fujishige, Matroid theory and its applications to system-engineering problems (in Japanese), *Systems and Control* 23 (1979) 11–20.

[20] S. Fujishige, Lexicographically optimal base of a polymatroid with respect to a weight vector, *Mathematics of Operations Research* 5 (1980) 186–196.

[21] S. Fujishige, Structures of polytopes determined by submodular functions on crossing families, Discussions Paper Series No. 121(81-22), Institute of Socio-Economic Planning, University of Tsukuba (August 1981).

[22] S. Fujishige, The theory of semimodular programs, Discussion Paper Series No. 135(82-2), Institute of Socio-Economic Planning, University of Tsukuba (January 1982).

[23] S. Fujishige, A note on Frank's generalized polymatroids, Discussion Paper Series No. 152(82-19), Institute of Socio-Economic Planning, University of Tsukuba (June 1982).

[24] S. Fujishige, Characterization of subdifferentials of submodular functions and its relation to Lovász's extension of submodular functions, Report No. 82241-OR, Institut für Ökonometrie und Operations Research, Universität Bonn (November 1982).

[25] S. Fujishige, Canonical decompositions of symmetric submodular systems, *Discrete Applied Mathematics* 5 (1983) 175–190.

[26] S. Fujishige and N. Tomizawa, An algorithm for finding a minimum-cost strongly connected reorientation of a directed graph, Discussion Paper Series No. 147(82-14), Institute of Socio-Economic Planning, University of Tsukuba (March 1982).

[27] S. Fujishige and N. Tomizawa, A note on submodular functions on distributive lattices, Discussion Paper Series No. 149(82-16), Institute of Socio-Economic Planning, University of Tsukuba (April 1982).
[28] R. Giles, Submodular Functions, Graphs and Integer Polyhedra (Ph. D. Thesis, Department of Combinatorics and Optimization, University of Waterloo, Waterloo, Ontario, Canada, 1975).
[29] M. Grötschel, L. Lovász and A. Schrijver, The ellipsoid method and its consequences in combinatorial optimization, *Combinatorica* 1 (1981) 169–197.
[30] R. Hassin, Minimum cost flow with set-constraints, *Networks* 12 (1982) 1–21.
[31] A.J. Hoffman, Some recent applications of the theory of linear inequalities to extremal combinatorial analysis, *Proceedings of Symposia in Applied Mathematics* 10 (1960) 113–127.
[32] M. Iri, A review of recent work in Japan on principal partitions of matroids and their applications, *Annals of the New York Academy of Sciences* 319 (1979) 306–319.
[33] M. Iri, Structural theory for the combinatorial systems characterized by submodular functions, Proceedings of the Silver Jubilee Conference on Combinatorics (Waterloo, 1982) (to appear).
[34] M. Iri, Applications of matroid theory, Proceedings of XI. International Symposium on Mathematical Programming (Bonn, 1982) (to appear).
[35] M. Iri and S. Fujishige, Use of matroid theory in operations research, circuits and systems theory, *International Journal of Systems Science* 12 (1981) 27–54.
[36] L.G. Khachiyan, A polynomial algorithm in linear programming (in Russian) *Doklady Akademii Nauk SSSR* 244 (1979) 1093–1096.
[37] M.K. Kozlov, S.P. Torosov and L.G. Khachiyan, Polynomial solvability of convex quadratic programming (in Russian), *Doklady Akademii Nauk SSSR* 248 (1979) 1051–1053.
[38] E.L. Lawler, Generalizations of the polymatroidal network flow model, Report BW 158/82, Department of Operations Research, Mathematical Centre, Amsterdam (April 1982).
[39] E.L. Lawler and C.U. Martel, Computing maximal 'polymatroidal' network flows, *Mathematics of Operations Research* 7 (1982) 334–347.
[40] E.L. Lawler and C.U. Martel, Flow network formulations of polymatroid optimization problems, *Annals of Discrete Mathematics* 16 (1982) 189–200.
[41] L. Lovász, Submodular functions and convexity, Proceedings of XI. International Symposium on Mathematical Programming (Bonn, 1982), to appear.
[42] L. Lovász and A. Recski, Selected topics on matroid theory and applications, Proceedings of the 10th Winter School, Supplemento ai Rendiconti del Circolo Matematico di Palermo, Italia, Serie II-numero 2 (1982) pp. 171–185.
[43] C.J.H. McDiarmid, Rado's theorem for polymatroids, *Mathematical Proceedings of the Cambridge Philosophical Society* 78 (1975) 263–281.
[44] M. Nakamura and M. Iri, A structural theory for submodular functions, polymatroids and polymatroid intersections, Research Memorandom RMI 81-06, Department of Mathematical Engineering and Instrumentation Physics, University of Tokyo (August 1981).
[45] R.T. Rockafellar, Convex analysis (Princeton University Press, Princeton, NJ, 1970).
[46] A. Schrijver, Total dual integrality from directed graphs, crossing families, and sub- and super modular functions, Proceedings of the Silver Jubilee Conference on Combinatorics (Waterloo, 1982), to appear.
[47] L.S. Shapley, Cores of convex games, *International Journal of Game Theory* 1 (1971) 11–26.
[48] J. Stoer and C. Witzgall, Convexity and optimization in finite dimensions I (Springer, Berlin, 1970).
[49] N. Tomizawa, Theory of hyperspaces (I) — supermodular functions and generalization of concept of 'bases' (in Japanese), Papers of the Technical Group on Circuits and Systems, Institute of Electronics and Communication Engineers of Japan, CAS80-72 (1980).
[50] N. Tomizawa, Theory of hyperspaces (III) — maximum deficiency = minimum residue theorem and its application (in Japanese), ibid., CAS80-74 (1980).
[51] N. Tomizawa, Theory of hyperspaces (X) — on some properties of polymatroids (in Japanese), ibid., CAS81-84 (1981).
[52] N. Tomizawa, Theory of hypermatroids, Applied Combinatorial Theory and Algorithms, RIMS Kokyuroku 427, Research Institute for Mathematical Sciences, Kyoto University (1981) pp. 43–50.
[53] N. Tomizawa, Quasimatroids and orientations of graphs, presented at the XI. International Symposium on Mathematical Programming (Bonn, 1982).

[54] N. Tomizawa, Theory of hedron space, Presented at the Szeged Conference on Matroids (Szeged, 1982).

[55] N. Tomizawa and S. Fujishige, Theory of hyperspaces (VIII) — on the structures of hypermatroids of network type (in Japanese), Papers of the Technical Group on Circuits and Systems, Institute of Electronics and Communication Engineers of Japan, CAS81-62 (1981).

[56] N. Tomizawa and S. Fujishige, Historical survey of extensions of the concept of principal partition and their unifying generalization to hypermatroids, Systems Science Research Report No. 5, Department of Systems Science, Tokyo Institute of Technology (April 1982).

[57] W.T. Tutte, A homotopy theorem for matroids, I, II, *Transactions of the American Mathematical Society* 88 (1958) 144–174.

[58] W.T. Tutte, Matroids and graphs, *Transactions of the American Mathematical Society* 90 (1959) 527–555.

[59] W.T. Tutte, Lectures on matroids, *Journal of Research of the National Bureau of Standards* 69B (1965) 1–48.

[60] D.J.A. Welsh, Matroid theory (Academic Press, London, 1976).

[61] H. Whitney, On the abstract properties of linear dependence, *American Journal of Mathematics* 57 (1935) 509–533.

[62] P. Wolfe, Finding the nearest point in a polytope, *Mathematical Programming* 11 (1976) 128–149.

[63] U. Zimmermann, Minimization on submodular flows, *Discrete Applied Mathematics* 4 (1982) 303–323.

[64] U. Zimmermann, Minimization of some nonlinear functions over polymatroidal flows, *Annals of Discrete Mathematics* 16 (1982) 287–309.

Mathematical Programming Study 22 (1984) 132-146
North-Holland

INNER ESTIMATES FOR THE GENERALIZED GRADIENT OF THE OPTIMAL VALUE FUNCTION IN NONLINEAR PROGRAMMING

Bernhard GOLLAN

Mathematisches Institut, Universität Würzburg, Am Hubland, D-8700 Würzburg, West Germany

Received 17 March 1983
Revised manuscript received 15 March 1984

Parameterized nonlinear programming problems and their optimal value functions $V = V(p)$ are considered near a reference parameter $\bar{p}$. It is known that under certain assumptions there exists an optimal solution of the reference problem whose Lagrange multiplier set intersects the generalized gradient of V at $\bar{p}$. We investigate such relationships in more detail and prove that this generalized gradient must meet any compact multiplier set (which is relevant when the optimal solution is not unique). This result can be extended to certain unbounded multiplier sets and be strengthened to hold also with (possibly smaller) multiplier sets which satisfy second order necessary conditions. As a novelty we give also inner estimates for the recession cone of $\partial V(\bar{p})$. Together with available outer estimates these inner estimates may lead to full information about the generalized gradient.

Key words: Mathematical Programming, Optimal Value Function, Multipliers, Generalized Gradients, Second Order Conditions.

1. Introduction

We consider constrained optimization problems whose feasible sets depend on a parameter p and investigate the optimal value as a function of p, denoted by $V(p)$. It is well known that even for nice data of the problem (such as differentiability of the constraint functions) the local behaviour of V may be quite bad. Therefore an appropriate local description of V is given by the generalized gradient $\partial V(\cdot)$ as it was defined by Rockafellar for arbitrary functions with values in $\mathbb{R} \cup \{\pm\infty\}$.

Some recent papers [5, 13, 14] are mainly concerned with outer estimates for $\partial V(\cdot)$. Under a weak regularity condition called tameness it is proved that $\partial V(\cdot)$ is contained in a set defined by the multiplier sets which belong to the respective optimal solutions of the reference problem. Another type of results may be called inner estimates. In a strong sense this term applies to statements of the following form. For some optimal solution x, the Lagrange multiplier set $\Lambda_1(x)$ is contained in $\partial V(\bar{p})$, where $\bar{p}$ is chosen as reference parameter. Such a result has been proved by Pomerol [8] for problems in Banach spaces. His assumptions practically amount to the following. The optimal solution of the reference problem is unique, the related Langrange multiplier set is weak *-compact and a stability condition holds. This last condition, however, requires detailed information about the feasible and optimal solutions of perturbed problems. The main topic of this paper are inner estimates in a weaker sense, namely statements of the following type. There is some optimal solution x such that

$$\partial V(\bar{p}) \cap \Lambda_1(x) \neq \emptyset, \tag{1.1}$$

where again $\Lambda_1(x)$ denotes the set of Lagrange multipliers. Of course, when $\Lambda_1(x)$ contains exactly one element, then (1.1) implies that $\Lambda_1(x) \subseteq \partial V(\bar{p})$ and thus (1.1) may result in inner estimates in the strong sense. Aubin [1] and Rockafellar [13] proved that under certain assumptions there is some optimal solution x with property (1.1). Aubin's result holds for problems in Banach spaces, but the constraints are assumed linear. Rockafellar treats a general class of nonsmooth optimization problems in finite dimensional spaces and assumes the problem to be tame (cf. (2.8)) and that $\partial V(\bar{p}) \neq \emptyset$. The main feature of our results is to show that (1.1) must hold not only for some optimal x, but rather for a whole class of optimal solutions. Mainly this is the class of all optimal solutions x for which $\Lambda_1(x)$ is nonempty and (weak *-) compact. In particular, if this class is nonempty, then $\partial V(\bar{p}) \neq \emptyset$. For problems with inequality constraints only, such results are stated in Theorem 2.1 and Corollaries 2.2 and 2.4; for general problems they are given in Theorem 4.4 and Corollary 4.5. As a novelty, we prove statements of form (1.1) also for the recession cone of $\partial V(\bar{p})$, which must intersect singular multiplier sets $\Lambda_0(x)$, cf. Theorems 2.3 and 4.4.

These results can be improved to hold also with the possibly smaller multiplier sets which satisfy second order necessary optimality conditions. This topic is pursued in Section 3. At different points of the paper we provide simple examples in order to illustrate the improvements made by our results. A broad class of problems where these results can be applied is found in the study of parameterized eigenvalue problems of the form $A(p)x = \lambda(p)B(p)x$, with symmetric matrices $A(\cdot)$ and $B(\cdot)$. However, these applications are far from being simple reformulations of these theorems and are worked out in a separate paper.

The crucial point of our approach is that we do not attack $\partial V(\bar{p})$ directly via its support function $V^{\uparrow}(\bar{p}; \cdot)$, cf. (1.3), as it was done in [5] in order to obtain the outer estimates. Rather we prove estimates for a similar notion of a generalized directional derivative also introduced by Rockafellar and denoted by $V^{\downarrow}(\bar{p}; \cdot)$, cf. (1.2). This is related to another generalized gradient $\tilde{\partial} V(\bar{p})$. But by results of Rockafellar we actually have $\tilde{\partial} V(\bar{p}) = \partial V(\bar{p})$ in the cases under consideration. We conclude this section with a collection of notations which will be used in the sequel.

For a real Banach space X the topological dual of all continuous linear functionals on X is denoted by X^*; it is endowed with the weak * topology.

For a closed subset $C \subseteq X$ and some $\bar{x} \in C$ the tangent cone $T_C(\bar{x})$ and the normal cone $N_C(\bar{x})$ are understood as in [12], i.e., $T_C(\bar{x})$ is the Clarke tangent cone.

For an arbitrary function $F: X \to \mathbb{R} \cup \{\pm\infty\}$ which is finite at some $\bar{x} \in X$, the lower subderivative of F at $\bar{x}$ has been defined by Rockafellar as follows:

$$F^{\downarrow}(\bar{x}; d) := \lim_{\varepsilon \downarrow 0} \left[\liminf_{\substack{(\eta, \alpha) \uparrow_F \bar{x} \\ t \downarrow 0}} \left(\sup_{\|d' - d\| \leq \varepsilon} \frac{F(\eta + td') - \alpha}{t} \right) \right], \tag{1.2}$$

where $(\eta, \alpha) \uparrow_F \bar{x}$ iff $(\eta, \alpha) \to (\bar{x}, F(\bar{x}))$ and $\alpha \leq F(\eta)$, cf. [12, (4.13)].

The upper subderivative $F^{\uparrow}(\bar{x}; d)$ is defined similarly, with inf and sup interchanged, and with $(\eta, \alpha) \downarrow^F \bar{x}$, which means that $(\eta, \alpha) \to (\bar{x}, F(\bar{x}))$ and $\alpha \geq F(\eta)$, cf. [12, (4.2)]. $F^{\uparrow}(\bar{x}; \cdot)$ is a l.s.c., convex, positively homogeneous function, not identically $+\infty$. $F^{\downarrow}(\bar{x}; \cdot)$ is an u.s.c., concave, positively homogeneous function, not identically $-\infty$. The generalized gradients are defined by

$$\partial F(\bar{x}) := \{z \in X^* \mid zd \leq F^{\uparrow}(\bar{x}; d) \text{ for all } d \in X\},$$

$$\tilde{\partial} F(\bar{x}) := \{z \in X^* \mid zd \geq F^{\downarrow}(\bar{x}; d) \text{ for all } d \in X\}.$$

They are discussed in detail in [12]. As mentioned there, one always has the relation $\tilde{\partial} F(\bar{x}) = -\partial(-F)(\bar{x})$; in 'most' cases also $\tilde{\partial} F(\bar{x}) = \partial F(\bar{x})$ holds, but without further assumptions this need not be true. If $\partial F(\bar{x}) \neq \emptyset$, then

$$F^{\uparrow}(\bar{x}; d) = \sup\{zd \mid z \in \partial F(\bar{x})\} \quad \text{for all } d \in X, \tag{1.3}$$

cf. [12, Theorem 4].

Note that when F is locally Lipschitzian then $F^{\uparrow}(\bar{x}; \cdot)$ coincides with the generalized directional derivative $F^0(\bar{x}; \cdot)$ defined by Clarke in [2].

Given a convex set C, the polar cone C^0 and the recession cone 0^+C are understood as in [10]. The closure of C is denoted by cl C, and for an arbitrary set A the convex hull is denoted by conv A. Furthermore we agree to set $\inf \emptyset = +\infty$ and $\sup \emptyset = -\infty$. Finally set $U_\rho(\bar{x}) = \{x \in X \mid \|x - \bar{x}\| \leq \rho\}$.

2. First order estimates

In this section we consider optimization problems of the following form.

$$V(p) := \inf\{f(x) \mid x \in S(p)\}, \qquad S(p) := \{x \in C \mid g(x) + p \in B\}, \tag{P1}$$

with the following assumptions:

$C \subseteq X$ and $B \subseteq Y$ are closed subsets of real Banach spaces X, Y, B is convex with int $B \neq \emptyset$. Furthermore we suppose that (2.1) or (2.2) holds:

$Y = \mathbb{R}^m$, B is the nonnegative orthant in Y, and f and the components g_i of g are locally Lipschitzian. (2.1)

f and g are continuously differentiable. (2.2)

Next we define the multiplier sets which will be used below. In all that follows we assume that $\bar{x} \in M$, where M denotes the set of all optimal solutions in $S(0)$ (i.e., in this and the next section we chose $\bar{p}=0$ as reference parameter). When (2.1) applies then define

$$\Lambda(\bar{x}) := \{(\lambda_0, \lambda) \in \mathbb{R} \times \mathbb{R}^m \mid 0 \in \lambda_0 \partial f(\bar{x}) + \sum_{i=1}^{m} \lambda_i \partial g_i(\bar{x}) + N_C(\bar{x}),$$

$$\lambda_i \geq 0,\ 0 \leq i \leq m,\ \lambda_i g_i(\bar{x}) = 0,\ 1 \leq i \leq m\}.$$

By Clarke's multiplier rule, cf. [2], $\Lambda(\bar{x})$ contains more than the trivial element 0.

When (2.2) applies let

$$\Lambda(\bar{x}) := \{(\lambda_0, \lambda) \in \mathbb{R} \times Y^* \mid \lambda_0 f'(\bar{x})v + \lambda(g(\bar{x}) - b + g'(\bar{x})v) \geq 0$$

$$\text{for all } b \in B \text{ and } v \in C - \bar{x};\ \lambda_0 \geq 0\}.$$

Again, the assumptions in (2.2) imply that $\Lambda(\bar{x}) \neq \{0\}$, cf. [5, Section 4].

Both in case (2.1) and (2.2) we derive the following multiplier sets from $\Lambda(\bar{x})$:

$$\Lambda_0(\bar{x}) := \{\lambda \mid (0, \lambda) \in \Lambda(\bar{x})\}, \qquad \Lambda_1(\bar{x}) := \{\lambda \mid (1, \lambda) \in \Lambda(\bar{x})\}. \tag{2.3}$$

Conditions for $\Lambda_1(\bar{x})$ to be nonempty and (weak *-) compact can be found in [15] and [6]. Define

$$s(p, \bar{x}) := \sup\{\lambda p \mid \lambda \in \Lambda_1(\bar{x})\},$$

the support function of $\Lambda_1(\bar{x})$, and

$$M_1 := \{x \in M \mid \Lambda_1(x) \text{ is nonempty and compact}\}.$$

We give a first result about inner estimates.

Theorem 2.1. *$\partial V(0) \cap \Lambda_1(x) \neq \emptyset$ for all $x \in M_1$.*

For the following extension let

$$G(x) := \{p \mid \lambda p < 0 \text{ for all nonzero } \lambda \in \Lambda_0(x)\}, \quad x \in M. \tag{2.4}$$

Note that $\emptyset \neq \operatorname{int} B - g(\bar{x}) \subseteq G(\bar{x}) = \operatorname{int} G(\bar{x})$. Actually $G(\bar{x}) = \operatorname{int} (\Lambda_0(x))^0$.

Corollary 2.2. *Let $M_1 \neq \emptyset$ and $\bar{x} \in M$ with $\Lambda_1(\bar{x}) \neq \emptyset$. If*

$$s(p, \bar{x}) \geq \inf\{s(p, x) \mid x \in M_1\} \quad \text{for all } p \in \partial G(\bar{x}),$$

then $\partial V(0) \cap \Lambda_1(\bar{x}) \neq \emptyset$.

Proof of Theorem 2.1. First we prove the theorem in case that (2.1) holds. Choose $x \in M_1$ and $p \in \mathbb{R}^m$. From [5, Proposition 3.1 and Theorem 3.2] we obtain the following.

We have $s(p, x) < \infty$ and for any $s > s(p, x)$ there exist $h \in T_C(x)$ and $\varepsilon > 0$ with

$$f^0(x; h) < s, \qquad g_i^0(x; h) + p_i + \varepsilon < 0, \quad i \in I(x), \tag{2.5}$$

where $I(x)$ denotes the index set of all active constraints at x. This implies that there exist $\tau_0 > 0$ and points $x(\tau)$ such that $x(\tau) \in S(\tau p')$ for all $0 \leq \tau \leq \tau_0$, $|p' - p| \leq \varepsilon$; furthermore $f(x(\tau)) < \tau s$. Hence $V(\tau p') - V(0) < \tau s$ for all τ, p' as before, and thus

$$\limsup_{t \downarrow 0} [\sup_{|p'-p| \leq \varepsilon} (V(tp') - V(0))/t] \leq s.$$

This holds for arbitrary $s > s(p, x)$, and hence we have proved that

$$V^{\downarrow}(0; p) \leq s(p, x) \quad \text{for all } p. \tag{2.6}$$

Furthermore, the convex function $p \to s(p, x)$ is finite and continuous on $\mathbb{R}^m$ which implies that its epigraph has nonempty interior. Therefore by a separation argument there exists some $\mu \in Y^*$ $(= \mathbb{R}^m)$ with

$$V^{\downarrow}(0; p) \leq \mu p \leq s(p, x) \quad \text{for all } p. \tag{2.7}$$

Here the first inequality is equivalent to $\mu \in \tilde{\partial} V(0)$, and the second is equivalent to $\mu \in \Lambda_1(x)$. Therefore

$$\tilde{\partial} V(0) \cap \Lambda_1(x) \neq \emptyset \quad \text{for all } x \in M_1.$$

Since Problem (P1) is given mainly by inequality constraints (with int $B \neq \emptyset$), the assumptions of [12, Proposition 4] are met by V, and hence V is directionally Lipschitzian at 0. By [12, Theorem 6] this in turn implies that $\tilde{\partial} V(0) = \partial V(0)$. Hence

$$\partial V(0) \cap \Lambda_1(x) \neq \emptyset \quad \text{for all } x \in M_1.$$

When (2.2) applies then the arguments are similar. We indicate the modifications which have to be made. Arguing as in the proof of [5, Theorem 4.2] we obtain the following analogue to (2.5). For all p we have $s(p, x) < \infty$, and for any $s > s(p, x)$ there are $v \in C - x$, α, β, $\varepsilon > 0$, $b \in$ int B, such that for all $\|p' - p\| \leq \varepsilon$ we have

$$f'(x)v < \alpha s, \qquad g'(x)v + \alpha p' = b' - g(x),$$

where $\|b' - b\| < \beta$ and $b + \{y \mid \|y\| \leq 2\beta\} \subset \text{int } B$.

As before this results in (2.6) and (2.7), with some $\mu \in Y^*$. Then continue as above.

Proof of Corollary 2.2. Let $\bar{x}$ meet the assumptions of the corollary. Then (2.6) holds for $x = \bar{x}$, for the following reasons. If $p \notin \text{cl } G(\bar{x})$ then $\lambda p > 0$ for some $\lambda \in \Lambda_0(\bar{x})$. This implies that $s(p, \bar{x}) = +\infty$ and hence $V^{\downarrow}(0; p) \leq s(p, \bar{x})$ holds trivially. If $p \in \partial G(\bar{x})$, then this inequality follows from (2.6) and the assumptions about $\bar{x}$.

It remains to consider $p \in G(\bar{x})$. If (2.1) applies then $s(p, \bar{x}) < \infty$ and exactly the same arguments as in the proof of Theorem 2.1 apply. If (2.2) applies then either $s(p, \bar{x}) = +\infty$ or $-\infty < s(p, \bar{x}) < +\infty$. In the first case the desired inequality is trivially true. In the second case the arguments of the preceding theorem apply.

Hence (2.6) holds for all p, with $x = \bar{x}$. Note that $s(p, \bar{x}) \leq 0$ for all $p \in B - g(\bar{x})$ which is a set with nonempty interior. Therefore (2.7) follows and one can continue as in the proof of Theorem 2.1.

So far we considered such optimal solutions $x \in M$ for which Lagrange multipliers exist, i.e., $\Lambda_1(x) \neq \emptyset$. But it was never ruled out that for some $\bar{x} \in M$ the case $\Lambda_1(\bar{x}) = \emptyset$ may happen. Below we discuss the further information which is obtained when this case occurs. $G(x)$ is the set defined in (2.4).

Theorem 2.3. *Let the constraints have form* (2.1) *and let* $\bar{x} \in M$ *with* $\Lambda_1(\bar{x}) = \emptyset$.

(i) *If* $p \in -G(\bar{x})$, *then* $V^{\uparrow}(0; p) = +\infty$. *Either* $\partial V(0) = \emptyset$ *or* $\partial V(0)$ *is unbounded and* $0^+\partial V(0) \neq \{0\}$.

(ii) $0^+\partial V(0) \cap \Lambda_0(\bar{x}) \neq \{0\}$, *if* $\partial V(0) \neq \emptyset$.

(iii) *If* $\partial V(0) \neq \emptyset$ *and* $\Lambda_0(\bar{x}) = \{\alpha\bar{\lambda} \mid \alpha \geq 0\}$, *i.e.*, $\Lambda_0(\bar{x})$ *is a halfray, then*

$$\Lambda_0(\bar{x}) \subseteq 0^+\partial V(0).$$

Remark. If (2.2) applies then one can show that $V^{\uparrow}(0; p) = +\infty$ for all $p \in g(\bar{x}) - \operatorname{int} B$, and hence $\partial V(0)$ is unbounded or empty.

Proof of Theorem 2.3. (i) Choose $-p \in G(\bar{x})$ arbitrarily. Then [5, Proposition 3.1] applies with $s(-p, \bar{x}) = -\infty$: Hence for any $s \in \mathbb{R}$ there exist $h \in T_C(\bar{x})$, $\varepsilon > 0$, such that relations (2.5) hold, with $x = \bar{x}$ and p replaced by $-p$. As analogue to (2.6) we obtain that

$$V^{\downarrow}(0; -p) = -\infty \quad \text{for all } -p \in G(\bar{x}).$$

Since V is directionally Lipschitzian (cf. the proof of Theorem 2.1), by [12, (5.8)] this is equivalent to

$$V^{\uparrow}(0; p) = +\infty \quad \text{for all } p \in -G(\bar{x}).$$

If $\partial V(0) \neq \emptyset$, this implies that $\partial V(0)$ is unbounded, cf. (1.3), and in a finite dimensional space ($p \in \mathbb{R}^m$) this is equivalent to $0^+\partial V(0) \neq \{0\}$.

(ii) By (1.3) and [10, Theorems 14.2 and 8.7] we have $(0^+\partial V(0))^\circ = \operatorname{cl} \operatorname{dom} V^{\uparrow}(0; \cdot)$. By (i) we obtain $(0^+\partial V(0))^\circ \cap (-G(\bar{x})) = \emptyset$. Hence $A_1 := -(\Lambda_0(\bar{x}))^\circ$ and $A_2 := (0^+\partial V(0))^\circ$ are separable, i.e., there exists some nonzero $\mu \in \mathbb{R}^m$ with $\mu A_1 \geq 0$ and $\mu A_2 \leq 0$. But

$$\mu A_1 \geq 0 \iff \mu(\Lambda_0(\bar{x}))^\circ \leq 0 \iff \mu \in \Lambda_0(\bar{x}),$$

and

$$\mu A_2 \leq 0 \iff \mu \in 0^+\partial V(0).$$

This proves the claim.

(iii) This is an immediate consequence of (ii), since both $\Lambda_0(\bar{x})$ and $0^+\partial V(0)$ are cones.

In order to draw further conclusions from Theorem 2.1 (by combining this result with the outer estimates in [5]) we need the notion of tameness. This has been used to derive outer estimates in [13] and [5].

Problem (P1) is called tame at $\bar{p}$ if

there exist $\delta_1 > V(\bar{p})$, $\delta_2 > 0$, such that for any sequence $\eta_k \to \bar{p}$ with $V(\eta_k) < \delta_1$ and $\|\eta_k - \bar{p}\| \leq \delta_2$, there is an optimal solution $x_k \in S(\eta_k)$ such that (x_k) converges to some $\bar{x} \in M$ (which may depend on the respective sequences). (2.8)

Corollary 2.4. *Let* (P1) *be tame at* $p = 0$.

(i) *If* $\Lambda_1(x) = \{\lambda_x\}$ *for all* $x \in M$ *(i.e., for each* $x \in M$ *there exists a unique Lagrange multiplier), then*

$$\partial V(0) = \text{cl conv}\{\lambda_x | x \in M\}.$$

(ii) *Let* $\dim Y < \infty$.

a. *If there is some* $\bar{\lambda}$ *with* $\Lambda_1(x) = \{\bar{\lambda}\}$ *for all* $x \in M$, *then* V *is strictly differentiable at* $p = 0$ *with* $V'(0) = \bar{\lambda}$. *(for the definition of strict differentiability cf.* [11, p. 341]).

b. *Conversely, if there are* $x_1 \neq x_2$ *in* M_1 *with* $\Lambda_1(x_1) \cap \Lambda_1(x_2) = \emptyset$, *then* V *is not strictly differentiable at* 0.

c. *If there exists some* $\bar{x} \in M$ *with* $\Lambda_1(\bar{x}) = \emptyset$, *then* V *is not locally Lipschitzian around* $p = 0$.

Proof. (i) By Theorem 2.1 we have $\lambda_x \in \partial V(0)$ for every $x \in M$. Since $\partial V(0)$ is a closed convex set, the inclusion

$$\text{cl conv}\{\lambda_x | x \in M\} \subseteq \partial V(0)$$

follows. Under the tameness assumption, [5, Theorem 3.6] in case (2.1) or [5, Theorem 4.4] in case (2.2) yield the reverse inclusion.

(ii) a. By [11, Corollary 2, p. 341] this follows directly from (i). b. Since $x_1, x_2 \in M_1$, $\partial V(0)$ must intersect both $\Lambda_1(x_1)$ and $\Lambda_1(x_2)$, and thus $\partial V(0)$ contains more than one element. c. By Theorem 2.3, $\partial V(0)$ is unbounded. This yields the claim, cf. [12, 13].

Remarks 2.5. Theorem 2.1 complements and extends some results by Rockafellar in [13, Corollary 4, p. 48]. There tameness of the problem and $\partial V(0) \neq \emptyset$ are assumed and it is proved that there is some $\bar{x} \in M$ with $\Lambda_1(\bar{x}) \cap \partial V(0) \neq \emptyset$. Here we treat problems with inequality constraints in infinite dimensional spaces and prove: If $M_1 \neq \emptyset$, then $\partial V(0) \neq \emptyset$ and $\partial V(0)$ intersects any set $\Lambda_1(x)$, $x \in M_1$.

Theorem 2.3 is a new type of result. Complementing the outer estimates for $0^+\partial V(0)$ given in [13, (5.9)] we prove inner estimates. Their combination may sometimes result in an exact determination of $0^+\partial V(0)$ (which Rockafellar calls the 'singular subgradient of V').

Example 2.6. For a simple problem with one constraint we determine $\partial V(0)$ exactly even though the tameness property fails.

Let $V(p) := \inf\{yx^2 \mid p - y \leq 0\}$. For $p = 0$ we get $V(0) = 0$ and $M = \{(x, y) \mid y = 0$ or $[y \geq 0$ and $x = 0]\}$.

If $y > 0$ then $\lambda = 0$, by the complementary slackness condition.

If $y = 0$ then $\lambda = x^2$. Hence $M = M_1$, and by Theorem 2.1 we have

$$[0, \infty) \subseteq \partial V(0).$$

Obviously V is nondecreasing and hence $V^{\uparrow}(0; -1) \leq 0$. Thus $\partial V(0)$ contains no negative real number and therefore

$$[0, \infty) = \partial V(0).$$

Of course, one could also check directly that $V(p) = 0$, if $p \geq 0$, and $V(p) = -\infty$, if $p < 0$, and find that $\partial V(0) = [0, \infty)$. Since V is not l.s.c. at 0, the problem cannot be tame at $p = 0$. Note that $\partial V(0)$ is unbounded even though all Lagrange multiplier sets are compact.

Example 2.7. In this example all the main results of Section 2 are needed in order to obtain full information about $\partial V(\bar{p})$. Below the unperturbed problem has two optimal solutions z_1 and z_2. At z_1 we have a unique Lagrange multiplier whereas at z_2 only singular multipliers exist. But Theorems 2.1 and 2.3 together with the outer estimates determine $\partial V(\bar{p})$ exactly. Let $z = (x, y) \in \mathbb{R}^2$ and

$$V(p) = \inf\{y^2 + (x-1)^2 \mid (x^2 + y^2)(2 - x) + p \leq 0\}.$$

Choose $\bar{p} = 0$, then obviously $V(0) = 1$ and $z_1 = (2, 0)$ and $z_2 = (0, 0)$ represent all optimal solutions. We have $z_1 \in M_1$ with $\Lambda_1(z_1) = \{0.5\}$, but $\Lambda_1(z_2) = \emptyset$ and $\Lambda_0(z_2) = [0, \infty)$. The tameness assumption is satisfied (to see this add the restriction $|z| < R$ with some $R > 2$) and hence the outer estimates of [5, 13] result in

$$\partial V(0) \subseteq \operatorname{conv}(\{0.5\} + [0, \infty)) = [0.5, \infty).$$

For z_1 Theorem 2.1 holds and gives $0.5 \in \partial V(0)$; in particular, $\partial V(0) \neq \emptyset$ and hence Theorem 2.3, (iii) gives $\Lambda_0(z_2) = [0, \infty) \subseteq 0^+ \partial V(0)$. Therefore the inner estimates result in $[0.5, \infty) \subseteq \partial V(0)$, and together with the outer estimate we obtain that $\partial V(0) = [0.5, \infty)$.

3. Second order estimates

It may happen that first order multiplier sets give no or only little insight into the behaviour of the value function, cf. [5, Example 5.5]. When the functions which define the problem are smooth enough then higher order necessary conditions may help to provide more precise information. In [5, Section 5] this program is carried out for several notions of directional derivatives and a lower subdifferential $\underset{\cdot}{\partial} V(0)$. In this section we show that Theorems 2.1 and 2.3 can be refined in a similar way. As in [5] we limit ourselves to conditions of second order, but everything carries over easily to higher order conditions as stated in [7]. For convenience we assume throughout this section that Y is of finite dimension. Some of the results remain true also when $\dim Y = +\infty$. We consider the problem

$$V(p) := \inf\{f(x) \mid x \in S(p)\}, \qquad S(p) := \{x \in X \mid g(x) + p \in B\}, \tag{P2}$$

where X is a real Banach space, $Y = \mathbb{R}^m$, $B \subseteq Y$ is a closed convex cone with nonempty interior, and f, g are twice continuously differentiable (in the Frechet sense).

The multiplier sets depend on so-called critical directions which are collected in the set

$$D(x) := \{d \in X \mid f'(x)d = 0,\ g'(x)d \in B + \mathbb{R}g(x)\},$$

which is a closed convex cone. Set $F := (f, g)$ and define

$$K(x, d) := F'(x)X + \{F''(x)(\beta d, \beta d) \mid \beta > 0\},$$

$$L(x, d) := \{(t_0, t) \mid t_0 < 0; \text{ there are } \gamma_1, \gamma_2 \in \mathbb{R} \text{ with }$$
$$t \in \operatorname{int} B + \gamma_1 g(x) + \gamma_2 g'(x)d\}.$$

The second order multiplier set for $x \in M$, $d \in D(x)$ is given by

$$\Lambda(x, d) := \{(\lambda_0, \lambda) \in \mathbb{R}^{1+m} \mid (\lambda_0, \lambda)(K(x, d) - L(x, d)) \geq 0\}.$$

As before we consider also

$$\Lambda_i(x, d) := \{\lambda \in \mathbb{R}^m \mid (i, \lambda) \in \Lambda(x, d)\}, \quad i = 0, 1,$$

and

$$s(p, x, d) := \sup\{\lambda p \mid \lambda \in \Lambda_1(x, d)\}.$$

As remarked in [5], we have $\Lambda(x, d) \neq \{0\}$ for all $x \in M$, $d \in D(x)$, and trivially $\Lambda(x, d) \subseteq \Lambda(x)$ holds. Define

$$G(x, d) := \{p \mid \lambda p < 0 \text{ for all nonzero } \lambda \in \Lambda_0(x, d)\},$$

$$\tilde{M}_1 := \{(x, d) \mid x \in M,\ d \in D(x),\ \Lambda_1(x, d) \text{ is nonempty and compact}\},$$

$$\tilde{M}_2 := \{(x, d) \mid x \in M,\ d \in D(x),\ \Lambda_1(x, d) = \emptyset\}.$$

Theorem 3.1. (i) $\partial V(0) \cap \Lambda_1(x, d) \neq \emptyset$ *for all* $(x, d) \in \tilde{M}_1$.
(ii) *If* $\partial V(0) \neq \emptyset$ *then*

$$0^+\partial V(0) \cap \Lambda_0(x, d) \neq \{0\} \textit{ for all } (x, d) \in \tilde{M}_2.$$

In particular: *If* $\tilde{M}_2 \neq \emptyset$ *then* $0^+\partial V(0) \neq \{0\}$, $\partial V(0)$ *is unbounded and* $V^{\uparrow}(0; p) = +\infty$ *for all* $p \in -G(x, d)$, $(x, d) \in \tilde{M}_2$.

Of course, part (i) can be refined similarly as in Corollary 2.2.

Proof. (i) Choose $(x, d) \in \tilde{M}_1$ and $p \in \mathbb{R}^m$. Then $s(p, x, d) < \infty$ and the arguments made in the proof of [5, Theorem 5.2] can be sharpened slightly. We obtain that for $s > s(p, x, d)$ there exist $v \in X$, $b \in \text{int } B$, β, $\varepsilon > 0$, γ_1, $\gamma_2 \in \mathbb{R}$, such that for all $|p' - p| \leq \varepsilon$ there is some $b' \in \text{int } B \cap U_\beta(b)$ with

$$f'(x)v + \tfrac{1}{2}f''(x)(d, d) < s,$$

$$g'(x)v + \tfrac{1}{2}g''(x)(d, d) + p' = b' + \gamma_1 g(x) + \gamma_2 g'(x)d,$$

and furthermore $b + U_{2\beta}(c) \subseteq \text{int } B$.

As in [5] this implies that $x(\tau) := x + \tau^{1/2}d + \tau v \in S(\tau p')$ for $0 < \tau \leq \tau_0$, $|p' - p| \leq \varepsilon$, and that

$$V(\tau p') \leq f(x(\tau)) < f(x) + \tau s = V(0) + \tau s.$$

Analogously to (2.6) this yields

$$V^{\downarrow}(0; p) \leq s(p, x, d) \tag{3.1}$$

for any $p \in \mathbb{R}^m$. By the same arguments as in the proof of Theorem 2.1 this implies that $\partial V(0) \cap \Lambda_1(x, d) \neq \emptyset$ for $(x, d) \in \tilde{M}_1$.

(ii) For $-p \in G(x, d)$ we have $s(-p, x, d) = -\infty$. As in the proof of Theorem 2.3 (with modifications as above) we obtain $V^{\downarrow}(0; -p) = -\infty$. Then continue as in the proof of Theorem 2.3 (note that $(\Lambda_0(x, d))^\circ = \text{cl } G(x, d)$).

Remarks. So far it is not known under which assumptions second order outer estimates of similar form as the first order outer estimates in [5] hold true. Therefore at this stage it is not possible to formulate consequences of Theorem 3.1 analogous to the facts stated in Corollary 2.4.

Example 5.5 in [5] is a problem where Theorem 3.1 gives additional information about V. In [5, Example 5.5] nothing is stated about $\partial V(0)$, but it is shown that $\tilde{M}_1 \neq \emptyset \neq \tilde{M}_2$ (whereas $M_1 = \emptyset$). Therefore the preceding second order results imply that $\partial V(0)$ is nonempty and unbounded, and the multiplier sets (cf. [5]) give some hints for details about $\partial V(0)$. Here we give another example.

Example 3.2. Minimize y subject to $-y + x^2 + p_1 \leq 0$ and $-y - x^2 + p_2 \leq 0$. We have

$$M = \{(0, 0)\} \quad \text{and} \quad \Lambda_1 = \{(\lambda_1, \lambda_2) | \lambda_1 \in [0, 1], \lambda_2 = 1 - \lambda_1\}.$$

We may choose $d = (1, 0)$ which leads to the smaller second order multiplier set

$$\Lambda_1(d) = \{(\lambda_1, \lambda_2) | \lambda_1 \in [\tfrac{1}{2}, 1], \lambda_2 = 1 - \lambda_1\}.$$

Thus, by Theorem 3.1, $\partial V(0)$ must intersect the set $\Lambda_1(d)$, which is a refined result compared to the first order information that $\emptyset \neq \partial V(0) \subseteq \Lambda_1$ (obtained from outer estimates).

4. Extensions

In the first part of this section we consider the following Problem (P3) and prove outer and inner estimates for the value function. The results of Lemmata 4.1 and 4.2 are regarded as auxiliary results for the subsequent considerations. The main topic of this section is the treatment of Problem (P4) which contains both inequality and equality constraints and parameters which enter the problem in a rather general form.

$$V(p) := \inf\{f(x) \mid x \in S(p)\},$$
$$S(p) := \{x \in X \mid g(x) \in B,\ h_1(x) = 0,\ h_2(x) + p = 0\}, \tag{P3}$$

where X, Y are real Banach spaces, $Z_1 = \mathbb{R}^k$, $Z_2 = \mathbb{R}^n$, $f: X \to \mathbb{R}$, $g: X \to Y$, $h_1: X \to Z_1$, $h_2: X \to Z_2$ are continuously differentiable mappings, $B \subseteq Y$ is a closed convex set with int $B \neq \emptyset$. Let tameness of this problem be understood as in (2.8). In this section we choose some $\bar{p}$ as reference parameter and let M denote the set of all optimal solutions in $S(\bar{p})$. It is obvious how the multiplier sets $\Lambda(x)$, $\Lambda_0(x)$, $\Lambda_1(x)$ defined in Section 2 have to be modified in order to include the equality constraints; for details see [5]. Note that $\Lambda_0(x)$, $\Lambda_1(x)$ are subsets of $Y^* \times \mathbb{R}^k \times \mathbb{R}^n$.

For $x \in M$ define also

$$\theta_1(x) := \{\nu \in \mathbb{R}^n \mid (\lambda, \mu, \nu) \in \Lambda_1(x)\}, \qquad \theta_0(x) := \{\nu \in \mathbb{R}^n \mid (\lambda, \mu, \nu) \in \Lambda_0(x)\}. \tag{4.1}$$

Lemma 4.1. *Let* (P3) *be tame at* $\bar{p}$.

(i) *If* $\nu p \leq 0$ *for all* $\nu \in \theta_0(x)$ *and all* $x \in M$, *then*

$$V^{\uparrow}(\bar{p}; p) \leq \sup_{x \in M} \sup\{\nu p \mid \nu \in \theta_1(x)\}.$$

(ii) $$\partial V(\bar{p}) \subseteq \text{cl conv}\Big[\bigcup_{x \in M} \theta_1(x) + \bigcup_{x \in M} \theta_0(x)\Big].$$

This is an immediate consequence of [5, Theorem 4.4].

Now we turn to inner estimates. Let

$$G(x) := \{p \mid \nu p \leq 0 \text{ for all } \nu \in \theta_0(x)\},$$
$$M_0 := \{x \in M \mid 0 \in \text{int}[(h_1'(x), h_2'(x))X] \subseteq Z_1 \times Z_2\},$$
$$M_1 := \{x \in M_0 \mid \theta_1(x) \text{ is nonempty and compact}\}, \tag{4.2}$$
$$M_2 := \{x \in M_0 \mid \theta_1(x) = \emptyset,\ \theta_0(x) \text{ is pointed}\}.$$

Lemma 4.2. (i) $-\partial(-V)(\bar{p}) \cap \theta_1(x) \neq \emptyset$ *for all* $x \in M_1$.

(ii) *If* ($P3$) *is tame at* $\bar{p}$ *and if* int$[\bigcup_{x \in M} G(x)] \neq \emptyset$, *then*

$$\partial V(\bar{p}) \cap \theta_1(x) \neq \emptyset \textit{ for all } x \in M_1.$$

(iii) *If furthermore* $\partial V(\bar{p}) \neq \emptyset$, *then*

$$0^+\partial V(\bar{p}) \cap \theta_0(x) \neq \{0\} \text{ for all } x \in M_2.$$

If $\theta_0(x)$ *is a halfray then* $\theta_0(x) \subseteq 0^+\partial V(\bar{p})$.

Proof. (i) Following the lines of the proof of Theorem 2.1 with slight modifications, one can conclude as there that for $x \in M_1$, $\tilde{\partial} V(\bar{p}) \cap \theta_1(x) \neq \emptyset$. But $\tilde{\partial} V(\bar{p}) = -\partial(-V)(\bar{p})$, and hence the claim follows.

(ii) Now [12, Proposition 4] does not apply any more. However, by our assumptions [13, Corollary 3, p. 48] is applicable (the $\cup$ there is a misprint and must read $\cap$). Hence V is directionally Lipschitzian and $\partial V(\bar{p}) = -\partial(-V)(\bar{p})$.

(iii) Choose $x \in M_2$. Then $\text{int } G(x) \neq \emptyset$, and for any $p \in -\text{int } G(x)$ it can be proved as in Theorem 2.3 that $V^{\downarrow}(\bar{p}; -p) = -\infty$. But V is directionally Lipschitzian and hence $V^{\uparrow}(\bar{p}; p) = +\infty$. Then continue analogously to the proof of Theorem 2.3, with the modification that here $G(x) = (\theta_0(x))^\circ$.

The main topic of this section is the discussion of the value function of Problem (P4):

$$V(p) := \inf\{f(x, p) \mid x \in S(p)\}, \qquad S(p) := \{x \in X \mid g(x, p) \in B, h(x, p) = 0\}. \tag{P4}$$

Here $B \subseteq Y$ is a closed convex set with nonempty interior, and f, g, h are continuously differentiable mappings $X \times P$ into $\mathbb{R}$, Y, Z, with real Banach spaces X, Y and finite dimensional spaces $Z = \mathbb{R}^k$, $P = \mathbb{R}^n$.

Choose some $\bar{p} \in P$ as reference parameter. Derivatives with respect to x, p are denoted by D_x, D_p.

Let M be the set of all optimal solutions in $S(\bar{p})$. Mainly we consider such optimal solutions $\bar{x}$ which satisfy the following regularity condition with respect to the equality constraints (if equality constraints are actually present in the model),

$$0 \in \text{int}[D_x h(\bar{x}, \bar{p})X]. \tag{4.3}$$

For $\bar{x} \in M$ the multiplier sets are given by

$$\Lambda(\bar{x}) := \{(\lambda_0, \lambda, \mu) \mid (\lambda_0 D_x f + \lambda D_x g + \mu D_x h)(\bar{x}, \bar{p}) = 0, \lambda_0 \geq 0,$$
$$\lambda(b - g(\bar{x}, \bar{p})) \leq 0 \text{ for all } b \in B\}, \tag{4.4}$$

$$\Lambda_i(\bar{x}) := (\lambda, \mu) \mid (i, \lambda, \mu) \in \Lambda(\bar{x})\}, \quad i = 0.1.$$

These induce other sets which are also convex and play the main role below. For reasons which become clear in the proof of Theorems 4.3 and 4.4, we denote them by θ as we did before when dealing with Problem (P3).

$$\theta(\bar{x}) := \{(\lambda_0, \nu) \in \mathbb{R}^{1+n} \mid \nu = (\lambda_0 D_p f + \lambda D_p g + \mu D_p h)(\bar{x}, \bar{p}), (\lambda_0, \lambda, \mu) \in \Lambda(x)\}, \tag{4.5}$$

$$\theta_i(x) := \{\nu \in \mathbb{R}^n \mid (i, \nu) \in \theta(x)\}, \quad i = 0, 1.$$

When f, g, h are given as above and when all spaces are of finite dimension, then the sets θ_0, θ_1 are the same as the sets K_0, K considered by Rockafellar, cf. [13, (4.6)]. Furthermore, θ_1 agrees with the set K defined by Gauvin and Dubeau, cf. [4, (2.2)].

Clearly θ_1 is compact when Λ_1 is compact, but not vice versa. As shown in Example 4.6, it is possible to have an unbounded set $\Lambda_1(x)$ and at the same time $\theta_1(x)$ not only compact but even unique.

First we state outer estimates.

Theorem 4.3. *Let* $(P4)$ *be tame at* $\bar{p}$.

(i) *If* $\nu p \leq 0$ *for all* $\nu \in \theta_0(x)$ *and all* $x \in M$, *then*

$$V^{\uparrow}(\bar{p}; p) \leq \sup_{x \in M} \sup\{\nu p \mid \nu \in \theta_1(x)\}.$$

(ii) $\partial V(\bar{p}) \subseteq \operatorname{cl}\operatorname{conv}\{\bigcup_{x \in M} \theta_1(x) + \bigcup_{x \in M} \theta_0(x)\}$.

This result will be proved together with the next theorem which provides inner estimates.

Let $G(x)$, M_1, M_2, be defined as in (4.2), but with $\theta_0(x)$, $\theta_1(x)$ given by (4.5) and $M_0 := \{\bar{x} \in M \mid (4.3) \text{ holds}\}$.

Theorem 4.4. (i) $-\partial(-V)(\bar{p}) \cap \theta_1(x) \neq \emptyset$ *for all* $x \in M_1$.

(ii) *If* $(P4)$ *is tame at* $\bar{p}$ *and if* $\operatorname{int}[\bigcap_{x \in M} G(x)] \neq \emptyset$, *then*

$$\partial V(\bar{p}) \cap \theta_1(x) \neq \emptyset \quad \textit{for all } x \in M_1.$$

(iii) *If furthermore* $\partial V(\bar{p}) \neq \emptyset$ *then*

$$0^+\partial V(\bar{p}) \cap \theta_0(x) \neq \{0\} \quad \textit{for all } x \in M_2.$$

If $\theta_0(x)$ *is a halfray then* $\theta_0(x) \subseteq 0^+\partial V(\bar{p})$.

Corollary 4.5. *Let* (P4) *be tame at* $\bar{p}$ *and* $M = M_1$.

(i) *Let* $\theta_1(x) = \{\nu_x\}$ *for every* $x \in M$, *i.e., for every optimal solution a Lagrange Multiplier exists and is unique). Then* $\partial V(\bar{p}) = \operatorname{cl}\operatorname{conv}\{\nu_x \mid x \in M\}$.

(ii) *If there is some* $\bar{\nu}$ *with* $\theta_1(x) = \{\bar{\nu}\}$ *for all* $x \in M$, *then* V *is strictly differentiable at* $\bar{p}$ *with* $V'(\bar{p}) = \bar{\nu}$.

Remarks. For the moment regard all spaces to be finite dimensional. Then Theorem 4.3 agrees with the results by Rockafellar in [13] for the differentiable case. A consequence of Theorem 4.3 is the following (cf. also [13, Corollary 2, p. 47]):

$$\begin{aligned}&\text{If } \theta_1(x) \text{ is nonempty and compact for all } x \in M \text{ then } V \text{ is}\\ &\text{locally Lipschitzian and } \partial V(\bar{p}) \subseteq \operatorname{cl}\operatorname{conv}[\textstyle\bigcup_{x \in M} \theta_1(x)].\end{aligned} \tag{4.6}$$

Theorems 5.1 and 5.3 in [4] come to the same conclusion as in (4.6), but under the (too restrictive) assumption that $\Lambda_1(x)$ is nonempty and compact for all $x \in M$.

The same holds with respect to [4, Corollary 5.4] which is closely related to Corollary 4.5 above.

The significant difference between these two assumptions can be seen in Example 4.6 below. Of course things have been arranged there in order to obtain this extreme behaviour. But in a natural way such effects come up already in problems with standard perturbations when the parameter under consideration enters only some of the constraints whereas the others are held fixed (as it is done in Problem (P3)).

Example 4.6. Let $X = Y = \mathbb{R}^2$, $P = \mathbb{R}$. Minimize x_1^2 subject to $px_2 \leq 0$ and $-px_2 - x_1^2 \leq 0$. For all $p > 0$ we have $M = \{(0, 0)\}$ and hence $V(p) = 0$ for all $p > 0$. Choose some $\bar{p} > 0$.

Then $\Lambda_1 = \Lambda_0 = \{(\lambda_1, \lambda_2) \mid \lambda_1 = \lambda_2 \geq 0\}$; this is unbounded and hence the results of [4] do not apply. However, we have $\theta_1 = \{0\}$, and Corollary 4.5 yields the right result, namely that $V'(\bar{p}) = 0$.

Proof of Theorems 4.3 and 4.4. We replace Problem (P4) by an equivalent formulation:

$$V(p) = \inf\{f(x, y) \mid (x, y) \in S(p)\},$$
$$S(p) := \{(x, y) \in X \times P \mid g(x, y) \in B,\ h(x, y) = 0,\ -y + p = 0\}. \tag{P4'}$$

Note that V really agrees with the value function of (P4), but that the reformulated problem has the formal structure of (P3).

The multiplier set at the optimal solution $(\bar{x}, \bar{y}) = (\bar{x}, \bar{p}) \in M \times \{\bar{p}\}$ is the following:

$$\Lambda(\bar{x}, \bar{p}) = \{(\lambda_0, \lambda, \mu, \nu) \mid (\lambda_0 D_x f + \lambda D_x g + \mu D_x h)(\bar{x}, \bar{p}) = 0,\ \lambda_0 \geq 0,$$
$$(\lambda_0 D_p f + \lambda D_p g + \mu D_p h)(\bar{x}, \bar{p}) = \nu,\ \lambda(b - g(x, p)) \leq 0 \text{ for all } b \in B\}.$$

Obviously $(\lambda_0, \lambda, \mu, \nu) \in \Lambda(\bar{x}, \bar{p})$ is equivalent to $(\lambda_0, \lambda, \mu) \in \Lambda(\bar{x})$ (the set defined in (4.4)) and ν given by the above representation. In particular, the sets $\theta_1(\bar{x}, \bar{p})$ and $\theta_0(\bar{x}, \bar{p})$ which appear when Lemmata 4.1 and 4.2 are applied to Problem (P4'), are the following (according to (4.1)):

$$\theta_1(\bar{x}, \bar{p}) = \{\nu \in \mathbb{R}^n \mid \nu = (D_p f + \lambda D_p g + \mu D_p h)(\bar{x}, \bar{p}),\ (\lambda, \mu) \in \Lambda_1(\bar{x})\},$$
$$\theta_0(\bar{x}, \bar{p}) = \{\nu \in \mathbb{R}^n \mid \nu = (\lambda D_p g + \mu D_p h)(x, \bar{p}),\ (\lambda, \mu) \in \Lambda_0(\bar{x})\},$$

where $\Lambda_0(\bar{x})$, $\Lambda_1(\bar{x})$ are the sets defined in (4.4). Thus $\theta_1(\bar{x}, \bar{p}) = \theta_1(\bar{x})$, $\theta_0(\bar{x}, \bar{p}) = \theta_0(\bar{x})$, cf. (4.5). Furthermore, (4.3) is equivalent to the linear independence of the gradients of all equality constraints in (P4'). It remains to remark that (P4') is tame at $\bar{p}$ exactly when (P4) has this property. Then by the above relations Theorems 4.3 and 4.4 are immediate consequences of Lemmata 4.1 and 4.2, applied to Problem (P4'). Corollary 4.5 follows from Theorems 4.3 and 4.4.

Remark. When the functions in Problem (P4) are at least twice continuously differentiable then the results of Theorem 4.4 can be refined similarly as in Section 3.

References

[1] J.-P. Aubin, "Further properties of Lagrange multipliers in nonsmooth optimization", *Applied Mathematics and Optimization* 6 (1980) 79–90.
[2] F.H. Clarke, " A new approach to Lagrange multipliers", *Mathematics of Operations Research* 1 (1976) 165–174.
[3] J. Gauvin, "The generalized gradients of a marginal function in mathematical programming", *Mathematics of Operations Research* 4 (1979) 458–463.
[4] J. Gauvin and F. Dubeau, "Differential properties of the marginal function in mathematical programming", *Mathematical Programming Study* 19 (1982) 101–119.
[5] B. Gollan, "On the marginal function in nonlinear programming", *Mathematics of Operations Research* 9 (1984) 208–221.
[6] F. Lempio and H. Maurer, "Differential stability in infinite dimensional nonlinear programming", *Applied Mathematics and Optimization* 6 (1980) 139–152.
[7] F. Lempio and J. Zowe, "Higher order optimality conditions", in: B. Korte, ed., Studies on Operations Research (North Holland, Amsterdam, 1981) pp. 145–189.
[8] J.-Ch. Pomerol, "The Lagrange multiplier set and the generalized gradient set of the marginal function of a differentiable program in a Banach space", *Journal of Optimization Theory and Applications* 38 (1982) 307–318.
[9] S.M. Robinson, "Stability theory for systems of inequalities, part II: Differentiable nonlinear systems", *SIAM Journal on Numerical Analysis* 13 (1976) 497–513.
[10] R.T. Rockafellar, *Convex analysis* (Princeton University Press, Princeton, NJ, 1970).
[11] R.T. Rockafellar, "Directionally Lipschitzian functions and subdifferential calculus", *Proceedings of the London Mathematical Society* 39 (1979) 331–355.
[12] R.T. Rockafellar, "Generalized directional derivatives and subgradients of nonconvex functions", *Canadian Journal of Mathematics* 32 (1980) 257–280.
[13] R.T. Rockafellar, "Langrange multipliers and subderivatives of optimal value functions in nonlinear programming", *Mathematical Programming Study* 17 (1982) 28–66.
[14] R.T. Rockafellar, "Marginal values and second-order necessary conditions for optimality", *Mathematical Programming* 26 (1983) 245–286.
[15] J.-J. Strodiot, V.H. Nguyen and R. Mifflin, "On conditions to have bounded multipliers in locally Lipschitz Programming", *Mathematical Programming* 18 (1980) 100–106.

Mathematical Programming Study 22 (1984) 147–166
North-Holland

NUMERICAL METHODS FOR BASIC SOLUTIONS OF GENERALIZED FLOW NETWORKS

Michael D. GRIGORIADIS and Tau HSU
Rutgers University, New Brunswick, NJ 08903, USA

Received 27 May 1983
Revised manuscript received 26 January 1984

A central problem in implementing specializations of the simplex method for solving large minimum-cost generalized network flow problems is the accurate and efficient computation of basic solutions. Bases for such problems are block-diagonal, with each block corresponding to a spanning subtree or 'one-tree', a connected subgraph which contains exactly one cycle. This paper reviews two previously proposed techniques for solving such one-tree systems and develops a new algorithm based on Gaussian elimination for use within a generalized network flow code. In addition to error bounds, the paper presents numerical results and a computational comparison of these methods.

Key words: Networks with Gains and Losses, One-Trees, Quasi-Trees, Cycle Systems, Gaussian Elimination, Round-Off Error.

1. Introduction

In this paper, we will examine numerical methods for obtaining basic solutions to specially structured linear programs of the form:

$$\text{GN:} \quad \min\{cx \mid Ax = b, 0 \leq x \leq h\} \tag{1}$$

where A is a real (m, n) matrix with at most two nonzero elements per column, b is a real m-vector and c, h and x are real n-vectors. Correspondence with the *minimum-cost generalized network flow problem* is established by associating the rows and columns of A with the vertices and edges of a finite undirected graph $G(V, E)$, the vector b with *exogenous flows* at the vertices V, and the vectors c, h and x with *costs, capacities* and *flows* on the edges E. The two nonzero elements of a column a_j of A, $j \in E$, to be denoted here as p_j and q_j, occuring, say, at rows r and $s \in V$, respectively, are referred-to as *flow multipliers* for the flow x_j. The *flow direction* in edge j depends upon the signs of p_j and q_j as follows:

1. If $p_j > 0$ and $q_j < 0$, then x_j is the flow *from r to s*. Such an edge is *flow-transforming* since a flow of $p_j x_j$ *out of* vertex r is transformed to a flow of $-q_j x_j$ *into* vertex s.

2. If $p_j > 0$ and $q_j > 0$, then x_j is the flow *from r and from s*. Such an edge is *flow-absorbing* (i.e. it acts as a *sink*) for a flow of $p_j x_j$ *out of r and* $q_j x_j$ *out of s*.

This research was supported by the National Science Foundation under Grant No. MCS-8113503.

Similarly, if $p_j < 0$ and $q_j < 0$, then edge j is *flow-generating* (i.e. it acts as a *source*) for a flow of $p_j x_j$ *into* r *and* $q_j x_j$ *into* s.

3. If $p_j > 0$ is the only nonzero element in column a_j, then edge j behaves as a *sink* which absorbs a flow of $p_j x_j$ *out of* r. Similarly, if $p_j < 0$, then j acts as a *source* which provides a flow of $p_j x_j$ *into* r. Such an edge may also be viewed as a self-loop at vertex r.

The problem of generalized network flows, often referred-to as *network flows with losses and gains*, was first introduced in 1958 by Jewell who extended the Ford–Fulkerson labeling algorithm for its solution [21]. Dantzig studied the problem under the title of *weighted distribution* [8]. E.L. Johnson observed that the Ford–Fulkerson labeling procedure may fail to converge unless all solutions remain basic and proposed data structures which ensure this property [22]. The first simplex-based code was developed by Maurras [24], followed by implementations using various data structures due to Glover et al. [13], Elam et al. [11], Brown and McBride [5], and a dual method due to Jensen and Bhaumik [19]. Jarvis and Jezior [18] and Grinold [16] studied the maximum flow problem. Wollmer discussed the generalized multicommodity flow problem [31]. Most of this work deals with the case of positive gains or losses (cases 1 and 3 above). Reviews of simplex-based methods may be found in books by Kennington and Helgason [23] and Jensen and Barnes [20]. It is interesting to note that the (0, 1)-integer case of GN with only flow-absorbing edges, unit flow multipliers and right-hand side elements, is the weighted matching problem for which Edmonds gave a polynomial-time algorithm in 1965 [9, 10].

The present state of algorithmic research involves the selection of appropriate data structures and experimentation with various implementations of the primal, dual or primal-dual simplex methods. Since such implementations are substantially faster than ordinary LP codes, they are already considered suitable for solving large, practical GN problems. Even though the manner in which basic solutions are computed within such codes may strongly affect numerical accuracy and the solution times, very little attention has been paid on analyzing the numerical behavior of such algorithms.

We are interested in solving two basic systems of the form $Bx = b$ and $uB = c_B$ where the basis B of A is block-diagonal, i.e. $B = \text{diag}(B_1, \ldots, B_p)$, $1 \leq p \leq m$. Each (nonsingular) *basis block* B_i corresponds to a subgraph $G_i(V_i, E_i)$ of G which may be a *spanning tree* or *spanning one-tree*, a connected graph with exactly one cycle [17]. Thus, the two basic systems to be solved for block i exhibit the additional structure

$$Bx \equiv \begin{pmatrix} C & D \\ 0 & T^* \end{pmatrix} \begin{pmatrix} x_C \\ x_F \end{pmatrix} = \begin{pmatrix} b_C \\ b_F \end{pmatrix} \equiv b, \tag{2}$$

$$uC \equiv (u_C, u_F) \begin{pmatrix} C & D \\ 0 & T^* \end{pmatrix} = (c_C, c_F) \equiv c_B \tag{3}$$

where we have omitted the block index i, and where the *cycle matrix* C is a special case of a k-order upper Hessenberg matrix, i.e.

$$C = \begin{pmatrix} p_1 & & & & q_k \\ q_1 & p_2 & & & \\ & q_2 & \ddots & & \\ & & \ddots & p_{k-1} & \\ & & & q_{k-1} & p_k \end{pmatrix} \tag{4}$$

and necessarily p_i, $q_i \neq 0$ for all i. The submatrix T^*, referred-to as a *forest matrix*, is upper triangular and consists of a set of triangular submatrices T_j along the diagonal. D is a matrix with at most one nonzero entry in each column. The row index of such a nonzero of D corresponds to a cycle vertex which is regarded as the *root node* of some subtree T_j. Figure 1.1 depicts a basis block for a one-tree which consists of one cycle and a forest of two subtrees T_1 and T_2. If G is a spanning tree, then C and D are vacuous.

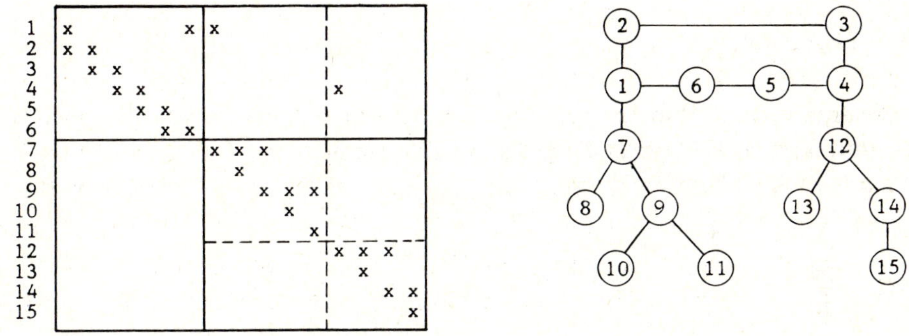

Figure 1.1. Typical basis block B_i and one-tree $G_i(V_i, E_i)$.

Now, to solve (2) we proceed in the obvious manner:

1. Solve the upper-triangular *forest subsystem* $T^*x_F = b_F$ for x_F by back substitution which, in practice, is known to result in a relatively accurate solution for most problems [26]. This task is straightforward and it is usually implemented via appropriate list functions similar to those used for minimum-cost network flow problems (see e.g. [13, 14, 23]).

2. Solve the *cycle subsystem* $Cx_C = b_C^* \equiv b_C - Dx_F$ for x_C. This step is considerably more involved due to the different levels of accuracy obtained by specializations of direct methods for solving linear systems. Two such specializations which have been widely used for this purpose are *direct substitution*, originally proposed by Dantzig [8, p. 423] and later used by Maurras [24], and *superposition*, proposed by Glover et al. [13], Kennington [23] and perhaps others.

The second system (3) is solved in a similar fashion, except that the cycle subsystem $u_C C = c_C$ is solved first for u_C and then the resulting forest subsystem $u_F T^* = c_F^* \equiv c_F - u_C D$ is solved for u_F.

We shall concentrate on solving the cycle subsystem $Cx_C = b_C^*$ and, by inference, the subsystem $u_C C = c_C$. In Section 2 and Section 3, we review direct substitution and superposition, where we generally argue that the former is efficient but numerically unstable whereas the latter is sufficiently stable but relatively inefficient, in particular for dense right hand sides. In Section 4, we propose the use of Gaussian elimination and we characterize the structure of the triangular decomposition of C. In Section 5 we investigate the numerical stability of the proposed method. In Section 6 we define appropriate data structures on the cycle vertices and state detailed algorithms for solving the two cycle subsystems mentioned above by first constructing the triangular factors of C and then by solving the resulting two triangular systems. Finally, in Section 7 we present some numerical results and we compare the computational behavior of all methods. With no loss of generality, we assume that the matrix A of GN is well-scaled and that C is a nonsingular cycle submatrix of A. For notational convenience, we will henceforth denote the two subsystems as $Cx = b$ and $uC = c$ by dropping the subscripts and superscript.

2. Direct substitution

To solve $Cx = b$, this method proceeds in two stages. First, forward elimination, with pivots along the diagonal, is used to obtain a *computed* value of x_k, say x_k^*, from

$$x_k = [b_k + \lambda_{k-1}(b_{k-1} + \cdots + \lambda_2(b_2 + \lambda_1 b_1) \cdots)]/p_k(1-\omega) \tag{5}$$

where $\omega = \prod_{i=1}^{k} \lambda_i$ with $\lambda_i = -q_i/p_i$. Second, x_k^* is substituted into $Cx = b$ and $(x_1^*, \ldots, x_{k-1}^*)$ are obtained by forward substitution in the (reduced) lower-triangular system

$$C^R x^R = b^R - q_k x_k^* e_1 \tag{6}$$

where C^R consists of the first $k-1$ rows and columns of C, b^R and x^R consist of the first $k-1$ elements of b and x respectively, and e_j denotes the j-th unit vector. Solving again for x_k from the k-th equation of $Cx = b$, i.e.

$$q_{k-1}x_{k-1}^* + p_k x_k = b_k \tag{7}$$

gives a *new* computed x_k^{**} where one expects to have $x_k^{**} = x_k^*$ as a computational check.

The implementation may be organized conveniently along two cycle traversals, starting from an arbitrary cycle vertex k, using a *predecessor* list function. The first traversal implements (5) to produce x_k^* and the second one computes the remaining x_i^* from (6). A total of $5k-2$ multiplications or divisions and $2k-1$ additions are thus required. Since the predecessor list is included in all GN codes, no additional storage is needed. Unfortunately, there are two drawbacks to such an implementation.

First, the computation of the products $\lambda_1 \cdots \lambda_p$ for $p=2,\ldots,k$ in (5) is prone to *exponent overflow* in both single and double precision arithmetic, whenever sufficiently many $|\lambda_i|>1$, a condition which depends entirely upon the given elements of C. Equilibrating or scaling C (or the matrix A, as it is sometimes done in GN codes) cannot alleviate this difficulty since it does not alter the value of ω.

Second, the forward substitution step in solving (6) tends to amplify the error included in x_k^*. To see this, consider $x_k^* = x_k + \xi_k$, as the sum of the exact value x_k and the error due to (5). Substituting x_k^* into (6) gives:

$$C^{\mathrm{R}}x^{\mathrm{R}} = b^{\mathrm{R}} - q_k x_k e_1 - q_k \xi_k e_1. \tag{8}$$

Now, the solution to (8) may be viewed as $x^{\mathrm{R}*} = x^{\mathrm{R}} + \xi^{\mathrm{R}}$ where x^{R} exactly solves $C^{\mathrm{R}}x^{\mathrm{R}} = b^{\mathrm{R}} - q_k x_k e_1$ and where $\xi^{\mathrm{R}} = (\xi_1,\ldots,\xi_{k-1})$ exactly solves $C^{\mathrm{R}}\xi^{\mathrm{R}} = -q_k\xi_k e_1$. The solution of the latter system is easily seen to be

$$\xi_j = -(q_k/p_j)\xi_k \prod_{i=1}^{j-1} \lambda_i, \quad j=1,\ldots,k-1, \tag{9}$$

which is a function of the initial error ξ_k. Finally, applying (7),

$$x_k^{**} = (b_k - q_{k-1}x_{k-1}^*)/p_k = x_k - (q_{k-1}/p_k)\xi_{k-1} \tag{10}$$

and substituting (9) into (10) yields

$$x_k^{**} = x_k + \omega\xi_k. \tag{11}$$

Thus, a small initial error ξ_k is amplified with no bound whenever $|\omega| \gg 1$, which may well be the case in the absence of any reasonable bound on the $|\lambda_i|$ for arbitrary cycle matrices C. To illustrate this consider the solution of the cycle subsystem (based on [3, p. 290])

$$\begin{pmatrix} -0.003 & 0 & 59.14 \\ -5.291 & -0.070 & 0 \\ 0 & 6.318 & -6.130 \end{pmatrix} \begin{pmatrix} x_1 \\ x_2 \\ x_3 \end{pmatrix} = \begin{pmatrix} 59.11 \\ -53.61 \\ 57.05 \end{pmatrix} \tag{12}$$

using four-digit arithmetic. The exact solution is $x_1 = x_2 = 10$ and $x_3 = 1$. The system is well-conditioned with respect to this arithmetic (the condition number $\kappa(C) = 11.32$; see e.g. [6], p. 181). Now, from (5) we obtain $x_3^* = 1.001$ with $\lambda_1 = -1764$, $\lambda_2 = 90.26$, $\lambda_3 = 9.648$ and $\omega = -1\,536\,000$. Forward substitution in (12) results in $x_1^* = 30$ and $x_2^* = -1501$. The error in x_3^* is $\xi_3 = 0.001$. Using (11) we see that $x_3^{**} = x_3 + \omega\xi_3 = 1 - 1\,536\,000 * 0.001 = -1535$, due to the enormous amplification of the relatively small initial error ξ_3.

However, using $x_3^* = 1.001$ in the third equation (instead of the first) gives $x_2^* = 10$ and then using x_2^* in the second equation gives $x_1^* = 10$, which illustrates the need for deciding upon a 'good' order of substitution. The triangular decomposition of C with partial pivoting, discussed in Section 4, provides a method for selecting such an order.

Finally, consider again the system (12) with the right-hand side (−59.17, −53.51, 69.31) for which $x=(10, 10, -1)$ is the exact solution. Performing (row) equilibration, the system becomes:

$$\begin{pmatrix} -0.00005073 & 0 & 1 \\ -1 & -0.01323 & 0 \\ 0 & 1 & -0.9702 \end{pmatrix}\begin{pmatrix} x_1 \\ x_2 \\ x_3 \end{pmatrix}=\begin{pmatrix} -1.001 \\ -10.13 \\ 10.97 \end{pmatrix} \tag{13}$$

for which $\lambda_1=-19\,710$, $\lambda_2=75.59$, $\lambda_3=1.031$ and $\omega=-1\,536\,000$ and from (5) $x_3^*=-1.001$. Then, forward substitution gives $x_1^*=0$, $x_2^*=765.7$, and from (11) $x_3^{**}=1535$. As expected, equilibration does not improve the numerical behavior of direct substitution.

3. Superposition

This method computes the solution x^* to $Cx=b$ by first solving the subsystems $Cx^{(j)}=b_je_j$, $j=1,\ldots,k$, and then by *superposing* their solution vectors $x^*=x^{(1)}+\cdots+x^{(k)}$. Thus, the method consists of forming the linear combination (with weights b_j) of those columns of C^{-1} which correspond to the nonzero elements of b.

Surprisingly, the method behaves better than direct substitution. We argue as follows. In solving the j-th subsystem for $x^{(j)}$, we first compute $x_j^{(j)*}$ by (5) which introduces a negligible initial error $\xi_j^{(j)}$ since the numerator is now simply b_j and no cancellation of terms is involved. Then, starting with $x_j^{(j)*}$, forward substitution gives $x_s^{(j)*}$, sequentially for $s=j+1,\ldots,k,\ 1,\ldots,j-1$. Recomputation of $x_j^{(j)**}$ from the j-th equation (analogous to (11)) induces an amplification of the initial error $\xi_j^{(j)}$ but this is seldom of any consequence since $\xi_j^{(j)}$ is very small in comparison to $x_j^{(j)}$. Finally, superposing the solution vectors of each subsystem, as mentioned above, *does* cause cancellation errors, but their effect is not amplified since there are no multiplications or divisions involved.

As an illustration, consider the previous example (12) with the same arithmetic. We solve the first subsystem: $x_1^{(1)*}=b_1/p_1(1-\omega)=-59.11/0.003(1+1\,536\,000)=-0.01283$, and by forward substitution in (12) $x_2^{(1)*}=0.9697$ and $x_3^{(1)*}=0.9995$. Then, we solve the second subsystem, starting with $x_2^{(2)*}=0.0004986$ to obtain $x_3^{(2)*}=0.0005139$ and $x_1^{(2)*}=10.13$. Similarly, we solve the third subsystem starting with $x_3^{(3)*}$ to obtain $x^{(3)*}=(-0.1194,\ 9.024,\ -0.000006059)$. Superposition yields $x^*=x^{(1)*}+x^{(2)*}+x^{(3)*}=(10.00, 9.994, 1.000)$, well within the acceptable accuracy of 3 digits.

The method requires k_0+1 traversals of the cycle, $2kk_0$ multiplications or divisions and $k_0(k-1)+1$ additions, where k_0 is the number of nonzeroes in b. It also requires space for an m-long vector *in addition* to the space used by the underlying GN code. We see that for $k_0=2$ superposition is slightly more efficient than direct substitution. Such is the case during each iteration of the revised simplex method where $b \equiv b_C$ (see Section 1) has at most two nonzeroes. However, superposition is more time consuming in other cases for which $k_0 \simeq k$, e.g. in the recomputation of the primal and dual variables within a GN code and also in basis partitioning algorithms for structured linear programs with GN subproblems, even though it has been used for this purpose [25]. Furthermore, the difficulty with exponent overflows persists just as in direct substitution.

4. Triangular decomposition of cycle matrices

Here we propose to solve the cycle subsystem $Cx = b$ by Gaussian elimination with partial (row) pivoting to first obtain a triangular decomposition of C so that $C = L^{-1}U$, and then to solve $L^{-1}y = b$ and $Ux = y$. A modification of the simplex method which uses such decompositions and a procedure for updating them at every iteration was first proposed by Bartels and Golub [1, 2]. In our case, the updating step is not worthwhile and is thus omitted in favor of recomputing the decomposition for each new C.

The triangular decomposition of C is computed by multiplying on the left by a sequence of $k-1$ *permutation* matrices P^t and *elementary elimination* matrices L^t (see e.g. [12, 30]), according to the recursion:

$$C^1 \equiv C, \qquad C^{t+1} = L^t P^t C^t \quad \text{for } 1 \leq t \leq k-1, \tag{14}$$
$$U \equiv C^k = L^{k-1}P^{k-1} \cdots L^1 P^1 C^1 \equiv LC.$$

Since C is upper Hessenberg, after $t-1$ pivots the t-th column of C^t has exactly one nonzero element below the t-th row, which occurs at row $t+1$. Thus, the t-th pivot step involves only rows t and $t+1$, and the row permutation P^t which is dictated by the choice of the pivot row, is limited to the possible interchange of these two rows of C^t. Similarly, only one L^t is required for each pivot step, where L^t is the identity matrix with the exception of one off-diagonal nonzero element λ_t at position $(t+1, t)$. The row operation $L^t(P^tC^t)$ clearly annihilates the $(t+1, t)$-th element of P^tC^t. Furthermore, since C has the special structure exhibited in (4), it is easy to see from (14) by induction that the t-th row of C^t has exactly two nonzeroes c^t_{tt} and c^t_{tk}, and that $c^t_{tk} = q_k$ whenever $P^1, \ldots, P^{t-1} \neq I$. It is also clear that $c^t_{t+1,t} = q_t$ and $c^t_{t+1,t+1} = p_{t+1}$, i.e., the $(t+1)$-st row of C^t is the original $(t+1)$-st row of C.

The *partial pivoting* strategy consists of selecting appropriate row permutations so that $|\lambda_t| \leq 1$ for all t, which reduces the magnitude of the elements of C^t. We thus distinguish two cases, depending upon the relative magnitude of c^t_{tt} and q_t:

Case 1. If the pivot element is c^t_{tt} then $P^t = I$, $\lambda_t = -q_t/c^t_{tt}$ and C^{t+1} differs from C^t only in its $(t+1)$-st row which, for $2 \leq t \leq k-2$, obviously has exactly two nonzeros $c^{t+1}_{t+1,t+1} = p_{t+1}$ and $c^{t+1}_{t+1,k} = c^t_{tk}\lambda_t$. We refer to the latter element as a *newly computed nonzero.* For $t = k-1$, the element $c^k_{kk} = p_k + c^{k-1}_{k-1,k}\lambda_{k-1}$ is the only nonzero in row k and it is newly computed.

Case 2. If the pivot element is $c^t_{t+1,t}$ $(=q_t)$ then $P^t \neq I$, $\lambda_t = -c^t_{tt}/q_t$ and C^{t+1} differs from C^t only in its t-th and $(t+1)$-st rows: The t-th row of C^{t+1} is the $(t+1)$-st row of C^t and the $(t+1)$-st row of C^{t+1} has two nonzero elements: $c^{t+1}_{t+1,t+1} = p_{t+1}\lambda_t$ and $c^{t+1}_{t+1,k} = c^t_{tk}$. So that for $2 \leq t \leq k-2$, the $(t+1)$-st row has a newly computed nonzero at position $t+1$, and a *previously computed nonzero* at position k unless $P^1, \ldots, P^{t-1} \neq I$ as mentioned earlier. For $t = k-1$, element $c^k_{kk} = c^{k-1}_{k-1,k} + p_k\lambda_{k-1}$ is the *only* nonzero in row k and it is newly computed.

Thus, U has at least one computed nonzero and, as argued earlier, each pivot operation may generate at most one newly computed nonzero. We summarize these observations below.

Lemma 1. *Let U be an upper triangular matrix computed from a nonsingular cycle matrix C of order k by applying Gaussian elimination with partial pivoting and let P^t be the row permutation matrix associated with the t-th pivot, $1 \leq t \leq k-1$. Then,*

1. *U has exactly two nonzeroes per row except the last which has one. Furthermore, denoting by u the t-th row of U after the t-th pivot operation, these nonzeroes are:*

a. *For $t = 1$:*

$$u_1 = p_1, \quad u_k = q_k \quad \text{if } P^1 = I,$$

$$u_1 = q_1, \quad u_2 = p_2 \quad \text{otherwise}$$

b. *For $2 \leq t \leq k-1$:*

(i) *If $P^t \neq I$, then $u_t = q_t$ and $u_{t+1} = p_{t+1}$*

(ii) *If $P^t = P^{t-1} = I$, then $u_i = p_i$ and u_k is a newly computed nonzero.*

(iii) *Otherwise $(P^t = I \neq P^{t-1})$, u_t is a newly computed nonzero and u_k is a previously computed nonzero unless all previous $P^{t-1}, \ldots, P^1 \neq I$, in which case, $u_k = q_k$.*

c. *In the last row, u_k is a newly computed nonzero.*

2. *U contains at least one but not more than $k-1$ computed nonzeroes, or equivalently U contains at least k but not more than $2k-2$ nonzero elements of C.* □

Remarks. 1. If *no pivoting* is used $(P^t = I)$, then $\lambda_t = -q_t/p_t$ for *all t*, and thus L^{-1} has a diagonal of all 1's, a subdiagonal of $-\lambda_1, \ldots, -\lambda_{k-1}$ and zeroes elsewhere. For later reference, we note that[1] $\|L^{-1}\| \leq 1 + \lambda_*$ where $\lambda_* = \max\{1, \max_i\{|\lambda_i|\}\}$.

2. For *partial pivoting* with $P^t \neq I$ for *all t*, L^{-1} has a subdiagonal of all 1's, the vector $(-\lambda_1, \ldots, -\lambda_{k-1}, 1)$ as its first row and zeroes elsewhere. In this case, $\|L^{-1}\| \leq k$.

3. From Lemma 1, part 1b-ii and Case 1 above, it is relatively easy to see that if *no pivoting* is used then U has the structure U_N depicted in Figure 4.1 with all newly computed elements in the last column where $\lambda^*_j = \prod^j_{i=1} \lambda_i$. Here, $\|U\| \leq c_*(1 + \lambda^{k-1}_*)$ where c_* is the maximum element of C in absolute value.

4. Similarly, for *partial pivoting* with $P^t \neq I$ for *all t*, Lemma 1, part 1b-i and Case 2 above imply that U has the structure U_P as shown in Figure 4.1 and $\|U\| \leq 2c_*$.

[1] $\|\cdot\|$ denotes the l_∞-norm which is used throughout the paper.

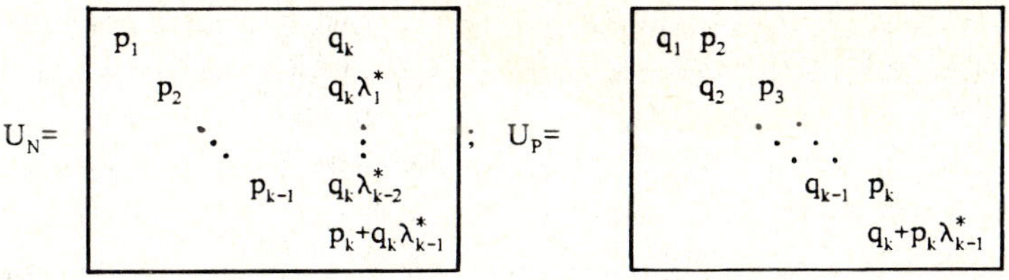

Figure 4.1. Two extreme structures of U.

5. In general, partial pivoting creates a U whose structure is a combination of (3) and (4) above. This is illustrated in Figure 4.2 for $k=8$, where the pivot row sequence of (1, 2, 4, 5, 3, 6, 7) produces U from C (‘x’ denotes an original element of C and ‘•’ denotes a computed nonzero element).

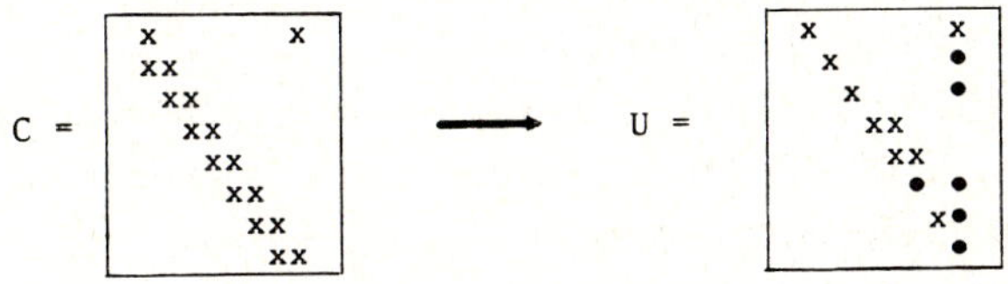

Figure 4.2. General structure of U obtained by partial pivoting.

6. In order for the partial pivoting strategy to select the best possible pivot sequence, the matrix C must be ‘well-scaled’. This is the case for ‘well-formulated’ practical GN problems in which the elements of the matrix A are comparable in size. In other instances, an appropriate procedure may be used to scale A, e.g. as suggested by Curtis and Reid [7] and Tomlin [27].

7. Using U_N from Figure 4.1 to solve for x_k yields equation (5). But the *order* of operations used in solving for $x_1, \ldots, x_{k-1}$ differs in Gaussian Elimination with no pivoting and in Direct Substitution (Section 2), even though these methods use the same fixed pivoting strategy. Thus, their numerical behavior may be different.

5. Error analysis

In this section, we study the effects of rounding errors in floating-point arithmetic when Gaussian elimination with partial pivoting is used to solve the cycle subsystem $Cx=b$. We assume that C is well-scaled and proceed along the lines originally suggested by Wilkinson (see e.g. [4, 12, 29]). This general approach was used by Bartels to analyze the error in updating triangular decompositions for the simplex method [2].

We consider the *inverse-error expression* $(C+\delta C)x=b$, where δC is a *perturbation matrix* and x is the exact solution to the so-perturbed system, and derive a bound for $\|\delta C\|$ in terms of the given elements of C. The perturbation matrix δC contains errors from computing the triangular decomposition of C *and* errors from solving the two resulting triangular systems, with corresponding inverse error expressions:

$$L^{-1}U = C + E \tag{15}$$

$$(L^{-1} + \delta L^{-1})(U + \delta U)x = b \tag{16}$$

where E is an *error matrix* for the triangular decomposition step, and δL^{-1} and δU are the error matrices associated with the solution of $L^{-1}y = b$ and $Ux = y$, respectively. Thus, $\delta C = E + U\delta L^{-1} + L^{-1}\delta U + \delta U \delta L^{-1}$ and since

$$\|\delta C\| \leq \|E\| + \|U\| \cdot \|\delta L^{-1}\| + \|L^{-1}\| \cdot \|\delta U\| + \|\delta U\| \cdot \|\delta L^{-1}\|, \tag{17}$$

$\|\delta C\|$ may be estimated by individually bounding the terms on the right-hand side, as described in the following paragraphs.

Triangular decomposition errors

We shall obtain an estimate for E in (15) by appropriately combining the error matrices E^t, $t = 1, \ldots, k-1$, where E^t represents the rounding errors caused by the t-th pivot operation. Denoting the elements of E^t by e^t_{ij}, we write from (14)

$$C^{t+1} = L^t P^t C^t + E^t. \tag{18}$$

At the conclusion of the t-th pivot step, E^t contains errors from two different sources:

1. From setting the $(t+1, t)$ element of C^{t+1} to zero: Considering the computation of λ_t (Cases 1 and 2, Section 5) whose error expression is[2] $\lambda_t = (-c^t_{t+1,t}/c^t_{t,t})(1 + \varepsilon^t_t)$, or equivalently $c^t_{t+1,t} + \lambda_t c^t_{tt} + \varepsilon^t_t c^t_{t+1,t} = 0$, we see that this error is:

$$e^t_{t+1,t} = c^t_{t+1,t}\varepsilon^t_t. \tag{19}$$

2. From the computation of a newly computed nonzero element: By Lemma 1, part 1b, such a nonzero of C^{t+1} is located either at $(t+1, k)$ if $P^t = I$, or at $(t+1, t+1)$ if $P^t \neq I$. Thus,

a. If, for some $t < k-1$, $P^t = I$, then $c^{t+1}_{t+1,k} = c^t_{t,k}\lambda_t\,(1 + \varepsilon^t_k)$ which gives the error:

$$e^t_{t+1,k} = c^t_{t,k}\lambda_t\varepsilon^t_k. \tag{20}$$

b. Otherwise $(P^t \neq I)$,

$$e^t_{t+1,t+1} = c^t_{t+1,t+1}\lambda_t\varepsilon^t_{t+1}. \tag{21}$$

c. For the last row $(t = k-1)$, if $P^{k-1} = I$, we have

$$c^k_{kk} = [c^{k-1}_{kk} + \lambda_{k-1}c^{k-1}_{k-1,k}(1 + \varepsilon^{k-1}_k)]/(1 + \varepsilon_k)$$

which gives the error:

$$e^{k-1}_{kk} = \lambda_{k-1}c^{k-1}_{k-1,k}\varepsilon^{k-1}_k - c^k_{kk}\varepsilon_k. \tag{22}$$

Otherwise $(P^{k-1} \neq I)$,

$$e^{k-1}_{kk} = \lambda_{k-1}c^{k-1}_{kk}\varepsilon^{k-1}_k - c^k_{kk}\varepsilon_k. \tag{23}$$

Now, let c_*^t be the maximum element of C^t in absolute value, and assume $c_{kk}^k \leqslant (|\lambda_{k-1}|+1)c_*^{k-1}$. By combining (19)–(23) and since[2] all $|\varepsilon_j^i| \leqslant \varepsilon$, we obtain

$$e_{ij}^t \leqslant \begin{cases} c_*^t \varepsilon & \text{for } i=t+1, j=t, \\ |\lambda_t| c_*^t \varepsilon & \text{for } i=t+1, j=k, t \neq k-1 \text{ and } P^t = I, \\ & \quad \textit{or } \text{for } i=j=t+1, t\neq k-1 \text{ and } P^t \neq I, \\ (2|\lambda_{k-1}|+1)c_*^{k-1}\varepsilon & \text{for } t=k-1, i=j=k, \\ 0 & \text{otherwise.} \end{cases} \tag{24}$$

Our next task is to get a bound for $\|E\|$. We first note that, due to the special structure of L^t and the location of the nonzero elements of E^t in (24),

$$(L^j)^{-1}E^t = E^t \quad \text{for } j \leqslant t \tag{25}$$

and similarly, since P^j interchanges only the j-th and $(j+1)$-st rows of E^t

$$(L^i)^{-1} \prod_{s=j}^{t} P^s E^t = \prod_{s=j}^{t} P^s E^t \quad \text{for } i<j\leqslant t. \tag{26}$$

Next, multiplying (18) by $(L^tP^t)^{-1}$ and using (25) we have

$$P^t(L^t)^{-1}C^{t+1} = C^t + P^tE^t. \tag{27}$$

Finally, (14) is equivalent to $L^{-1}U = P^1(L^1)^{-1} \cdots P^{k-1}(L^{k-1})^{-1}C^k$ which, after repeated applications of (27) and (26) and collection of terms, simplifies to

$$L^{-1}U = P^1(L^1)^{-1} \cdots P^{k-2}(L^{k-2})^{-1}(C^{k-1} + P^{k-1}E^{k-1}) = \cdots = C + E$$

where clearly

$$E = \sum_{j=1}^{k-1} \left(\prod_{i=1}^{j} P^iE^j \right). \tag{28}$$

In the special case of *no pivoting*, this reduces to

$$E = E^1 + \cdots + E^{k-1} \tag{29}$$

and by substituting (24) into (29)

$$e_{ij} \leqslant \begin{cases} c_*\varepsilon & \text{for } i=j+1, 2\leqslant i \leqslant k, \\ \lambda_*^{i-1} c_* \varepsilon & \text{for } 2 \leqslant i \leqslant k-1, j=k, \\ (2\lambda_*^{k-1}+1)c_*\varepsilon & \text{for } i=j=k, \\ 0 & \text{otherwise} \end{cases} \tag{30}$$

and

$$\|E\| \leqslant 2(\lambda_*^{k-1}+1)c_*\varepsilon. \tag{31}$$

[2] For rounded operations with normalized floating-point numbers, the *unit roundoff* is $\varepsilon = 1/2\beta^{1-t}$ where t is the number of digits in base β used for the mantissa. Denoting by $\oplus$ any of the operators +, −, ×, /, then a floating-point operation for two numbers a and b is $\text{fl}(a\oplus b) = (a\oplus b)(1+\varepsilon_{ab})$, or $\text{fl}(a\oplus b) = (a\oplus b)/(1+\varepsilon_{ab})$ where $|\varepsilon_{ab}| \leqslant \varepsilon$ (see e.g. [12] p. 90–97 and [28]).

For *partial pivoting*, $|\lambda^t| \leq 1$ for all t, and thus $c_*^t \leq c_*$ for $1 \leq t \leq k-1$. The worst case occurs when $P^t \neq I$ for *all t*. In this case, from (24) and (28), we have

$$e_{ij} \leq \begin{cases} c_* \varepsilon & \text{for } i=j=1, \\ 2c_* \varepsilon & \text{for } i=1, 2 \leq j \leq k-1, \\ 3c_* \varepsilon & \text{for } i=1, j=k, \\ 0 & \text{otherwise} \end{cases}$$

and

$$\|E\| \leq 2kc_* \varepsilon.$$

We have thus shown

Lemma 2. *The matrices L^{-1} and U, computed by Gaussian elimination using floating-point arithmetic with unit round-off ε, satisfy $L^{-1}U = C + E$, where*

$$\|E\| \leq \begin{cases} 2(\lambda_*^{k-1}+1)c_* \varepsilon & \textit{with no pivoting}, \\ 2kc_* \varepsilon & \textit{with partial pivoting}. \end{cases} \qquad \square$$

System solution errors

After completing the triangular decomposition, additional errors result from solving the two triangular systems: $L^{-1}y = b$ and $Ux = y$. We consider these separately.

1. Let y be the exact solution of the perturbed system $(L^{-1} + \delta L^{-1})y = b$. Then,

a. If $P^t = I$ for all t, then L^{-1} has the special structure described in Section 4, Remark 1. Thus $y_1 = b_1$ and $y_{i+1} = (-\lambda_i(1+\varepsilon_{ii})y_i + b_{i+1})/(1+\varepsilon_{i+1,i+1})$ for $2 \leq i \leq k-1$, which gives:

$$(\delta L^{-1})_{ij} \leq \begin{cases} \varepsilon & \text{for } i=j \neq 1, \\ \varepsilon|\lambda_j| & \text{for } i=j+1, \\ 0 & \text{otherwise} \end{cases}$$

and

$$\|\delta L^{-1}\| \leq \varepsilon(\lambda_* + 1). \tag{32}$$

b. If $P^t \neq I$ for all t, recalling Remark 2 of Section 4 and arguing as in (a) above with $|\lambda_t| \leq 1$ results in[3]

$$\|\delta L^{-1}\| \leq 1.01\,\varepsilon(k+2)(k-1)/2. \tag{33}$$

2. Now, let x be the exact solution to the perturbed upper triangular system $(U + \delta U)x = y$. Proceeding in the same fashion as in (a) and (b) above we obtain:

a. For the special structure of U described in Remark 3 of Section 4,

$$\|\delta U\| \leq 1.01\,\varepsilon c_*(2 + \lambda_*^{k-1}). \tag{34}$$

[3] Note that $1 - k\varepsilon \leq \prod_{i=1}^{k}(1+\varepsilon_i) \leq 1 + 1.01k\varepsilon$ provided that $|\varepsilon_i| \leq \varepsilon$ for $i = 1, \ldots, k$ and if $k\varepsilon \leq 0.01$ (see e.g. [12, p. 92]). For single precision arithmetic with mantissa lengths of at least 23 bits ($\varepsilon \simeq 10^{-7}$), the latter assumption is reasonable for cycle lengths of $k \leq 10\,000$.

b. For the structure of U given in Remark 4 of Section 4,

$$\|\delta U\| \leqslant 3.03 \varepsilon c_* . \tag{35}$$

Finally, to estimate $\|\delta C\|$ we sum up the bounds for each term of (17) by using Lemma 2, equations (32)–(35) and the bounds for $\|U\|$ and $\|L^{-1}\|$ given in Remarks 1–4 Section 4. This results in a complicated expression which, for $6.06\varepsilon \ll 1$, a perfectly reasonable assumption for any computing environment, may be simplified as stated below.

Theorem 1. *The solution x computed by Gaussian elimination exactly solves the perturbed system $(C+\delta C)x = b$, where*

$$\|\delta C\| \leqslant \begin{cases} 1.01\varepsilon C_*(7\lambda_*^k + 3\lambda_* + 5) & \text{with no pivoting,} \\ 5.05\varepsilon c_* k^2 & \text{with partial pivoting.} \end{cases} \qquad \square$$

As expected, the bound for no pivoting grows rapidly with increasing values of $\lambda_* > 1$, in contrast to the bound for partial pivoting which grows only in terms of the maximum element of C and the cycle length k. For $k > 2$ the bound for partial pivoting is sharper by a power of k than the corresponding bound for arbitrary C given in the literature (see e.g. [29, p. 108], [12, p. 106] and [4, p. 120]).

6. Data structures and algorithms

In order to efficiently implement the procedure outlined in Section 4–5 we introduce a number of list functions on the vertices of the graph G for computing, storing and retrieving the nonzero elements of L and U.

We recall that U has at most $k-1$ computed nonzeros (Lemma 1, part 2), which may be accommodated in an m-long array. The remaining nonzeroes of U, being elements of C, may be retrieved from the list functions for the flow multipliers associated with each edge of G. Now, treating q_k as a computed nonzero, the locations of the nonzero elements in each row (cycle vertex) of U may be easily identified by knowing whether an interchange of two adjacent rows was required during pivoting. This is recorded in a boolean *permutation indicator* list function on the cycle vertices. Thus, a sequential review of the elements of U may be implemented by traversing the vertices of the cycle clockwise, with the aid of the (circular) *predecessor* and the permutation indicator lists. For back-substitution, it is necessary to traverse the cycle counter-clockwise which is done via a *Successor* list function.

The data structure for L is straightforward since every off-diagonal element λ_t is always the $(t+1, t)$-element of L^t. Thus, a list function which maps a vertex t to λ_t suffices to store L.

We propose to use the following list functions on the cycle vertices V_C.

$F \;: (i \in V_C) \to (j \in V_C)$ – *Predecessor* (*Father*) of cycle vertex i, clockwise.

$F^* : (i \in V_C) \to (j \in V_C)$ – *Successor* of cycle vertex i, counterclockwise, i.e. $j = F(i) \Leftrightarrow F^*(j) = i$ for each i and $j \in V_C$.

L: $(i \in V_C) \to \{\text{Reals}\}$ – Off-diagonal element of L^i associated with vertex i.

U : $(i \in V_C) \to \{\text{Reals}\}$ – *Computed Nonzero* of U associated with vertex i. If there are two computed nonzeroes at i, then the *previously computed* nonzero is $U(F^*(i))$.

P : $(i \in V_C) \to \{\text{True, False}\}$ – *Permutation Indicator*, i.e.

$$P(i) = \begin{cases} \text{True} & \text{if rows } i \text{ and } i+1 \text{ are permuted,} \\ \text{False} & \text{otherwise.} \end{cases}$$

All traversals of a cycle must consistently start at some vertex, say θ, which we refer-to as the *origin* of the cycle. Since, in general, there is more than one cycle in a GN basis, the set of origin vertices must be stored. In GN codes which employ a *depth* list function [23, 5, 15], the origin vertex for each cycle may be stored within this list since the depth of a cycle is 0 and all other entries of the list are positive integers. The list P may be combined with any other positive-valued list such as F^*. Under these assumptions, the space complexity of the proposed method is $3m$ higher than existing GN codes which use direct substitution, and $2m$ higher than those which use superposition.

Three algorithms are of interest. The first (LUPP) computes a triangular decomposition of C by Gaussian elimination with partial pivoting and creates the lists L, U and P. Given these lists, the second algorithm (FTRAN) computes the solution to $Cx = b$ for a given b, and the third algorithm (BTRAN) solves $zC = c$ for a given c. In all cases, we assume that in addition to the list functions defined above, the origin vertex for the cycle, and the list functions $p(\cdot)$ and $q(\cdot)$, which describe the given flow multipliers, are available.

Algorithm LUPP. *Method*: LUPP performs partial pivoting according to Lemma 1 by traversing the cycle via F, starting at vertex $F(\theta)$ and terminating at vertex θ. For implementational simplicity, the algorithm treats q_k as a 'computed' nonzero.

Input: Origin vertex θ and lists F, F^*, p and q.

Output: Lists L, U and P.

0. *Initialization*: $i \leftarrow F(\theta)$, $\alpha \leftarrow p(i)$, $U(i) \leftarrow \beta \leftarrow q(\theta)$.

1. *Partial Pivoting*:

1.1. If $i = \theta$, go to 2. Else, if $|q(i)| \leq |\alpha|$, go to 1.3.

1.2. *Permutation and Elimination*: $L(i) \leftarrow -\alpha/q(i)$, $\alpha \leftarrow p(F(i)) \cdot L(i)$, $U(i) \leftarrow \beta$, $U(F(i)) \leftarrow \alpha$, $P(i) \leftarrow \text{True}$. Go to 1.4.

1.3. *Elimination*: $L(i) \leftarrow -q(i)/\alpha$, $\beta \leftarrow \beta \cdot L(i)$, $U(F(i)) \leftarrow \beta$, $\alpha \leftarrow p(F(i))$, $P(i) \leftarrow \text{False}$.

1.4. $i \leftarrow F(i)$, go to 1.

2. *Update Origin Vertex*:

$$U(\theta) \leftarrow \begin{cases} U(\theta) + \beta & \text{if } P(F^*(\theta)) = \text{True,} \\ U(\theta) + \alpha & \text{otherwise.} \end{cases}$$

$P(\theta) \leftarrow \text{False}$. STOP. □

i	deep	p	q	F	F^*	L	U	P
1	(1, 2)	p_1	q_1	2	5	$-q_1/p_1$	q_5	False
2	(2, 3)	p_2	q_2	3	1	$-q_2/p_2$	$q_5 L_1$	False
3	(3, 4)	p_3	q_3	4	2	$-p_3/q_3$	$q_5 L_1 L_2$	True
4	(4, 5)	p_4	q_4	5	3	$-q_4/p_4 L_3$	$p_4 L_3$	False
5	(5, 1)	p_5	q_5	1	4	—	$p_5 + q_5 L_1 L_2 L_4$	False

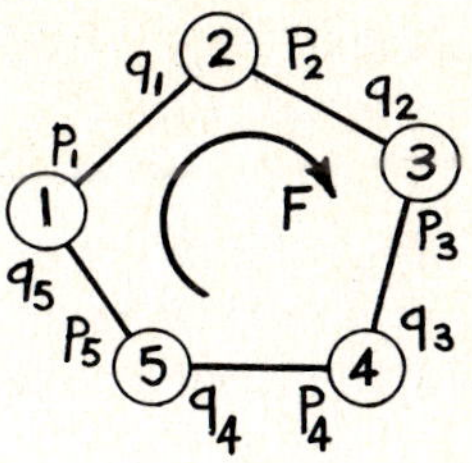

Figure 6.1. Example: List functions after applying LUPP.

Figure 6.1 provides an illustration of the algorithm and of the underlying data structures. In this 5-vertex example, $\theta = 5$ and the given list functions are p, q, F and F^*. Pivoting proceeds from edge (1, 2) or equivalently from vertex 1, and it is assumed to have produced the permutation sequence: False, False, True, False.

Algorithm FTRAN. *Method*: To solve $Cx = b$, FTRAN first traverses the cycle clockwise using list F to compute $b \leftarrow Lb$. Then, it traverses the cycle counterclockwise via F^* to solve $Ux = b$ by back substitution.

Input: Origin vertex θ and lists F, F^*, L, U, P, p, q and b.

Output: List x.

0. *Initialization*: Set $i \leftarrow F(\theta)$.

1. *Forward Elimination – Compute* $b \leftarrow Lb$:

1.1. If $i = \theta$, go to 2.

1.2. If $P(i) = \text{True}$, then $b^* \leftarrow b(i)$, $b(i) \leftarrow b(F(i))$, $b(F(i)) \leftarrow b^*$.

1.3. $b(F(i)) \leftarrow b(F(i)) + b(i)L(i)$.

1.4. $i \leftarrow F(i)$ and go to 1.1.

2. *Back-Substitution – Solve* $Ux = b$: Set $x(i) \leftarrow b(i)/U(i)$ and $j \leftarrow F^*(i)$.

2.1. If $j = \theta$, STOP.

2.2. If $P(j) = \text{True}$, then $x(j) \leftarrow (b(j) - p(i) \cdot x(i))/q(j)$. Else

$$x(j) \leftarrow \begin{cases} (b(j) - U(j) \cdot x(\theta))/p(j) & \text{if } P(F^*(j)) = \text{False} \\ (b(j) - U(F^*(j)) \cdot x(\theta))/U(j) & \text{otherwise.} \end{cases}$$

2.3. $i \leftarrow j$, $j \leftarrow F^*(i)$ and go to 2.1. □

For instance, to solve $Cx = b$ by FTRAN using the lists depicted in Figure 6.1, we first apply step 1 to obtain $b' = (b_1, b_2 + b_1 L_1, b_4, b_3 + b'_2 L_2 + b_4 L_3, b_5 + b'_4 L_4)$. Then, we solve $Ux = b'$ for x by back-substitution (step 2) which results in $x_5 = b'_5/U_5$, $x_4 = (b'_4 - U_3 x_5)/U_4$, $x_3 = (b'_3 - p_4 x_4)/q_3$, $x_2 = (b'_2 - U_2 x_5)/p_2$, and $x_1 = (b'_1 - U_1 x_5)/p_1$.

Algorithm BTRAN. *Method*: To solve $zC = c$, BTRAN first traverses the cycle clockwise using list F to solve $zU = c$ for z. Then, it traverses the cycle counterclockwise via F^* to compute $z \leftarrow zL$.

Input: Origin vertex θ and lists F, F^*, L, U, P, p, q and c.
Output: List z.
0. *Initialization*: $i \leftarrow F(\theta)$.
1. *Solve* $zU = c$:
1.1. If $i = \theta$, go to 1.4.
1.2. If $P(i) = $ True, then $z(i) \leftarrow c(i)/q(i)$, $c(F(i)) \leftarrow c(F(i)) - z(i) \cdot p(F(i))$.
Else, if $P(F^*(i)) = $ False, then $z(i) \leftarrow c(i)/p(i)$, $c(\theta) \leftarrow c(\theta) - z(i) \cdot U(i)$.
Else $z(i) \leftarrow c(i)/U(i)$, $c(\theta) \leftarrow c(\theta) - z(i) \cdot U(F^*(i))$.
1.3. $i \leftarrow F(i)$ and go to 1.1.
1.4. $z(\theta) \leftarrow c(\theta)/U(\theta)$.
2. *Apply L Backward on* z: $i \leftarrow F^*(\theta)$.
2.1. If $i = \theta$, STOP.
2.2. $z(i) \leftarrow z(i) + L(i) \cdot z(F(i))$.
If $P(i) = $ True, then $z^* \leftarrow z(i)$, $z(i) \leftarrow z(F(i))$, $z(F(i)) \leftarrow z^*$.
2.3. $i \leftarrow F^*(i)$ and go to 2.1. □

To illustrate the solution of $zC = c$ by BTRAN, we follow step 1 in Figure 6.1 and compute $z' = (c_1/p_1, c_2/p_2, c_3/q_3, (c_4 - p_4 z_3')/U_4, (c_5 - U_1 z_1' - U_2 z_2' - U_3 z_4')/U_5)$. Then, we use step 2 to apply L backward on z' which results in $z_5 = z_5'$, $z_3 = z_4' + L_4 z_5$, $z_4 = z_3' + L_3 z_3$, $z_2 = z_2' + L_2 z_3$, and $z_1 = z_1' + L_1 z_2$.

Algorithm LUPP requires one cycle traversal, $2k - 2$ multiplications or divisions and 1 addition. FTRAN requires two cycle traversals, $3k - 2$ multiplications or divisions and $2k - 2$ additions. Thus, a total of three cycle traversals, $5k - 4$ multiplications or divisions and $2k - 1$ additions are required to solve a *single* system $Cx = b$. The complexity of BTRAN is the same as that of FTRAN.

7. Computational results

We have developed experimental programs to test the computational behavior of all four methods discussed in the previous sections. These are Direct Substitution (DS), Superposition (SUP), Triangular Decomposition with no pivoting (TNP), and with partial pivoting (TPP). These programs have been coded for single precision floating-point arithmetic using ANSI-66 FORTRAN and have been executed for this test on a DEC-2060T computer which accommodates floating-point numbers of magnitude 1.4×10^{-39} to 3.4×10^{38}. In this system, exponent overflows (EOF) are handled by interrupts which produce a warning message for the first few times, the number in question is reset to the maximum floating-point number, and processing is resumed. Exponent underflows are handled similarly but the result is set to zero.

In order to test these programs we developed a random problem generator to create cycle subsystems $Cx = b$. This generator first creates a random cycle matrix C and a random *integer* solution x according to given input parameters and then it computes the right hand side b (usually b turns out to be 100% dense), so that x is the *exact* solution to the so-generated system. The input parameters are: The cycle length k, ranges for p_i, the multiplier ratios $\gamma_i = |q_i/p_i|$, the flows x_i, and the percentage of γ_i's which are greater than 1. Pseudorandom numbers from a uniform distribution are used throughout the program. The parameters and characteristics of our test problems are listed in Table 7.1 where each group includes four problems with cycle lengths of $k = 100$, 200, 300 and 400 vertices.

Table 7.1
Numerical characteristics of generated test problems

Problem Group	Flow multipliers		Ratios γ_i		RHS $\lvert b_i\rvert$	Exact sol'n $\lvert x_i\rvert$
	$\lvert p_i\rvert$	$\lvert q_i\rvert$	>1	Range		
A	$\leqslant 10$	$\leqslant 15$	50%	[0.5, 1.5]	$\leqslant 250$	$\leqslant 10$
B	$\leqslant 10$	$\leqslant 15$	70%	[0.5, 1.5]	$\leqslant 250$	$\leqslant 10$
C	$\leqslant 100$	$\leqslant 1000$	50%	[0.1, 10.0]	$\leqslant 110\,000$	$\leqslant 100$

Our comparison of the computed solution x^* versus the generated integer-valued exact solution x is based on the maximum *relative error* r, defined as

$$\max_i\{|(x_i - x_i^*)/x_i| \text{ if } |x_i| > 1 \text{ or } |x_i - x_i^*| \text{ otherwise}\}.$$

It is well known that if a computer system represents floating-point numbers with a mantissa of s significant decimal digits, and if the condition number of a given matrix is approximately 10^d, then at least $(s-d)$ digits of accuracy should be provided by the method used to solve a linear system associated with such a matrix ([6, p. 179]). In our case $s \simeq 8$ for single precision. Now, comparing the condition numbers $\kappa(C)$ (which are based on the l_∞-norm) with the relative errors for the problems listed in Table 7.2, we see that our TPP algorithm uniformly achieves this accuracy for our computing environment *and* test data.

All methods achieved acceptable solutions for the first group of problems. For the second group, where the percent of $\gamma_i > 1$ was increased to 70%, the solutions by DS and TNP are no longer acceptable, even though no exponent overflow occurred. TNP is slightly more accurate than DS although this may not always be the case. Both DS and TNP failed for the third group of problems with or without an exponent overflow. SUP produced reasonably accurate results, but by no means the best, except when it suffered exponent overflows.

The execution times given in Table 7.2 pertain to the solution of a *single* system $Cx = b$, and are *normalized* with respect to the execution times for the TPP method which executes all 100, 200, 300 and 400-node problems in 5.7, 11.3, 17.0 and 22.6 milliseconds, respectively. DS is the fastest of all methods. Its execution time is 2/3 of the time for TPP, except for problems C2–C4 for which DS encountered several exponent overflow interrupts. TNP is slightly faster than TPP since it does not use

Table 7.2
Performance of methods for solving cycle systems $Cx = b$ with fully dense b

Pr. ID.	k	ω	Cond. No. $\kappa(C)$	Maximum Relative Error, r				Normal'd Exec. Times		
				r_{DS}	r_{SUP}	r_{TNP}	r_{TPP}	t_{DS}	t_{SUP}	t_{TNP}
A1	100	-5.3×10^{-2}	1.2×10^{2}	6.9×10^{-7}	1.0×10^{-6}	5.1×10^{-7}	9.1×10^{-7}	0.68	33.1	0.94
A2	200	-7.7×10^{-2}	2.1×10^{3}	4.5×10^{-5}	4.9×10^{-4}	4.6×10^{-5}	1.7×10^{-5}	0.68	66.3	0.94
A3	300	4.9×10^{-3}	9.6×10^{2}	1.0×10^{-5}	9.5×10^{-5}	1.4×10^{-4}	8.9×10^{-5}	0.67	98.6	0.94
A4	400	1.3×10^{-3}	3.0×10^{2}	6.6×10^{-5}	2.2×10^{-4}	6.5×10^{-5}	6.4×10^{-5}	0.67	130.3	0.94
B1	100	-1.7×10^{3}	1.9×10^{2}	4.4×10^{-4}	2.7×10^{-6}	3.8×10^{-6}	1.0×10^{-7}	0.68	32.8	0.94
B2	200	-1.3×10^{8}	8.0×10^{1}	3.0×10^{1}	1.6×10^{-5}	1.5×10^{1}	4.3×10^{-7}	0.68	66.5	0.95
B3	300	1.5×10^{11}	5.2×10^{2}	1.0×10^{5}	1.2×10^{-5}	3.2×10^{1}	6.0×10^{-7}	0.68	98.7	0.94
B4	400	-4.9×10^{14}	3.1×10^{2}	1.4×10^{8}	2.5×10^{-5}	5.0×10^{7}	1.8×10^{-6}	0.68	131.2	0.95
C1	100	-2.5×10^{21}	7.0×10^{2}	1.2×10^{15}	1.8×10^{-5}	1.8×10^{14}	1.8×10^{-6}	0.67	32.9	0.94
C2	200	EOF	2.6×10^{4}	8.5×10^{36}	4.2×10^{10}	7.9×10^{35}	1.5×10^{-6}	0.87	58.4	1.00
C3	300	EOF	4.0×10^{4}	8.5×10^{36}	6.7×10^{29}	4.1×10^{36}	4.2×10^{-4}	1.20	99.8	1.29
C4	400	EOF	8.0×10^{3}	8.5×10^{36}	5.0×10^{37}	1.1×10^{35}	1.1×10^{-5}	1.33	162.1	1.37

the permutation indicator list. We have verified experimentally that our FTRAN and BTRAN algorithms consistently execute within a fixed time ratio of approximately 1.0 (for FTRAN) to 1.15 (for BTRAN).

For fully dense b, SUP is the least efficient of the four methods. Such high densities do arise in practice, e.g. in recomputing the flows (to reduce numerical errors), the duals (when new cycles form), and in basis partitioning applications. Clearly, the time for computing the duals (solving $uC = c$ via SUP) is comparable to t_{SUP} for solving $Cx = b$, given in Table 7.2. Existing GN codes use DS for this task (see e.g. [23, p. 112]) apparently in order to avoid this heavy computational burden. But, the inherent numerical inaccuracy of DS may well affect the accuracy of the dual solution and consequently that of the optimality conditions, particularly when sufficiently large cycles appear in the basis.

As mentioned in Section 3, SUP would solve $Cx = b$ slightly faster than DS whenever b has two nonzero elements ($k_0 = 2$). Such is the case of 'updating' the incoming column in the simplex method. This is also clear from the t_{SUP} column of Table 7.2. The time required to solve *two* (instead of k) systems via SUP is approximately $2t_{SUP}/k$, a 3% advantage over DS.

Even though DS and SUP are faster than TPP for the case of at most two nonzeroes in b, in practice, TPP may turn out to be more efficient than these two methods. We reason as follows. The execution times for TPP include two components: Time spent for creating the triangular decomposition via LUPP and time for applying FTRAN to solve the two resulting triangular systems. We have measured these two components to be approximately $\frac{2}{5}$ for LUPP and $\frac{3}{5}$ for FTRAN. Now, within the setting of a GN code, it is reasonable to expect that the cycle of a particular one-tree may remain unaltered over a pivot subsequence which involves only the restructuring of subtrees. Thus, the cost of creating the triangular decomposition by LUPP is 'amortized' over a sequence of subsequent FTRANs involving this cycle, until the next tear-down or restructuring of the cycle. With this observation, TPP would not only be more accurate but also more efficient than DS and SUP whenever such pivot subsequences have a length of three or more. Such is the case when all GN bases have one component. For multiple components, the LUPP cost for a newly-formed cycle is amortized over those subsequent pivots that 'access' its triangular decomposition. Computational results for GN Problems of 200 to 2000 vertices indicate that a cycle is accessed an average of 3 to 13 times [15].

It is relatively easy to integrate the LUPP, FTRAN and BTRAN algorithms of the previous section into an existing GN code. The subroutine LUPP is simply appended to the code since triangular decomposition is applied only to cycle nodes. If no cycles change, the updating procedures for GN bases remain the same and nothing is affected. Whenever a new cycle is formed or an old cycle is changed, LUPP is 'called' to compute the list functions L, U and P for the newly formed (or updated) cycle. Our FTRAN and BTRAN subroutines would usually replace the corresponding subroutines in an existing code. In some GN codes, particularly in those that use direct substitution, the FTRAN and BTRAN functions are embedded within other subroutines. In such instances, one would first have to identify the relevant sequence of statements in the code and then replace these with the corresponding steps of the algorithms given in Section 6. An implementation along these lines is described in [15].

Acknowledgements

The authors thank Professor M.L. Balinski for his stimulating communication on the direct substitution method, and Professor A. Gerasoulis for his valuable suggestions on some aspects of this work.

References

[1] R.H. Bartels and G.H. Golub, "The simplex method of linear programming using LU decomposition", *Communications of the ACM* 12 (1969) 266–268.

[2] R.H. Bartels, "A stabilization of the simplex method, *Numerische Mathematik* 16 (1971) 414–434.

[3] R.L. Burden, J.D. Faires and A.C. Reynolds, *Numerical analysis*, 2nd ed. (Prindle, Weber and Schmidt, Boston, MA, 1981).

[4] E.K. Blum, *Numerical analysis and computation: Theory and practice* (Addison-Wesley, Reading, MA, 1972).

[5] G. Brown and R.D. McBride, "Solving generalized networks", presented at ORSA/TIMS meeting, Detroit, April 1982.

[6] S.D. Conte and C. de Boor, *Elementary numerical analysis, An algorithmic approach*, 3rd ed. (McGraw-Hill, New York, 1980).

[7] A.R. Curtis and J.K. Reid, "On the automatic scaling of matrices for Gaussian elimination", *Journal of the Institute of Mathematics and its Applications* 1 (1972) 118–124.

[8] G.B. Dantzig, *Linear programming and extensions* (Princeton University Press, NJ, 1963).

[9] J. Edmonds, "Maximum matching and a polyhedron with (0, 1) vertices", *Journal of Research National Bureau of Standards* 69B, 1, 2 (1965) 125–130.

[10] J. Edmonds and E.L. Johnson, "Matching: A well-solved class of integer linear programs" in: R. Guy ed., *Combinatorial Structures and Their Applications* (Gordon and Breach, New York, 1970), pp. 89–92.

[11] J. Elam, F. Glover and D. Klingman, "A strongly convergent primal-simplex algorithm for generalized networks", *Mathematics of Operations Research* 4, 1 (1979) 39–59.

[12] G.E. Forsythe and C.B. Moler, *Computer solution of linear algebraic systems*, (Prentice-Hall, Englewood, NJ, 1967).

[13] F. Glover, D. Klingam and J. Stutz, "Extensions of the augmented predecessor index method to generalized network problems", *Transportation Science* 7 (1974) 377-384.

[14] M.D. Grigoriadis, "Minimum-cost network flows, Part I: An implementation of the network simplex method", LCSR-TR-37, Laboratory for Computer Science Research, Rutgers University (New Brunswick, NJ, 08903, 1982).

[15] M.D. Grigoriadis and T. Hsu, "A numerically robust adaptation of the simplex method for generalized network flows", presented at ORSA/TIMS meeting, San Francisco, 1984. Also, LCSR-TR-56, Laboratory for Computer Science Research, Rutgers University (New Brunswick, NJ 08903, 1984).

[16] R.C. Grinold, "Calculating maximal flows in a network with positive gains", *Operations Research* 21 (1973) 528-541.

[17] M. Held and R. Karp, "The traveling saleman problem and minimum spanning trees", *Operations Research* 18 (1970) 1138-1162.

[18] J.J. Jarvis and A.M. Jezior, "Maximal flow with gains through a special network", *Operations Research* 20 (1972) 678-688.

[19] P.A. Jensen and G. Bhaumik, "A flow augmentation approach to the network with gains minimum cost flow problem", *Management Science* 23(6) (1977) 631-643.

[20] P.A. Jensen and J.W. Barnes, *Network flow programming* (Wiley, New York, 1980).

[21] W.S. Jewell, "Optimal flow through network with gains", *Operations Research* 10 (1962) 476–499.

[22] E.L. Johnson, "Networks and basic solutions", *Operations Research* 14 (1966) 619–623.

[23] J.L. Kennington and R.V. Helgason, *Algorithms for network programming* (Wiley, New York, 1980).

[24] J.F. Maurras, "Optimization of the flow through networks with gains", *Mathematical Programming* 3 (1972) 135–144.

[25] R.D. McBride and A. Gupta, "Solving embedded generalized network problems", Working paper, University of Southern California (Los Angeles, October, 1982).

[26] G.W. Stewart, *Introduction to matrix computation* (Academic Press, New York, 1973).

[27] J.A. Tomlin, "On scaling linear programming problems", *Mathematical Programming Study* 4 (1975) 146–166.

[28] J.H. Wilkinson, "Error analysis of floating-point computation", *Numerische Mathematik* 2 (1960) 319–340.

[29] J.H. Wilkinson, *Rounding errors in algebraic processes* (Prentice-Hall, Englewood Cliffs, NJ, 1963).

[30] J.H. Wilkinson, *The algebraic eigenvalue problem* (Oxford University Press, London, 1965).

[31] R.D. Wollmer, "Multicommodity networks with resource constraints: The generalized multicommodity flow problem", *Networks* 1 (1972) 245–263.

Mathematical Programming Study 22 (1984) 167–184
North-Holland

MODELLING WITH INTEGER VARIABLES

R.G. JEROSLOW
College of Management, Georgia Institute of Technology, Atlanta, GA, USA

J.K. LOWE
U.S. Air Force, Maxwell Air Force Base, Montgomery, AL, USA

Received 18 December 1983
Revised manuscript received 10 April 1984

We give several results which characterize MIP-representability of sets and functions, and the finite union of MIP-representable sets. These results extend earlier ones due to R.R. Meyer. We provide representations whose linear relaxations have an efficient formulation, while retaining maximum accuracy to the problem modelled. Under mild hypotheses, these properties are retained as integer variables are set to values in nodes of branch-and-bound trees.

Key words: Integer Programming, MIMM, Modelling.

Introduction

The interest in modelling practical and mathematical problems as integer and mixed-integer programs begins with a paper by Dantzig (1963). Since that paper, many 'standard problem formulations' are routinely available in elementary textbooks. However, not until the more recent work of Meyer (1974, 1975, 1976, 1981) has there been thorough exploration, or, perhaps, even awareness, of the 'limits' of integer modelling. Our aim in this report is to elucidate the subleties of integer modelling. In the course of doing so, we develop modelling techniques which appear to be very effective.

For example, the fixed-charge problem can be modelled without a minimum usage constraint. On the other hand, the fixed 'benefit-problem' must contain a minimum usage stipulation (which must be met before the benefit is received) before it can be modelled as an integer problem. Our results explain why this is so.

In Section 1 of our report, we find that, for bounded sets, or functions defined on bounded domains (Corollary 1.2), the subtleties are neither very complex, nor restrictive. However, for a few unbounded sets the restrictions are so severe that, unless the unboundedness can be removed, one would naturally seek a different mathematical formulation of the sets. For example, many unbounded functions can be modelled as generalized linear complementarity problems, which have no integer modelling (the set $\{(x, y) \geq 0 | x \cdot y = 0\}$ is one such; see Theorem 1.7).

This author's research has been partially supported by grant ECS8001763 of the National Science Foundation, USA and completed while the author was an Alexander-von-Humboldt Senior Scientist at Institute for Ökonometrie and Operations Research of the University of Bonn.

It is important to know that a bound must be used. This kind of information occurred in Meyer's earliest studies of integer modelling (1976) in connection with the fixed-charge problem (which is representable only in its bounded form, when rational data are stipulated). This choice of a suitable bound can be a major issue in algorithmic implementation.

In Section 1, we develop basic definitions and many corollaries which lead to the heart of our investigations into set representability. Theorem 1.7 describes necessary and sufficient conditions for the union of representable sets to a MIP-representable. In Section 2, we explore Meyer's concept of bounded representability, and develop necessary and sufficient conditions for bounded MIP-representability (Theorem 2.1). For bounded representable sets, our results guarantee one representation in which all integer variables are binary variables; furthermore, these binary variables occur in a common set-partitioning constraint. The practical applications of our work involve primarily bounded-representability.

We then focus upon the implementation of MIP-representations, and ask how 'accurate' the representations are after the integer variables are 'relaxed' to continuous (rational) variables, as in branch-and-bound algorithms (Section 3). We easily show that the best possible relaxed MIP-representation is exactly the convex hull of the original set. When the specific MIP-representation is exactly the convex hull, it is called a 'sharp' representation. We provide sharp representations for many models. Many representations in common use are sharp, but some important ones are not.

Our problem formulations are either derived from or motivated by disjunctive programming as in Balas (1974) and Jeroslow (1977) and a problem studied by Beale (1979, Section 4).

While Meyer's theory permits representations which involve irrational data, we have restricted ourselves to rational representation only. Some of our results do depend upon the field used, and we make a few comments on this issue below, as seems appropriate. In other respects, the reader should understand that all matrices and vectors are rational below.

Throughout this paper, we denote the convex span respectively the closure of set S as conv(S) respectively cl(S), and clconv(S) is the closure of conv(S) (see Rockafellar (1970) or Stoer and Witzgall (1970)). Similarly cone(S) denotes the cone generated by S. The convex span and cone operations are taken only with rational multipliers. Below, Q denotes the rationals and Z denotes the integers.

We assume that the reader is familiar with the intended use of MIP-representability in problem formulation, as illustrated in Meyer's papers.

1. Representability of the finite union of representable sets

We define a set $S \subseteq Q^n$ as MIP-representable if there are rational matrices A_1, A_2, A_3, and vector b with the property that: $x \in S$, if and only if, for some u, $v \geq 0$, with u integer, we have

$$A_1x + A_2u + A_3v = b. \tag{1.1}$$

We note that our definition extends to the case where x is constrained to be integer. For example, let $x = (x_1, \ldots, x_n)$ and let I be any subset of $\{1, \ldots, n\}$; then if S is MIP-representable so is the set $\{x \in S | x_i \in Z$ (integers) for all $i \in I\}$ (this is an easy exercise). Our definition of MIP-representation can be extended to functions in different ways, via the MIP-representability of the graph $\text{gr}(f) = \{(z, x) | z = f(x)\}$ of f, or of the epigraph $\text{epi}(f) = \{(z, x) | z \geq f(x)\}$ of f, etc. Meyer's definition in Meyer (1975) of a rational MIMM for a function f corresponds to MIP-representability of epi(f), and is the appropriate notion if f appears only in a minimizing objective function. When f appears in the constraints of a program, generally one needs MIP-representability of gr(f). We do not care to go in detail into these matters here, but only note that care must be taken to specify the exact representability used, since e.g. possibly epi(f) is representable but not gr(f) (see Jeroslow (1980)). In what follows, we do not explicitly discuss function representability.

Our first theorem is an extension of the finite basis theorem for polyhedra (Rockafellar (1970)).

Theorem 1.1. *Suppose that $S \neq \emptyset$ is MIP-representable. There are constants a, $b \geq 0$ such that for all integers $p \geq 1$, $S_{a+bp} \neq \emptyset$ and*

$$S = S_{a+bp} + pM, \tag{1.2}$$

where

$$S_\alpha = \left\{ x \in S \,\middle|\, \sum_{i=1}^{n} |x_i| \leq \alpha \right\}, \tag{1.3}$$

and

$$M = \{x \in Z^n | \textit{for some } u, v \geq 0, \textit{ with } u, v \textit{ integer}, A_1x + A_2u + A_3v = 0\}. \tag{1.4}$$

Proof. The containment $S \supseteq S_{a+bp} + pM$ is immediate once we show that $S_{a+bp} \neq \emptyset$. In fact, if $x \in S$ and $x' \in M$ then by a direct computation, $x + px' \in S$.

We note that the nonempty set P, defined as

$$P = \{(x, u, v) | A_1x + A_2u + A_3v = b;\ u, v \geq 0\} \tag{1.5}$$

is a polyhedron. Therefore, from the finite basis theorem, there are disjoint finite sets J and K, points (x^t, u^t, v^t) of P for $t \in J$, and recession directions of P for $t \in K$, with

$$P = \text{conv}(\{(x^j, u^j, v^j) | j \in J\}) + \text{cone}(\{(x^k, u^k, v^k) | k \in K\}). \tag{1.6}$$

We define $a' = \max_{j \in J} \|x^j\|$, and $b = \sum_{k \in K} \|x^k\|$, and $a = \max\{a', \|x^0\|\}$, where x^0 is a point in S. (Note, $\|x\| := |x_1| + \cdots + |x_n|$). Since $b \geq 0$ for $p \geq 1$ we have $x^0 \in S_{a+pb}$. Thus $S_{a+bp} \neq \emptyset$ and we now must establish the containment $S \subseteq S_{a+bp} + pM$.

We may assume in what follows, that (x^k, u^k, v^k) is an integer vector for $k \in K$ (recall that A_1, A_2, A_3, b are rational). In particular, $x^k \in M$ for $k \in K$.

Let $x \in S$. Then for some $u, v \geq 0$ with u integer, we have $(x, u, v) \in P$. Therefore there is a solution to

$$(x, u, v) = \sum_{r \in J} \lambda_j (x^j, u^j, v^j) + \sum_{k \in K} \tau_k p(x^k, u^k, v^k),$$
$$1 = \sum_{j \in J} \lambda_j; \quad \lambda_j \geq 0, \ j \in J, \ \tau_k \geq 0, \ k \in K. \tag{1.7}$$

Let q_k denote the integer part of τ_k and let f_k be the fractional part, so that $\tau_k = q_k + f_k$, $f_k < 1$, and $q_k, f_k \geq 0$. Define

$$(x', u', v') = \sum_{j \in J} \lambda_j (x^j, u^j, v^j) + \sum_{k \in K} f_k p(x^k, u^k, v^k), \tag{1.8a}$$

$$(x^*, u^*, v^*) = \sum_{k \in K} q_k p(x^k, u^k, v^k). \tag{1.8b}$$

From (1.8a), so we have $x = x' + x^*$. We must show that $x' \in S_{a+bp}$ and $x^* \in pM$.

Since $(x, u, v) = (x', u', v') + (x^*, u^*, v^*)$ and u and u^* are integer vectors, u' is an integer vector. From (1.8a), $(x', u', v') \in P$. Hence $x' \in S$. Also from (1.5), and (1.8a), $\|x'\| \leq a' + p \sum_{k \in K} f_k \|x^k\| \leq a' + pb$. Thus $x' \in S_{a+bp}$. Since $x^k \in M$ and q_k is integer for $k \in K$, we have $x^* \in pM$ from (1.8b). □

We note that Theorem 1.1 defines any representable set S in terms of a bounded part (S_{a+bp}) and a discrete unbounded part (pM) as in the finite basis theorem for polyhedra. For $\alpha \in Q$, S_α is MIP-representable; hence S_{a+bp} is representable. In addition, M is an integral monoid with a finite basis (see Jeroslow (1978)).

Theorem 1.1 is false, if irrational data A_1, A_2, A_3, b are permitted. For example, $S = \{(x_1, x_2) \geq 0 \mid x_2 = x_1\sqrt{2}\}$ is representable with A_1 irrational, and $M = \{(0, 0)\}$. Then (1.2) is false for any p, since S_{a+bp} is bounded and S is unbounded. Our proof above fails, since the rationality of A_1, A_2, A_3 and b is needed to insure that recession directions (x^k, u^k, v^k) are integer. As a general heuristic rule – not intended to be a precise principle – we do need the rational field for our results on unbounded representable sets, but not for bounded ones. Indeed, the proof of the next result does not require integer recession directions, and reveals that the result actually does not use our assumption of rational data.

Corollary 1.2. *If S is both bounded and MIP-representable, S is a finite union of polytopes.*

Proof. It suffices to show that a bound can be placed on the vector u occurring in the representation.

Assume that $x \in S$. From the proof of Theorem 1.1, and since S has no recession directions, we have $x^k = 0$ for $k \in K$. Hence in (1.8a), $x' = x$ and $A_1 x + A_2 u' + A_3 v' = b$

with $u', v' \geq 0$ and u' integer (recall $x' \in S$). A bound on $\|u'\|$ is $\max_{j\in J} \|u^j\| + \sum_{k\in K} \|u^k\|$ directly from (1.8). $\square$

Corollary 1.3. *If S is representable,* conv(S) *is a polyhedron.*

Proof. Apply (1.2) for $p=1$. Since M is a monoid we have

$$\operatorname{conv}(S_{a+b}+M) = \operatorname{conv}(S_{a+b}) + \operatorname{cone}(M). \tag{1.9}$$

To establish (1.9), note that clearly

$$\operatorname{conv}(S_{a+b}+M) \subseteq \operatorname{conv}(S_{a+b}) + \operatorname{cone}(M), \tag{1.10}$$

is valid. To prove that

$$\operatorname{conv}(S_{a+b}+M) \supseteq \operatorname{conv}(S_{a+b}) + \operatorname{cone}(M), \tag{1.11}$$

we have for any $v \in \operatorname{conv}(S_{a+b}) + \operatorname{cone}(M)$,

$$v = \sum_j \lambda_j s^j + \sum_k u_k m_k, \tag{1.12}$$

where $\sum_j \lambda_j = 1$, $\lambda_j \geq 0$, $u_k \geq 0$, $m_k \in M$, and $s^j \in S_{a+b}$.

Let $\lambda_1 > 0$ (WLOG), then we have

$$m := \sum_k u_k m_k / \lambda_1 \in \operatorname{cone}(M), \tag{1.13}$$

and also

$$v := \lambda_1(s^1 + m) + \sum_{j\neq 1} \lambda_j(s^j + 0). \tag{1.14}$$

But since $(s^1+m) \in S_{a+b} + \operatorname{cone}(M)$, then $v \in \operatorname{conv}(S_{a+b} + \operatorname{cone}(M))$. Furthermore, if $w = s + w'$, where $w' \in \operatorname{cone}(M)$ and $s \in S_{a+b}$, then for some integer $D \geq 1$, $Dw' \in M$, and by factoring,

$$s + w' = 1/D(s + Dw') + 1/D(s+0) + \cdots + 1/D(s+0), \tag{1.15}$$

where $(s+Dw') \in S_{a+b} + M$ and $(s+0) \in S_{a+b} + M$. Therefore $(s+w') \in \operatorname{conv}(S_{a+b}+M)$, and

$$S_{a+b} + \operatorname{cone}(M) \subseteq \operatorname{conv}(S_{a+b}+M). \tag{1.16}$$

Finally,

$$\operatorname{conv}(S_{a+b}) + \operatorname{cone}(M) \subseteq \operatorname{conv}(S_{a+b} + \operatorname{cone}(M)) \subseteq \operatorname{conv}(S_{a+b}+M), \tag{1.17}$$

By Corollary 1.2, $\operatorname{conv}(S_{a+b})$ is a polytope. Since M has a finite basis, cone(M) is a polyhedral cone. By the converse to the Finite Basis Theorem (Rockafellar (1970)), conv(S) is a polyhedron. $\square$

Corollary 1.4. *S is an MIP-representable set if and only if there is a bounded MIP-representable set S' and an integral monoid M with a finite basis such that $S = S' + M$.*

Moreover, if S' is an MIP-representable set (bounded or not) and M is an integral monoid with finite basis, then $S = S' + M$ is MIP-representable.

Proof. The 'only if' is established by Theorem 1.1. If $S = S' + M$, let A_1, A_2, A_3, and b be rational vectors such that

$$x \in S' \Leftrightarrow \text{for some } u,\, v \geqslant 0,\ u \text{ integer},\ A_1x + A_2u + A_3v = b, \tag{1.18}$$

and let A_1 be an integral matrix such that

$$m \in M \Leftrightarrow \text{for some } w \geqslant 0 \text{ integer},\ m = A_4x. \tag{1.19}$$

Then we have,

$$x \in S \Leftrightarrow \text{there are } x_1, x_2;\ u,\ v,\ w \geqslant 0 \text{ with } u,\ w \text{ integer such that}$$

$$x = x_1 + x_2,\ A_1x_1 + A_2u + A_3v = b \text{ and } A_1w = x_2, \tag{1.20}$$

which is an MIP-representation of S. □

The next result is a technical corollary that produces a necessary condition for MIP-representability using recession directions.

Corollary 1.5. *Suppose S is MIP-representable and $x^* \in Q^n$ is a vector for which there exists some $x^0 \in S$ and a sequence $\tau_k \to +\infty$ with*

$$x^0 + \tau_k x^* \in S \quad \text{for all } k = 1, 2, \ldots. \tag{1.21}$$

There is an $\sigma > 0$, $\sigma \in Q$, with the property that for all $x \in S$ and all integers $p \geqslant 0$,

$$x + p\sigma x^* \in S \quad \text{and} \quad \sigma x^* \in Z^n \text{ (integers)}. \tag{1.22}$$

Remark. The conditions (1.21) are sometimes abbreviated by saying that x^* is *discrete recession direction* of S at x^0.

Proof. By Theorem 1.1 for each $k = 1, 2, \ldots,$ there exists $x^k \in S_{a+b}$ and $m^k \in$ cone(M) with $x^0 + \tau_k x^* = x^k + \tau_k m^k$, where $\tau_k m^k \in M$ (so that $m^k \in$ cone$((M))$. Then note

$$\|x^* - m^k\| \leqslant \|x^k - x^0\| / \tau_k \leqslant B / \tau_k, \tag{1.23}$$

where B is a bound independent of k.

Since $\tau_k \to +\infty$ we have $x^* \in$ clcone$(M) =$ cone(M), as cone(M) is finitely generated. Hence for some $\sigma > 0$, we have $\sigma x^* \in M$, which implies $\sigma x^* \in Z^n$. Finally, from Theorem 1.1, $x + p\sigma x^* \in S$ whenever $x \in S$ and $p \geqslant 0$ is integer. □

Example 1.1. Let $S \subseteq Q$ be defined by $S = \{x \in Z \mid x = 0 \text{ or } x \geqslant 2\}$. An MIP-representation of S is

$$x \in S \Leftrightarrow \text{there exist } u_1, u_2 \geqslant 0,\ u_1, u_2 \text{ integer such that } x - 2u_1 - 3u_2 = 0. \tag{1.24}$$

With $x^*=1$, $x^0=2$, the hypotheses of Corollary 1.5 are met by letting $\tau_k=k$. The conclusion holds if $\sigma=2$, but not if $\sigma=1$, i.e. $\sigma=1$ does not satisfy $x+\sigma x^* p\in S$ for $p=1$ and $x=0$.

Suppose we are given representable sets $S_1,\ldots,S_t$. Trivially, their intersection is representable, but what conditions must hold for their union $(S_1\cup S_2\cup\cdots\cup S_t)$ to be representable? The next fixed-charge example illustrates some complexities of this question.

Example 1.2. Let f be the fixed-charge function

$$f(x)=\begin{cases}1 & \text{if } x>0,\\ 0 & \text{if } x=0.\end{cases} \tag{1.25}$$

Note that $\text{epi}(f)=S_1\cup S_2$ are polyhedra (thus representable) defined by

$$S_1=\{(z,x)\mid z\geqslant 0,\ x=0\},\qquad S_2=\{(z,x)\mid z\geqslant 1,\ x\geqslant 0\}. \tag{1.26}$$

However, $S_1\cup S_2$ is *not* representable. In fact, with $(z^*,x^*)=(0,1)$ and $(z^0,x^0)=(1,0)$ the hypotheses (1.21) of Corollary 1.5 hold for $S_1\cup S_2$. But the conclusion (1.22) fails since $(0,0)\in\text{epi}(f)$ and $(0,0)+\gamma(0,1)=(0,\gamma)\notin\text{epi}(f)$ for any $\gamma>0$. The precise relative placement of S_1 with respect to S_2 (and not merely recession directions) also makes a difference. E.g., if $S_1'=\{(z,-1)\mid z\geqslant 1\}$ then $S_1'\cup S_2$ is MIP-representable (exercise).

The simple construction of the linear system (1.27) in our next result will play an important role in our results to follow. This construction derives from disjunctive methods as in Balas (1974), Jeroslow (1974 and 1977).

Lemma 1.6. *If $S_1,\ldots,S_t$ are bounded MIP-representable sets, then their union is MIP-representable.*

Proof. By Corollary 1.2, it suffices to show that any union of polytopes is MIP-representable, since each S_i is a union of polytopes. Let $P_i=\{x\mid A^i x\geqslant b^i\}$ be non-empty polytopes for $i=1,\ldots,s$. Note that the boundedness property implies that $A^i x\geqslant 0\Leftrightarrow x=0$. A representation of $P_1\cup\cdots\cup P_s$ is given by

$$\begin{gathered}x=\sum_{i=1}^{s} x^i;\quad A^i x^i-b^i\lambda_i\geqslant 0,\quad \lambda_i\geqslant 0 \text{ for } i=1,\ldots,s,\\ \sum_{i=1}^{s}\lambda_i=1 \text{ and } \lambda_i \text{ integer for } i=1,\ldots,s.\end{gathered} \tag{1.27}$$

□

Before presenting the main theorem of this section, we must formally define our use of recession directions. Our definition is restricted to discrete recession vectors and from this point on, any mention of recession directions implies discrete vectors. (Note this restriction has no effect upon Corollary 1.5.)

An integer vector x^* is a discrete global recession direction of a set S if, for all $x \in S$ and integers $p \geq 0$, we have $x + px^* \in S$. The set of all discrete global recession vectors is denoted $\mathrm{rec}(S)$.

Note that $\mathrm{rec}(S)$ is closed under addition, and integer multiples, and $0 \in \mathrm{rec}(S)$; thus $\mathrm{rec}(S)$ is an integer monoid.

Lemma 1.6 is still valid if each polyhedron P_i gives the very same continuous recession directions $\{x \mid A^i x \geq 0\}$, independent of i. In this case, the $P_i \neq \emptyset$ need not be bounded. The same proof of Lemma 1.6 justifies this claim. In particular, if each function f_i is polyhedral on a nonempty domain ($\mathrm{dom}(f_i) \neq \emptyset$) and the recession directions of all sets $\mathrm{dom}(f_i)$ are the same for $i \in I(i)$, then the function

$$f(x) = \min\{f_i(x) \mid x \in \mathrm{dom}(f_i)\} \tag{1.28}$$

is MIP-representable.

We next state our main result for the finite union of representable sets.

Theorem 1.7. *Suppose that $S_1, \ldots, S_t$ are MIP-representable nonempty sets. Then $S = S_1 \cup \cdots \cup S_t$ is MIP-representable if and only if every discrete recession direction of every S_i has a positive multiple which is a discrete recession direction of S.*

Proof. A recession direction x^* of $S_i \neq \emptyset$ satisfies the hypothesis of Corollary 1.5. The conclusion of Corollary 1.5 must hold if S is MIP-representable, hence our conclusion is necessary.

Next suppose our condition holds; we must show that the corresponding S is MIP-representable. Let the sets $S^i_{a_i+b_ip}$ and M^i and constants $a_i, b_i \geq 0$ be defined for S_i as in Theorem 1.1. From (1.4) each integer monoid M_i has a set of generators $a_{ij} \in \mathbb{R}^n$ for $j \in I(i)$, where $I(i)$ is an index set. Each a_{ij} is a discrete recession direction for S_i, and, from our condition, $\sigma_{ij} a_{ij} \in \mathrm{rec}(S)$ for some $\sigma_{ij} > 0$. Since $\sigma_{ij} \in Q$, we may take $\sigma_{ij} \in Z$ by clearing denominators. Then $\sigma_i = \Pi_{j \in I(i)} \sigma_{ij}$ is an integer and $\sigma_i M_i \subseteq \mathrm{rec}(S)$. Furthermore, from Theorem 1.1 we have

$$S_i = S^i_{a_i+b_i\sigma_i} + \sigma_i M_i, \quad i = 1, \ldots, t. \tag{1.29}$$

In fact, we have

$$S = \left(\bigcup_{i=1}^{t} S^i_{a_i+b_i\sigma_i} \right) + \left(\sum_{i=1}^{t} \sigma_i M_i \right). \tag{1.30}$$

The containment ($\subseteq$) in (1.30) is trivial, since if $x \in S$ we have $x \in S_i$ for some $i = 1, \ldots, t$, and (1.29) holds. As to the containment ($\supseteq$) in (1.30), let $x = x^0 + x^*$, with $x^0 \in \bigcup_{i=1} S^i_{a_i+b_i\sigma_i}$ and $x^* \in \sum_{i=1} \sigma_i M_i$. Since each $S^i_{a_i+b_i\sigma_i} \subseteq S_i$, we have $x^0 \in S$. Since $\sigma_i M_i \subseteq \mathrm{rec}(S)$, we have $x^* \in \mathrm{rec}(S)$. Thus $x \in S$.

Each set $S^i_{a_i+b_i\sigma_i}$ is a bounded set, and MIP-representable; thus by Lemma 1.6, the union is representable. Since each M_i is finitely generated, so is $\sum_{i=1}^{t} \sigma_i M_i$. By applying the sufficient condition of Corollary 1.4 to (1.30), S is MIP-representable. □

From Theorem 1.7, we may determine representability of any finite collection of representable sets using discrete recession directions. Moreover, from the proof of Theorem 1.7, it is not necessary that every discrete recession direction of S_i have a positive multiple that is in rec(S); we only need $\sigma_i M_i \subseteq \text{rec}(S)$ for a suitable integer $\sigma_i \geq 1$, where M_i is the integer monoid of (1.4). One easily proves that $M_i \subseteq \text{rec}(S_i)$, but the converse is false, as our next example shows. This example also illustrates the fact that M_i depends not only on S_i, but also on a specific representation for S_i.

Theorem 1.7 is false if real data A_1, A_2, A_3, b are used. In the case of real data, Meyer has shown in Meyer (1975) that the arbitrary finite union of representable sets is representable. Both polyhedra $P = \{(x_1, x_2) \geq 0 \mid x_1 = 0\}$ and $Q = \{(x_1, x_2) \geq 0 \mid x_2 = 0\}$ are representable with rationals. By Meyer's result, $P \cup Q$ is representable using reals. However, by Theorem 1.7, $P \cup Q$ is not representable using rationals, since the discrete recession direction (0, 1) of P has no multiple which is a recession direction for S at the point $(1, 0) \in Q \subseteq S$.

The next example illustrates Theorem 1.7.

Example 1.3. A representation of the set $S = \{x \mid x \geq 0\}$ is

$$x = x_1 + 2x_2, \quad 0 \leq x_1 \leq 2, \; x_2 \geq 0, \; x_2 \in Z. \tag{1.31}$$

Using (1.4) to determine M, we find that

$$M = \{x = 2x_2 \mid x_2 \geq 0, x_2 \in Z\}, \tag{1.32}$$

as $x_1 = 0$ is forced in M. We know that a recession direction of S is $x^* = 1$, but in this representation $x^* = 1 \notin M$. (However, whenever our MIP-representation results in a $x^* \in \text{rec}(S)$, but $x^* \notin M$, we will always have some $\sigma > 0$, with $\sigma x^* \in M$). If '2' is replaced by '1' as a coefficient in (1.31), the resulting M is indeed all of the discrete global recession directions.

An alternate form of Theorem 1.7 requires the cone of the recession directions of the set S to be equal to the union of the cones of recession directions of each S_i.

Corollary 1.8. *Let $S_1, \ldots, S_t$ be MIP-representable and put $S = S_1 \cup \cdots \cup S_t$. Then S is MIP-representable if and only if*

$$\text{cone}(\text{rec}(S)) = \bigcup_{i=1}^{t} \text{cone}(\text{rec}(S_i)) = \sum_{i=1}^{t} \text{cone}(\text{rec}(S_i)). \tag{1.33}$$

In particular, if S is MIP-representable, the right-hand side of (1.33) *is a cone.*

Proof. By Theorem 1.7, S is *MIP*-representable if and only if

$$\text{rec}(S_i) \subseteq \text{cone}\{\text{rec}(S)\} \quad \text{for } i = 1, \ldots, t. \tag{1.34}$$

(Recall that the cone operation is only with respect to rational multipliers).

Suppose S is representable and $x^* \in \text{rec}(S)$. Fix any $x^0 \in S$. Then there is some S_i such that $x^0 + qx^* \in S_i$ for infinitely many $q \geq 0$. By Corollary 1.5, we also have

$\sigma x^* \in \text{rec}(S_i)$ for at least one $\sigma > 0$ and $i \in \{1, \ldots, t\}$. Thus,

$$\text{rec}(S) \subseteq \bigcup_{i=1}^{t} \text{cone}\{\text{rec}(S_i)\}. \tag{1.35}$$

Upon taking the cone generated by each set S_i in (1.34) and applying (1.35), we find

$$\text{rec}(S) \subseteq \bigcup_{i=1}^{t} \text{cone}\{\text{rec}(S_i)\} \subseteq \text{cone}\{\text{rec}(S)\}. \tag{1.36}$$

We now show that $C = \bigcup_{i=1}^{t} \text{cone}\{\text{rec}(S_i)\}$ is a cone. This will also demonstrate that $C = \sum_{i=1}^{t} \text{cone}\{\text{rec}(S_i)\}$.

In fact, if $x, y \in C$ then for certain j, k we have $x \in \text{cone}\{\text{rec}(S_j)\}$ and $y \in \text{cone}\{\text{rec}(S_k)\}$. By (1.36), $x, y \in \text{cone}\{\text{rec}(S)\}$ so $x + y \in \text{cone}\{\text{rec}(S)\}$; i.e., for some $\sigma > 0$, $\sigma \in Q$, we have $\sigma(x+y) \in \text{rec}(S)$. By the first containment in (1.36), for some p we have $\sigma(x+y) \in \text{cone}\{\text{rec}(S_p)\}$ and hence $x + y \in \text{cone}\{\text{rec}(S_p)\}$. As C is closed under addition (and clearly closed under positive multiples), C is a cone.

By applying the cone operation to rec(S) in (1.36) we obtain (1.33) when S is representable.

Suppose (1.33) holds. It implies $\text{cone}\{\text{rec}(S_i)\} \subseteq \text{cone}\{\text{rec}(S)\}$ for $i = i = 1, \ldots, t$, in which case (1.34) follows, forcing S to be representable. □

Note that, as we saw in the fixed-charge example (Example 1.2), it is not sufficient for the r.h.s. of (1.33) to be a cone. But if the $\text{cone}\{\text{rec}(S_i)\}$ is independent of $i = 1, \ldots, t$, the result holds.

Corollary 1.9. *If* $\text{cone}\{\text{rec}(S_i)\}$ *is independent of* $i = 1, \ldots, t$, *and each* S_i *is representable, then* $S = S_1 \cup \cdots \cup S_t$ *is representable.*

Proof. Fix $i = 1, \ldots, t$. If $x^* \in \text{cone}\{\text{rec}(S_t)\}$, then for all $j = 1, \ldots, t$, we have $x^* \in \text{cone}(\text{rec}(S_j))$. Hence, for any j there is some integer $\sigma_j > 0$ for which we have $\sigma_j x^* \in \text{rec}(S_j)$.

Define $\sigma = \prod_{i=1} \sigma_i$. Then $\sigma x^* \in \text{rec}(S_j)$ for all $j = 1, \ldots, t$. Thus $\sigma x^* \in \text{rec}(S)$, and the necessary condition of Theorem 7 is verified. □

We note, from example (1.2), that the sufficiency condition of Corollary 1.9 is not necessary.

Proposition 1.10. *If S is MIP-representable, then S is closed.*

Proof. Suppose $x^n \in S$ for $n = 1, 2, \ldots$ and $x^n \to x^0$. With $\alpha = \|x^0\| + 1$, S_α is a bounded representable set. By Corollary 1.2, S_α is a finite union of polytopes, and thus closed. Since $x^n \in S_\alpha$ for large n, $x^0 \in S_\alpha \subseteq S$. □

Example 1.4. The 'fixed-benefit function' f given by

$$f(x)=\begin{cases} 0, & x=0,\\ -1, & x>0,\, x\leqslant M,\end{cases}$$

for each M finite or infinite, is not MIP-representable, since epi(f) is not closed and Proposition 1.10 applies. (Note that $(-1, 1/n)\in \text{epi}(f)$ for each n but $(-1,0)\notin \text{epi}(f)$. (N.B. In the 'minimizing format' we have assumed, cost is minimized, so that a benefit or profit shows up as negative cost. If one switches to a maximizing format, the difficulties will remain).

By the kind of reasoning as in the example, one easily establishes the known result that a representable function is lower semicontinuous.

Example 1.5. The 'fixed-benefit function with minimum usage level' δ is given by

$$f(x)=\begin{cases} 0, & x=0,\\ -1, & \delta\leqslant x\leqslant M.\end{cases}$$

For $M\geqslant\delta>0$, and M finite, it is MIP-representable. In fact, one easily shows that epi(f) is the union of two polyhedra which have $(1,0)$ as their sole recession direction, and our earlier remarks apply. In fact, by Theorem 2.1 and remarks to follow, f is representable using only binary integer variables in a set-partitioning constraint. (Actually, we can have M infinite and f will be representable; this is a somewhat harder exercise.)

We conclude this section with a necessary and sufficient condition for a set S to be MIP-representable.

Theorem 1.11. *A set S is MIP-representable if and only if S has the form*

$$S=\left(\bigcup_{i=1}^{t} P_t\right)+M \tag{1.37}$$

for a finite set of polytopes $P_1,\ldots,P_t$ and an integer monoid M with finite basis.

Proof. Necessity follows from Corollary 1.4 and the Corollary 1.2. Sufficiency follows by Corollary 1.4, since $S'=\bigcup_{i=1}^{t} P_i$ is MIP-representable ($\text{rec}(P_i)=\{0\}$ as each P_i is a polytope, and e.g. Corollary 1.9 can be used). □

2. Bounded-integer representability

In this section we explore Meyer's concept of bounded representability. We define S as *bounded* MIP-representable if there are rational matrices A_1, A_2, A_3 and a vector b, plus a vector bound u^0 such that

$$x \in S \Leftrightarrow \text{there are } u,\ v \geq 0,\ u \text{ integer and } u \leq u^0,$$
$$\text{and } A_1x + A_2u + A_3v = b. \tag{2.1}$$

Note that S need not necessarily be bounded as a set, in order to be bounded MIP-representable; e.g. we can have $S = P_1 \cup P_2$ where P_1 and P_2 are unbounded polyhedra with $\text{rec}(P_1) = \text{rec}(P_2)$.

The astute reader will note that we do not need our hypothesis of rational data in this section.

Theorem 2.1. *S is bounded MIP-representable if and only if:*

(i) *S is a finite union of polyhedra; and also*

(ii) *The following condition is satisfied for every $x^* \in Q^n$: If there exists $x^0 \in S$ and $\tau_k \to +\infty$ with $x^0 + \tau_k x^* \in S$ for $k = 1, 2, \ldots$, then for all $\tau \geq 0$ ($\tau \in Q$) and all $x \in S$, we have $x + \tau x^* \in S$.*

Remark. The condition in (ii) is, in words, that every local discrete recession direction of S is also a *continuous* global recession direction of S.

Proof. First suppose S is bounded MIP-representable. Clearly, S is a union of polyhedra.

When $x^* \neq 0$ is such that there exists an $x^0 \in S$ and $\tau_k \to +\infty$ with $x^0 + \tau_k x^* \in S$ for all $k = 1, 2, \ldots$, then there exist u_k, $v_k \geq 0$ with u_k integer such that

$$A_1(x^0 + \tau_k x^*) + A_2u_k + A_3v_k = b, \text{ and } u_k \leq u^0 \quad \text{for all } k. \tag{2.2}$$

Dividing both sides of (2.2) by τ_k we have $A_1x^* + A_3(v_k/\tau_k) = (b - A_2u_k - A_1x^0)/\tau_k$. Since the r.h.s. of this expression goes to 0 as $k \to \infty$, there is a bound $B > 0$ with $\|A_3(v_k/\tau_k)\| \leq B$. By compactness, a subsequence of $A_3(v_k/\tau_k)$ approaches a vector w^* in the closure of the cone $C = \{w \mid w = A_3v,\ v \geq 0\}$. Since C is a polyhedral, $w^* \in C$ and hence $w^* = Av^*$ for some $v^* \geq 0$. It then easily follows that

$$A_1x^* + A_3v^* = 0. \tag{2.3}$$

By rationality of A_1, A_3, x^*, we may assume v^* is rational. By a direct computation, $x + \tau x^* \in S$ whenever $x \in S$ and $\tau \geq 0$. (If data is not rational, we do not need to be concerned about the rationality of x^* or v^*.)

For the converse, suppose that S is a finite union of nonempty polyhedra $S = P_1' \cup \cdots \cup P_s'$ and the condition (ii) holds. If we utilize the notation $P_i' = \{x \mid A^i x \geq b_i\}$, then (1.27) provides a bounded representation of S.

In fact, let the set represented by (1.27) be denoted S'. For $i = 1, \ldots, s$ by putting $\lambda_i = 1$ and $\lambda_j = 0$ for $j \neq i$, $x^j = 0$ for $j \neq i$, we see that $P_i \subseteq S'$. Thus $S \subseteq S'$. Next, let $x \in S'$, so that (1.27) holds for certain x^i and λ_i. For some $k = 1, \ldots, s$ we have $\lambda_k = 1$ and $\lambda_i = 0$ for $i \neq k$. Without loss of generality, $k = 1$. If $i \neq 1$ and $A^i x^i \geq 0$, then $x^0 + \tau x^i \in P_i'$ for all $\tau \geq 0$ whenever $x^0 \in P_i'$. By the hypothesis (ii), $x^- + \tau x^i \in S$

whenever $x^- \in S$ and $\tau \geq 0$. We have $x^1 \in S$; hence $x^1 + x^2 \in S$; hence $x^1 + x^2 + x^3 \in S$; etc. In this manner, we establish that $x = \sum x^i \in S$. Hence $S' \subseteq S$. We conclude that $S = S'$, as desired. $\square$

Theorem 2.1 excludes nontrivial integer monoids, as well as many other unbounded representable sets, from being bounded MIP-representable.

The proof of the 'sufficiency' part of Theorem 2.1 reveals that, when S is bounded MIP-representable, it has at least one representation (1.27) in which all the integer variable occurring are binary variables that appear in the same set-partitioning constraint (i.e. $\sum_i \lambda_i = 1$).

Condition (ii) is not the same as requiring that all polyhedra occurring in the union $S = P_1 \cup \cdots \cup P_t$ have the same recession directions. For example, with $t = 2$, $P_1 = \{(x_1, x_2) \geq 0 \mid 0 \leq x_1, x_2 \leq 1\}$, $P_2 = \{(x_1, x_2) \geq 0 \mid 1 \leq x_1, 0 \leq x_2 \leq 2\}$, we see that condition (ii) holds even though P_1 has no recession directions while P_2 has one.

However, when S is bounded MIP-representable we may assume 'without loss of generality' that all P_i, $1 \leq i \leq t$, have the same recession directions. To see this, let C_i be the polyhedral cone of recession directions of P_1, where $S = P_1 \cup \cdots \cup P_t$, and let $C = C_1 + \cdots + C_t$. Then C is a polyhedral cone, so each set $P_i' = P_i + C$ is a polyhedron, and each P_i' has C as its cone of recession directions. Then it is easy to see, using (ii), that $S = P_1' \cup \cdots \cup P_t'$, since $S \supseteq P_i'$ for each $i = 1, \ldots, t$.

3. Sharpness of representations

Another property of MIP-representation involves the 'relaxation' of the nonnegative integral variables u of (1.1) to nonnegative rational values.

Our interest in this relaxation lies in the fact that, in branch-and-bound algorithms, this relaxation is utilized in the subproblems formed. Obviously, it is desirable for this relaxation to be 'as accurate as possible'. We next prove that the relaxed representation contains the convex hull of the original set S. This fact places a substantial limitation on the accuracy of the relaxation, although it is a simple result. We do not need rational data for results in this section, as our proofs show. However, some remarks require rational data.

Proposition 3.1. *Whenever matrices A_1, A_2, A_3, and a vector b exist such that*

$$x \in S \Rightarrow \text{there are } u, v \geq 0 \text{ with } u \text{ integer and } A_1 x + A_2 u + A_3 v = b, \tag{3.1}$$

then the following holds:

$$x \in \text{clconv}(S) \Rightarrow \text{there are } u, v \geq 0 \text{ such that } A_1 x + A_2 u + A_3 v = b. \tag{3.2}$$

Proof. $x \in \text{conv}(S)$ implies there exist $\lambda_i \geq 0$ for $i = 1, \ldots, t$ such that

$$x = \sum_{i=1}^{t} \lambda_i x^i \quad \text{where } \sum_{i=1}^{t} \lambda_i = 1 \text{ and } x^i \in S \text{ for } i = 1, \ldots, n. \tag{3.3}$$

By taking a summation of the following valid equalities,

$$A_1\lambda_i x^i + A_2\lambda_i u^i + A_3\lambda_i v^i = \lambda_i b_i, \quad u^i \text{ integer for } i=1,\dots,t$$

we get

$$A_1\left(\sum_{i=1}^{t} \lambda_i x_i\right) + A_2\left(\sum_{i=1}^{t} \lambda_i u_i\right) + A_3\left(\sum_{i=1}^{t} \lambda_i v_i\right) = b$$

which gives (3.2), with 'clconv(S)' replaced by conv(S)'. Then (3.2) follows since the set $\{b - A_2 u - A_3 v \mid u, v \geq 0\}$ is a polyhedron, and so is closed. □

A MIP-representation of S (1.1) in which (3.2) is bi-conditional ($\Leftrightarrow$) is called 'sharp'. A sharp representation is as accurate as possible in its linear relaxations. By Proposition 3.1, this may not be very accurate; but inappropriate MIP representations can be even less accurate.

By Corollary 1.3, clconv(S) = conv(S) when S is representable. Therefore, a specific representation of S (1.1) is sharp, precisely when the condition '$x \in \text{conv}(S)$' is equivalent to the right-hand-side condition of (3.2). Thus the concept of 'sharpness' defined here is the same as that in Meyer (1981) in the case of functions occurring only in the criterion function.

The concept of sharpness is in essence Meyer's concept of 'linear relaxation optimal'. We are using it in a broader context here of sets rather than functions, and prefer the shorter name.

Example 3.1. The bounded fixed charge function

$$f(x) = \begin{cases} 0, & x = 0, \\ 1, & 0 < x \leq 1, \\ +\infty, & x \notin [0, 1], \end{cases}$$

has an MIP representation of epi(f) as: '$z \geq \frac{1}{2}x$, $0 \leq x \leq 1$, z integer'. The LP relaxation defines the function $\frac{1}{2}x$ on [0, 1], and the epigraph of $\frac{1}{2}x$ is larger than conv(epi(f)). Of course, the MIP representation '$z \geq x$, $0 \leq x \leq 1$, z integer' is sharp.

In Lemma 1.6 we prove that a union of polytopes is representable. We now show that the corresponding representation depicted by (1.27) is sharp whenever all the polyhedra have the same recession directions, which is very mild requirement (by our remarks in Section 2).

Proposition 3.2. *If all the nonempty polyhedra P_i ($i = 1, \dots, s$) have the same directions of recession, then* (1.27) *is a sharp representation of $P_1 \cup \cdots \cup P_s$.*

Proof. We omit the proof that (1.27) is a representation of $P_1 \cup \cdots \cup P_s$. This proof is easily accomplished by, e.g., using our argument below for the sharpness of the representation, and then noting that (when the λ_i are all integer) we have $\lambda_1 = 1$ below.

To establish sharpness, we need only show one direction of the biconditional, and cite Proposition 3.1. (To establish a representation, we need only show the same direction of the bi-conditional in (1.1), as the converse direction is trivial for (1.27). I.e., for the set S defined in (1.27), one trivially has $S \supseteq P_1 \cup \cdots \cup P_s$).

In (1.27) relax 'λ_i integer' to '$\lambda_i \in Q$' for $i=1, \ldots, s$. Then if $\lambda_i > 0$, we have $x_i/\lambda_i = w_i \in P_i$. We find that $x = \sum_{\lambda_i>0} \lambda_i w^i + \sum_{\lambda_i=0} w^i$, where $w^i \in P_i$ if $\lambda_i > 0$ and w^i is a recession direction of P_i if $\lambda_i = 0$.

Without loss of generality, $\lambda_1 > 0$ (as $\sum \lambda_i = 1$). Then $\lambda_1 w^1 + \sum_{\lambda_i=0} w^i = \lambda_1(w^1 + \sum_{\lambda_i=0} w^i/\lambda_1)$. Since each w^i is a recession direction of P_1 also, $w' = w^1 + \sum_{\lambda_i=0} w^i/\lambda_1 \in P_1$. Thus, without loss of generality, we can assume $x = \sum_{\lambda_i>0} \lambda_i w^i$. Hence $x \in \operatorname{conv}(P_1 \cup \cdots \cup P_s)$. □

If the MIP-representation of S' is sharp in Corollary 1.4, the construction gives a sharp representation of S. Thus sharp representations always exist.

Note that as integer variables are set to zero or one (i.e., 'arbitrated') in a branch-and-bound algorithm, the representation (1.27) of the union of polyhedra $P_1 \cup \cdots \cup P_t$ as obtained following the idea of the previous paragraph will correspond to subunions, provided that all recession cones are identical. We call this property *heriditary sharpness.*

The juxtaposition of sharp representations for two sets S_1 and S_2, while a representation for $S_1 \cap S_2$, is typically not sharp. E.g. if S_1 is a polyhedron and $S_2 = Z^n$ is the n-dimensional integers, the linear relaxation of the juxtaposed representations gives the polyhedron S_1 (which is typically not the convex span of the integer points in S_1).

It is sometimes convenient to state the 'polyhedral' disjunctive representation (1.27) in an alternative form, called the 'extreme point' form.

Let $P_i = \operatorname{conv}\{x^{ij} \mid j \in I(i)\} + C$, where the index set $I(i)$ depends on $i = 1, \ldots, s$ and the polyhedral cone $C = \operatorname{cone}\{v^k \mid k \in K\}$ is independent of i. The 'extreme point' representation of $P_1 \cup \cdots \cup P_s$ is:

$$x = \sum_{i=1}^{s} \sum_{j \in I(i)} \lambda_{ij} x^{ij} + \sum_{k \in K} \sigma_k v^k, \quad \lambda_i = \sum_{j \in I(i)} \lambda_{ij},$$

$$1 = \sum_{i=1}^{s} \lambda_i, \quad \lambda_i \text{ integer}, \ \lambda_i, \lambda_{ij}, \sigma_k \geq 0. \tag{3.4}$$

Proposition 3.3. *If all the nonempty polyhedra P_i ($i = 1, \ldots, s$) have the same directions of recession, then* (3.4) *is a sharp representations of $P_1 \cup \cdots \cup P_s$.*

Proof. Let S' be the set defined by (3.4) in variables $x \in \mathbb{R}^n$, and let $S = P_1 \cup \cdots \cup P_s$.

If $x \in P_i$, let $\lambda_{ij}, \sigma_k \geq 0$ such that $\sum_{j \in I(i)} \lambda_{ij} = 1$ and $x = \sum_{j \in I(i)} \lambda_{ij} x^{ij} + \sum_{k \in K} \sigma_k v^k$. Then by putting $\lambda_i = 1$ and $\lambda_j = 0$ for $j \neq i$, (3.4) holds. Hence $S' \supseteq P_i$. As i was arbitrary, $S \subseteq S'$.

If $x \in S'$, suppose that $\lambda_i = 1$ in (3.4). Then $\lambda_j = 0$ for $j \neq i$ and so $x = \sum_{j \in I(i)} \lambda_{ij} x^{ij} + \sum_{k \in K} \sigma_k v^k \in P_i$. Thus $x \in S$, and so $S \supseteq S'$.

We have established that $S = S'$, i.e. that (3.4) is a representation of S.

To see that (3.4) is sharp, only one direction of the bi-conditional need be established (i.e. that if x solves the linear relaxation, then $x \in \text{conv}(S)$).

Put $I = \{i \mid \lambda_i > 0\}$; we have $I \neq \emptyset$. For $i \in I$, define $\lambda'_{ij} = \lambda_{ij}/\lambda_i$. Without loss of generality, $\lambda_1 > 0$. With $\sigma'_k = \sigma_k/\lambda_1$, we have:

$$x = \sum_{i \in I} \sum_{j \in I(i)} \lambda_{ij} x^{ij} + \sum_{k \in K} \sigma_k v^k = \sum_{i \in I} \lambda_i \left(\sum_{j \in I(i)} \lambda'_{ij} x^{ij} \right) + \sum_{k \in K} \sigma_k v^k$$

$$= \lambda_1 \left(\sum_{j \in I(1)} \lambda'_{ij} x^{ij} + \sum_{k \in K} \sigma'_k v^k \right) + \sum_{i \in I, i \neq 1} \lambda_i \sum_{j \in I(i)} \lambda'_{ij} x^{ij}.$$

We also have, for $i \in I$, $\sum_{j \in I(s)} \lambda'_{ij} = \lambda_i/\lambda_i = 1$.

Upon putting $x(1) = \sum_{j \in I(1)} \lambda'_{ij} x^{ij} + \sum_{k \in K} \sigma'_k v^k$, $x(i) = \sum_{j \in I(i)} \lambda'_{ij} x^{ij}$ for $i \in I$ and $i \neq 1$, we easily prove that $x(i) \in P_i \subseteq S$ for $i \in I$. Since $x = \sum_{i \in I} \lambda_i x(i)$ and $\sum_{i \in I} \lambda_i = 1$, $x \in \text{conv}(S)$, as desired. □

We remark that the representation (3.4) is also hereditarily sharp.

Note that the λ_i can be removed in the linear relaxation of (3.4), which becomes:

$$x = \sum_{i=1}^{s} \sum_{j \in I(i)} \lambda_{ij} x^{ij} + \sum_{k \in K} \sigma_k v^k, \quad 1 = \sum_{i=1}^{s} \sum_{j \in I(i)} \lambda_{ij}, \qquad \lambda_{ij}, \sigma_k \geq 0. \tag{3.5}$$

Only $(n+1)$ constraints appear in (3.5). The linear relaxation of (1.27) contains many more constraints. We were motivated to develop (3.4) at a means of generalizing Beale's construction in Beale (1979, Section 4) and placing that in our framework.

The 'common wisdom' regarding the Simplex Algorithm, in practical applications, is that its running time is linear in the number of constraints, and is affected by the number of variables only slightly. Clearly, this practical observation is based on experience where the number of variables is not exponential! For 'arbitrary' polyhedra P_i, the number of extreme points x^{ij} (i.e. the size of $I(i)$) is exponential, so the exact representation (3.4) is of no practical value.

However, a second observation from experience is that, when representations are called for, the P_i which arise are not 'arbitrary' at all, and have very few extreme points. Part of the reason for this is that the P_i thus occurring are of small dimension, but that is not the entire reason. In any event, for several common representations there are not many variables λ_{ij}, and the $(n+1)$ linear inequalities needed assumes primary importance.

4. Comments on applications

In an earlier version of this paper (Jeroslow and Lowe (1981)), we concluded with several examples of the use of the modelling techniques (1.27) and (3.4) which emerge in our proofs above.

At the present time, we have a series of computer experiments under way, and we next cite results from preliminary runs from two of these experiments.

One of the experiments is in the context of 'either/or' constraints. There, we discovered that the standard modelling suggested by many texts is very poor. For example, suppose that variables x_1, x_2 are constrained by $0 \leqslant x_1, x_2 \leqslant 200$, and in addition at least one of these constraints hold:

$$x_1 - x_2 \geqslant 100 \quad \text{or} \quad -x_1 + x_2 \geqslant 100. \tag{4.1}$$

A best possible lower bound on $x_1 - x_2$ or $-x_1 + x_2$ is -200, and using this bound, often (4.1) is modelled in this manner:

$$x_1 - x_2 \geqslant 100 - 300z, \qquad -x_1 + x_2 \geqslant 300z - 200, \tag{4.2}$$

where z is a new variable introduced, to take a value $z \in \{0, 1\}$. In the linear relaxation of (4.2), '$z \in \{0, 1\}$' is replaced by '$0 \leqslant z \leqslant 1$', and when that is done, (4.2) has no value at all. It is in fact an exercise to prove that, for any $0 \leqslant x_1, x_2 \leqslant 200$ we have (4.2) satisfied by setting $z = \max\{0, 1/3 - v/300\}$, where $v = x_1 - x_2$, and we have $z \in [0, 1]$. On the other hand, (4.1) implies $x_1 + x_2 \geqslant 100$, and a sharp representation would not lose this information.

In our experiment on 'either/or' constraints, we have observed run time advantages exceeding 20-to-1 (ard node count advantages exceeding 100-to-1) for a batch of random problems each involving no more than 45 binary variables. While these problems are small, the advantage clearly seems to increase with problem size. This occurs despite the fact, that our hereditarily sharp representations significantly increase the size of the linear program.

As it is common to test ideas empirically by means of worked examples, it may be worth while here to mention how misleading that practice can prove in this context. As the linear relaxations solved with these new representations are much larger than with those from the textbooks, in order to see the advantage of the newer formulations, one needs at least small-to-medium sized problems. Only on such problems does the branch-and-bound activity become sufficiently important, that it outweighs the disadvantage of more pivoting (with larger bases). On tiny problem, such as those solved by hand, the earlier formulations are almost always better in terms of CPU time.

We also are partly through an experiment involving fixed charges where a smaller advantage in run time (of the nature of 3-to-1) has emerged on very small problem, with only 15 fixed charges. In this case, our sharp representation decreased the number of integer variables, as well as the number of variables and constraints.

These experiments will be reported in forthcoming papers. In these papers, we will also touch upon modelling issues that go beyond sharpness and hereditary sharpness, and which emerge when one goes more deeply into the details of a practical implementation of the modelling techniques above.

Acknowledgement

We have benefitted from the careful proofing and helpful suggestions of the referees, including the comments above regarding the need for our hypotheses of rational data.

References

E. Balas, "Disjunctive programming: Facets of the convex hull of feasible points", Report 348, GSIA, Carnegie–Mellon University (Pittsburgh, PA, 1974).

M. Bazaraa and M. Shetty, *Nonlinear programming* (Wiley, New York, 1979).

E.M.L. Beale, "Branch-and-bound methods for mathematical programming", in: P.L. Hammer, E.L. Johnson and B.H. Korte, eds., Discrete optimization II (North-Holland, Amsterdam, 1979).

G. Dantzig, *Linear programming and extensions* (Princeton University Press, Princeton, NJ, 1963).

R. G. Jeroslow, "Representations of unbounded optimizations as integer programs", *Journal on Optimization Theory and its Applications* 30 (1980) 339–351.

R.G. Jeroslow, "Cutting-planes for relaxation of integer programs", MSSR NO. 347, GSIA, Carnegie–Mellon University (Pittsburgh, PA, 1974).

R.G. Jeroslow, "Cutting plane theory: Disjunctive methods", *Annals of Discrete Mathematics* 1 (1977) 293–330.

R.G. Jeroslow, "Some basis theorems for integral monoids", *Mathematics of Operations Research* 3 (1978) 145–154.

R.G. Jeroslow and J.K. Lowe, "Modelling with integer variables", College of Management, Georgia Institute of Technology (Atlanta, GA, 1981).

R.R. Meyer, "On the existence of optimal solutions to integer and mixed-integer programming problems", *Mathematical Programming* 7 (1974) 223–235.

R.R. Meyer, "Integer and mixed-integer programming models: General properties", *Journal of Optimization Theory and Applications* 16 (1975) 191–206.

R.R. Meyer, "Mixed integer minimization models for piecewise linear functions of a single variable", *Discrete Mathematics* 16 (1976) 163–171.

R.R. Meyer, "A theoretical and computational comparison of 'equivalent' mixed integer formulations", *Naval Research Logistics Quarterly* 28 (1981) 115–131.

R.R. Meyer, M.V. Thakkar and W.P. Hallman, "Rational mixed integer and polyhedral union minimization models", *Mathematics of Operations Research* 5 (1980) 135–146.

R.T. Rockafeller, *Convex analysis* (Princeton University Press, Princeton, NJ, 1970).

J. Stoer and C. Witzgall, *Convexity and optimization in finite dimensions I* (Springer-Verlag, New York, 1970).

Mathematical Programming Study 22 (1984) 185–205
North-Holland

MULTIPOINT METHODS FOR SEPARABLE NONLINEAR NETWORKS

P.V. KAMESAM

Department of Mathematics and Computer Science, University of Maryland, Baltimore County, Catonsville, MD 21228, USA

R.R. MEYER

Computer Sciences Department, University of Wisconsin-Madison, Madison, WI 53706, USA

Received 31 March 1983
Revised manuscript received 26 March 1984

Iterative piecewise-linear approximation methods are considered for separable, convex nonlinear network problems. A comparison is made between 'fixed grid' approximations of 2, 4, and 6 segments per variable and 'implicit grid' strategies that generate segments as needed, but store at most a 2-segment approximation at any time. It is shown that the implicit grid methods are linearly convergent, and this predicted behavior is confirmed by highly accurate solutions within 7 iterations of problems with up to 2238 variables. Since the computing time per iteration is only slightly more for implicit grids than for fixed grids, the numerical results presented show overall computing times are less for implicit grids. A lower bounding technique based on the error of approximation is also developed. This technique is highly useful if Lagrangian relaxations are difficult to solve.

Key words: Piecewise-Linear Approximation, Nonlinear Networks, Separable Problems.

1. Introduction

Throughout this paper we are concerned with algorithms for the following minimization problem:

$$\begin{aligned} \underset{x\in\mathbb{R}^n}{\text{minimize}} \quad & f(x)=\sum_{i=1}^{n} f_i(x_i) \\ \text{subject to} \quad & Ax=b, \\ & l\leq x\leq u, \end{aligned} \tag{1.1}$$

where $l\in\mathbb{R}^n$, $u\in\mathbb{R}^n$, $x=(x_1,\ldots,x_n)\in\mathbb{R}^n$, $b\in\mathbb{R}^m$, A is an $m\times n$ node-arc incidence matrix for a directed network, and each f_i is a continuous convex function on $[l_i, u_i]$.

There are many optimization problems that may be represented in the form (1.1). Examples are estimation of data elements in input–output tables (Bachem and Korte, 1981) determination of steady-state flows in electrical and water distribution

This material is based upon work supported by the National Science Foundation under Grant No. MCS-8200632.

networks (Collins et al. 1978), and reservoir control problems (Bertsekas, 1976). Details of these problems are given in Section 7.

Problem (1.1) has a simple structure and numerous nonlinear programming algorithms can be readily used or specialized for (1.1). Because of the network constraints, however, it is tempting to consider linearizing the objective function, particularly since in applications problems of type (1.1) tend to be fairly large and the speed of linear network flow codes can therefore be used to best advantage. The proposed algorithms do so by constructing piecewise-linear approximations and then solving a sequence of linear network subproblems.

2. Review of previous work

Several authors have proposed a variety of solution approaches. Collins et al. (1978) discusses the Frank–Wolfe method and piecewise linearization of the objective function. Dembo and Klincewicz (1978) and Beck et al. (1983) specialize the reduced gradient method to solve (1.1) and show how some second order information may be used to obtain an approximation to Newton's method. Bachem and Korte (1981) and Cottle and Duvall (1982) propose algorithms for a subclass wherein each f_i is a quadratic function and the constraints define a transportation polytope.

Non-iterative separable programming techniques can also be used. These standard approaches, though readily applicable, have the disadvantage that there is a trade-off between the accuracy of the solution and the size of the linearized subproblems to be solved. Thakur (1978) gives error bounds on the optimal objective value and the optimal solution. Bazaraa and Shetty (1979) discusses an iterative separable programming procedure based on generalized programming.

These separable programming methods use what amounts to a global approximation of the objective function. Meyer (1980) and Kao and Meyer (1981) on the other hand, develop local approximation methods and show these methods to be computationally effective.

The primary advantages of the local approximation methods are (1) the size of the linearized subproblem is kept small, (2) they avoid unnecessary function evaluations which would be required for a global piecewise-linear approximation of the objective function, and (3) in the small neighborhood of the current iterate in which it is used, the approximation is quite accurate.

3. Solution methodology

In what follows, we discuss two algorithms for (1.1). The first of these is an extension of the algorithms proposed by Kao and Meyer (1981) and uses a local piecewise-linear approximation with a fixed number of segments, which in the computational experiments reported in Section 7 was 2, 4, or 6.

The second algorithm uses an implicit global approximation (to the objective function) that is modified each iteration. The approximation is implicitly global in the sense that we initially use only two segments, and generate additional approximating segments only as needed. Computationally, the approximation utilizes at most two piecewise-linear segments per variable at any time. In essence, we try to retain the computational advantages of a local approximation method, while using (in theory) a global approximation to the objective function. A similar implicit grid approach is discussed in Jensen and Barnes (1980), but there a uniform grid size for all variables is considered and no results are given on lower bounds or convergence.

4. Local approximation methods in separable programming

4.1. Local piecewise-linear approximation

These local approximation methods use piecewise-linear approximations over appropriately chosen neighborhoods. In order to describe these methods clearly, we introduce the following notation, and define a local approximation around $\hat{m} \in \mathbb{R}^n$.

An ordered triple $(\hat{l}, \hat{m}, \hat{u})$ is said to be *admissible* if it satisfies the following conditions

$$l \leq \hat{l} \leq \hat{m} \leq \hat{u} \leq u, \tag{4.1}$$

$$\hat{l}_i < \hat{m}_i \quad \text{if } l_i < \hat{m}_i, \tag{4.2}$$

and

$$\hat{m}_i < \hat{u}_i \ \text{ if } \hat{m}_i < u_i, \quad i = 1, 2, \ldots, n. \tag{4.3}$$

An *approximating problem* $P(\hat{l}, \hat{m}, \hat{u})$ is defined as

$P(\hat{l}, \hat{m}, \hat{u})$:

$$\min_x \quad \sum \hat{f}_i(x_i)$$

$$\text{s.t.} \quad x \in \{x \mid Ax = b\} \cap [\hat{l}, \hat{u}]$$

where $\hat{f}_i(x_i)$ is a piecewise-linear approximation (with $2s$ segments) of $f_i(x_i)$ over $[\hat{l}_i, \hat{u}_i]$ that agrees with f_i at $\hat{m}_i \pm k\beta$, $k = 0, 1, \ldots s$, where $\hat{m}_i + s\beta = \hat{u}_i$ and $\hat{m}_i - s\beta = \hat{l}_i$ ($\beta > 0$ is the length of the subdivision interval except (perhaps) for the end segments and may depend on i; for simplicity of notation, however, we avoid reference to these dependencies). It is well-known (Kao and Meyer, 1981) that the problem $P(\hat{l}, \hat{m}, \hat{u})$ may be converted into an equivalent *linear* network flow problem under the convexity assumption on f.

Algebraically, we have

$$\hat{f}_i(x_i)=\begin{cases} f_i(\hat{m}_i) & \text{for } x_i=\hat{m}_i,\\ f_i(\hat{m}_i+(k-1)\beta)+c_i^{+k}\{x_i-(\hat{m}_i+(k-1)\beta)\} & \text{for } (k-1)\beta\leqslant x_i-\hat{m}_i\leqslant k\beta,\\ f_i(\hat{m}_i-(k-1)\beta)+c_i^{-k}\{x_i-(\hat{m}_i-(k-1)\beta)\} & \text{for } -k\beta\leqslant x_i-\hat{m}_i\leqslant -(k-1)\beta\\ \quad (k=1,2,\ldots,s) \end{cases} \tag{4.1.1}$$

where

$$\begin{aligned} c_i^{+k}&=\{f_i(\hat{m}_i+k\beta)-f_i(\hat{m}_i+(k-1)\beta)\}/\beta,\\ -c_i^{-k}&=\{f_i(\hat{m}_i-k\beta)-f_i(\hat{m}_i-(k-1)\beta)\}/\beta \end{aligned} \tag{4.1.2}$$

with the understanding that if $\hat{m}_i=\hat{l}_i=l_i$, then the approximation is defined only on the interval $[\hat{m}_i, \hat{u}_i]$. Similar comments hold if $\hat{m}_i=\hat{u}_i=u_i$. Because of the convexity of f_i, we have $\hat{f}_i(x_i)\geqslant f_i(x_i)$ for $x_i\in[\hat{l}_i, \hat{u}_i]$. Letting $\hat{S}$ be the feasible region of $P(\hat{l}, \hat{m}, \hat{u})$, this implies that if $\tilde{x}\in\hat{S}$, and $\hat{f}(\tilde{x})<\hat{f}(\hat{m})$, then $f(\tilde{x})<f(\hat{m})$. We thus have strict monotonicity of the iterates, since in the algorithms proposed, $\hat{m}$ is the most recently generated feasible solution and a new iterate must yield an improved value of $\hat{f}$. From this viewpoint, $\hat{f}$ is a local piecewise-linear approximation in the neighborhood $[\hat{l}, \hat{u}]$ of the most recent iterate and $\hat{S}$ is not empty, since it contains $\hat{m}$.

4.2. Error bounds and optimality conditions

Let $\hat{E}_i\equiv\max_{\hat{l}_i\leqslant x_i\leqslant\hat{u}_i}[\hat{f}_i(x_i)-f_i(x_i)]$. Then $\hat{E}_i$ satisfies

$$\hat{E}_i\leqslant\max_{[\hat{l}_i,\hat{u}_i]}\hat{f}_i(x_i)-\min_{[\hat{l}_i,\hat{u}_i]}f_i(x_i)\leqslant\max_{[\hat{l}_i,\hat{u}_i]}f_i(x_i)-\min_{[\hat{l}_i,\hat{u}_i]}f_i(x_i).$$

Thus, as the length of the interval $[\hat{l}_i, \hat{u}_i]$ goes to zero, $\hat{E}_i$ must tend to zero. Letting $E(\hat{l}, \hat{m}, \hat{u})\equiv\max_{[\hat{l},\hat{u}]}[\hat{f}(x)-f(x)]=\sum_{i=1}^n\hat{E}_i$, we have

4.2.1. Lemma. *If* (l^k, m^k, u^k) *is a sequence of admissible triples such that* $l^k\to\bar{m}$, $m^k\to\bar{m}$, $u^k\to\bar{m}$ *then* $E(l^k, m^k, u^k)\to 0$.

The following lemma relating the optimal values of $P(\hat{l}, \hat{m}, \hat{u})$ and (1.1) where x^* is optimal for $P(\hat{l}, \hat{m}, \hat{u})$ and x^{**} is optimal for (1.1), is taken from Kao and Meyer (1981).

4.2.2. Lemma. *If* x^* *is an optimal solution of an approximating problem* $P(\hat{l}, \hat{m}, \hat{u})$, *where* $(\hat{l}, x^*, \hat{u})$ *is admissible, then the following lower and upper bounds hold for the optimal value* z^{**} *of* (1.1).

$$f(x^*)-\sum_{i=1}^n\hat{e}_i\leqslant z^{**}\leqslant f(x^*)\quad\text{where } \hat{e}_i\geqslant\max_{\hat{l}_i\leqslant x_i\leqslant\hat{u}_i}(\hat{f}_i(x_i)-f_i(x_i)). \tag{4.2.1}$$

The following theorem from Kao and Meyer (1981) give conditions under which $\bar{x} \in S := \{x \mid Ax = b, l \leq x \leq u\}$ is an optimal solution of (1.1).

4.2.3. Theorem. *Let* $\{(\bar{l}^k, \bar{x}, \bar{u}^k)\}$ *be a sequence of admissible triples such that* $\bar{l}^k \to \bar{x}$, $\bar{u}^k \to \bar{x}$, *where* $\bar{x} \in S$. *Then* $\bar{x}$ *is optimal for* (1.1) *iff* $\bar{x}$ *is optimal for each of the problems* $P(\bar{l}^k, \bar{x}, \bar{u}^k)$, $k = 1, 2, \ldots$.

The first algorithm given below starts with an $\bar{x} \in S$ and conducts a search procedure at $\bar{x}$ by considering a sequence of approximating problems of the type described in Theorem 4.2.3. If $\bar{x}$ is optimal for (1.1), then the search procedure establishes the optimality of $\bar{x}$ by the above optimality conditions. Otherwise, the search procedure is terminated as soon as an improved feasible solution is obtained. This improved feasible solution is used in the next iteration, where again a search procedure is carried out. This algorithm is very similar to algorithm 2 of Kao and Meyer (1981) except that in their algorithm these authors use an approximating function consisting of at most two piecewise-linear segments. In algorithm (5.1) below, the approximating function may have up to $2s$ piecewise-linear segments. Thus, algorithm (5.1) is a straightforward extension of algorithm 2 of Kao and Meyer (1981).

5. Algorithm based on local piecewise-linear approximation

5.1. Algorithm. (a) Let $\alpha \in (0, 1)$, let $\bar{x}$ be a feasible solution available at the start of the current iteration, and let $\hat{\lambda}$ be a grid size vector (determined by a procedure described below) for the current iteration.

(b) Denoting by $\bar{x}(\delta)$ an optimal solution of $P(\bar{l}(\delta), \bar{x}, \bar{u}(\delta))$, where $\bar{l}_i(\delta) \equiv \max\{l_i, \bar{x}_i - s\delta\lambda_i\}$ and $\bar{u}_i(\delta) \equiv \min\{u_i, \bar{x}_i + s\delta\lambda_i\}$, (with the understanding that $\bar{x}(\delta)$ is taken to be $\bar{x}$ if $\bar{x}$ is optimal for $P(\bar{l}(\delta), \bar{x}, \bar{u}(\delta))$), then $\bar{x}$ is optimal for (1.1) if $\bar{x}(\delta) = \bar{x}$ for $\delta = \alpha^0, \alpha^1, \alpha^2, \ldots$.

(c) Otherwise, let $\hat{\alpha}$ be the first power of α such that $f(\bar{x}(\hat{\alpha})) < f(\bar{x})$ and use $\bar{x}(\hat{\alpha})$ as a starting feasible solution for the next iteration, for which a new grid size vector is also generated.

In step (b) of the algorithm, we consider an infinite sequence of approximating problems to test the optimality of $\bar{x}$ for (1.1). In practice, this is avoided by computing a lower bound on the optimal value of (1.1) and terminating the computation if the current solution agrees with the lower bound to a desired accuracy. Theoretically, in order to establish a convergence theorem, step (c) of Algorithm 5.1 must be modified and grid size vector generation must obey certain rules (see Kao and Meyer, 1981). Computationally, step (c) is quite adequate, and a good approach to the determination of $\hat{\lambda}$ is described in the next section.

5.2. Determination of the bounds

From a theoretical point of view, it is necessary to ensure that the search procedure in step (c) be started with bounds $\hat{l}$ and $\hat{u}$ sufficiently far away from $\bar{x}$. On the other hand, if $\hat{l}$ and $\hat{u}$ are far from $\bar{x}$, then step (c) is time-consuming. The choices of $\hat{l}$ and $\hat{u}$ thus have a significant impact on the computational efficiency of the algorithm. We now describe a strategy for determining $\hat{l}$ and $\hat{u}$ which proved to be quite satisfactory.

At the beginning of the jth iteration, we have available a feasible solution x^{*j-1}, which is the optimal solution of the $(j-1)$st iteration along with a set of optimal multipliers μ^* associated with the constraints $\{Ax=b\}$. Consider the following Lagrangian relaxation problem

$$\min_{l\leq x\leq u} \{f(x)-\mu^{*\mathrm{T}}(Ax-b)\}. \tag{5.2.1}$$

Because f is separable, (5.2.1) reduces to

$$\sum_{i=1}^{n} \min_{l_i\leq x_i\leq u_i} \{f_i(x_i)-\mu^{*\mathrm{T}}A^i x_i\}+\mu^{*\mathrm{T}}b, \tag{5.2.2}$$

where A^i is the ith column of A.

It is well known that the minimum value z of (5.2.1) gives us a lower bound on the optimal value of (1.1). Let $\hat{x}$ solve (5.2.1). Unless $\hat{x}$ is optimal for (1.1), $\hat{x}$ is not feasible for (1.1). Nevertheless, $\hat{x}$ can be used to determine the initial bounds, $\hat{l}$, $\hat{u}$ for the jth iteration. We have used the following rule and found it to be quite satisfactory:

Let λ_i^{j-1} be the grid size for $\hat{f}_i$ in the $(j-1)$st iteration. Set $\lambda_i^j = \min[\max\{\frac{1}{2}\lambda_i^{j-1}, |x_i^{*j-1}-\hat{x}_i|\}, (f(x^{*j-1})-z)^{1/2}]$. (The motivation for this rule is that we wish to combine the observations that the errors in the values of the variables are bounded by a multiple of the square root of the function value error (see Appendix for details) and that a large change in the value of a variable in a particular iteration is often followed by a relatively large change in the variable in the succeeding iteration.) In Section 7, we present computational results with Algorithm 5.1 using this heuristic rule.

In the next section, we describe a different algorithm based on implicit global approximation.

6. Implicit global approximations in separable programming

Separable programming traditionally consists of choosing a fixed grid of points in an interval $[l_i, u_i]$ and then approximating the function f_i on $[l_i, u_i]$ by the piecewise-linear segments determined by the function evaluations at the grid points. Unless a very fine grid of points in $[l_i, u_i]$ is chosen, the resulting error of approximation is likely to be high. Collins et al. (1978) considers the following problem: for a specified number of grid points, determine a grid so that the error of approximation is minimized. After determining the grid, an LP is solved to get an approximate solution to (1.1). They, however, report that determining the grid took more computational effort, than solving the LP (with 8 segments per variable). Even with such strategies, the error of approximation is still likely to be high, unless a very fine grid is used.

We propose an alternative to these traditional approaches. Initially start with a crude approximation defined by a coarse grid. After solving the resulting approximation, set up another approximating problem around the current iterate, this time with a finer grid. Proceeding iteratively, we stop when a solution to a desired accuracy is obtained.

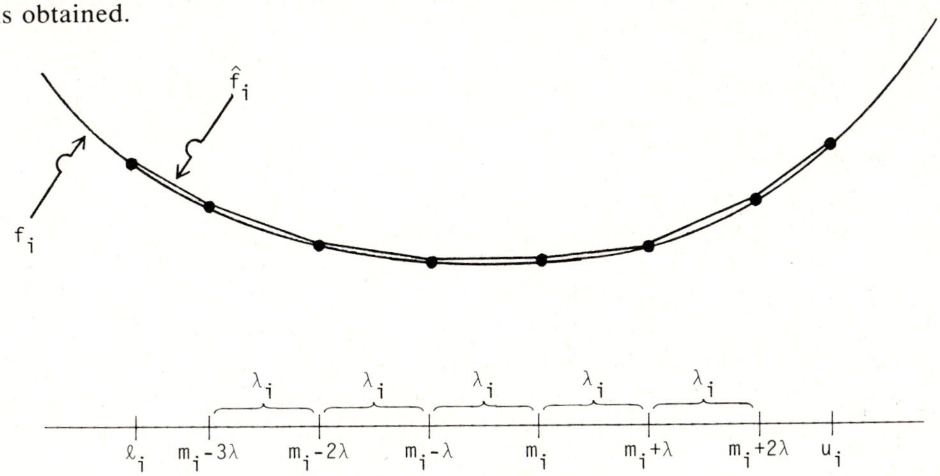

Fig. 1. A global piecewise linear approximation $\hat{f}_i$ of f_i.

The approximating function used is similar to (4.1.1) except that the approximation is (in theory) not restricted to a neighborhood of the current iterate. Thus $\hat{f}_i$ below is a piecewise-linear approximation of f_i over the interval $[l_i, u_i]$. The approximating function $\hat{f}_i$ agrees with f_i at $m_i \pm k\lambda_i$, $k = 0, 1, 2, \ldots$ and also at the end points l_i and u_i (see Figure 1). λ_i is the length of the subdivision interval except (perhaps) the end segments. For simplicity of notation, however, we assume that the length of each segment is equal to λ_i. The difference between this approximating function and (4.1.1) is that the latter is a local approximation restricted to the chosen neighborhood.

Let λ^j be the vector of subdivision interval lengths in the jth iteration. The corresponding approximating problem is

$\mathrm{AP}^j(m^j, \lambda^j)$:

$$\min_x \quad \hat{f}^j(x) = \sum_{i=1}^{n} \hat{f}_i^j(x_i)$$

$$\text{s.t.} \quad Ax = b, \quad l \leq x \leq u.$$

6.2. Proposition. *Let $\hat{f}_i^j$ be an approximation of f_i over $[l_i, u_i]$. Let*

$$\varepsilon_i^j \equiv \max_{x_i \in [l_i, u_i]} \{\hat{f}_i^j(x_i) - f(x_i)\}.$$

Then

$$\varepsilon_i^j \to 0 \quad \text{as } \lambda^j \to 0.$$

Further, if

$$\varepsilon^j = \sum_{i=1}^{n} \varepsilon_i^j = \max_{x \in [l, u]} \{\hat{f}^j(x) - f(x)\},$$

then

$$\varepsilon^j \to 0 \quad \text{as } \lambda^j \to 0.$$

Proof. Follows from the continuity of f_i.

6.3. Proposition. *Let x^* be an optimal solution of $\mathrm{AP}^j(m^j, \lambda^j)$ and x^* an optimal solution of (1.1). Then the following lower and upper bounds hold on the optimal value of (1.1).*

$$\hat{f}^j(x^*) - \varepsilon^j \leq f(x^{**}) \leq f(x^*).$$

Proof. $\hat{f}^j(x^*) \leq \hat{f}^j(x^{**}) = f(x^{**}) + [\hat{f}^j(x^{**}) - f(x^{**})] \leq f(x^{**}) + \varepsilon^j$. Since x^* is feasible for (1.1), the upper bound holds.

6.4. Lemma. *Let $\{\lambda^j\}$ be such that $\lambda^j > 0$, $j = 1, 2, \ldots$ and $\{\lambda^j\} \to 0$. Then a point x^* is an optimal solution of (1.1) iff x^* is an optimal solution of each of the approximating problems $\mathrm{AP}^j(x^*, \lambda^j)$, $j = 1, 2, \ldots$.*

Proof. ($\Leftarrow$) By Proposition 6.3,

$$f(x^*) - \varepsilon^j \leq f(x^{**}) \leq f(x^*) \quad \forall j = 1, 2, \ldots.$$

But, by Proposition 6.2, $\varepsilon^j \to 0$ as $\lambda^j \to 0$, and therefore $f(x^*) = f(x^{**})$.

($\Rightarrow$) Since the objective function $\hat{f}^j$ of $\mathrm{AP}^j(x^*, \lambda^j)$ satisfies $\hat{f}^j(x) \geq f(x)$ $\forall x \in [l, u]$ and agrees with f at x^*, x^* is an optimal solution of each of the problems $\mathrm{AP}^j(x^*, \lambda^j)$, $j = 1, 2, \ldots$.

We are now ready to state our second algorithm to solve (1.1). At the jth iteration, we solve the approximating problem $\mathrm{AP}^j(x^{*j-1}, \lambda^j)$, where x^{*j-1} is the optimal solution of the previous iteration. (Both our Algorithms 5.1 and 6.5 assume that a starting feasible solution is available. Various methods can be used to find an $x \in S$. In all our experiments, however, we solved an approximating problem of type (AP), (with a small number of segments per function) to obtain the initial feasible solution. If the original problem is infeasible, this is detected at the initial stage.)

6.5. Algorithm. Step 0: Let $\alpha \in (0, 1)$ and let x^{*0} be a starting feasible solution. Let $0 < \lambda_i^0 \leq u_i - l_i \; \forall i = 1, 2, \ldots, n$. Set $j \leftarrow 1$.

Step 1: Let $\lambda_i^j = \alpha \lambda_i^{j-1}$ $(i = 1, 2, \ldots, n)$.

Step 2: Determine $x^{*j}(\lambda^j)$, an optimal solution of $\mathrm{AP}^j(x^{*j-1}, \lambda^j)$. Set $j \leftarrow j+1$ and go to Step 1.

6.6. Theorem. *Let Algorithm* 6.5 *generate a sequence of points* $\{x^{*j}\}$ *and let the corresponding sequence of dual multipliers associated with the constraints* $Ax = b$ *be* $\{\mu^{*j}\}$. *Then* $f(x^{*j}) \to z^{**}$, *the optimal objective value of* (1.1). *Each accumulation point of* $\{x^{*j}\}$ *is an optimal solution of* (1.1). *If, in addition, the right and left derivatives* $f'_{i+}(l_i)$ *and* $f'_{i-}(u_i)$, $i = 1, 2, \ldots, n$, *are finite, then the sequence* $\{\mu^{*j}\}$ *is bounded and each accumulation point of* $\{\mu^{*j}\}$ *is an optimal set of dual multipliers for* (1.1).

Proof. See Appendix.

6.7. Theorem. *Let Algorithm* 6.5 *generate a sequence of points* $\{x^{*j}\}$ *and let* x^{**} *be an optimal solution of* (1.1). *If*

(i) $f(x)$ *is strongly convex on* $[l, u]$, *and*

(ii) $f(x)$ *is boundedly convex on* $[l, u]$ *or* $\nabla f(x)$ *is Lipschitz continuous on* $[l, u]$, *then* $\{x^{*j}\}$ *converges to* x^{**} *linearly.*

Proof. See Appendix.

6.8. Computational aspects

While stating Algorithm 6.5, we have not mentioned how the subproblems $(\mathrm{AP})^j$ are to be solved. It is well-known that $(\mathrm{AP})^j$ can be converted to an equivalent linear program. Specifically, $\mathrm{AP}^j(x^{*j-1}, \lambda^j)$ is equivalent to the following problem.

$(\mathrm{LPAP})^j$:

$$\min_{x_i^{+k}, x_i^{-l}} \quad \sum_{i=1}^{n} \sum_{k} c_i^{+k} x_i^{+k} - \sum_{i=1}^{n} \sum_{l} c_i^{-l} x_i^{-l}$$

$$\text{s.t.} \quad \sum_{k} A x^{+k} - \sum_{l} A x^{-l} = 0,$$

$$0 \leq x^{+k} \leq \lambda^j,$$

$$0 \leq x^{-l} \leq \lambda^j,$$

where c_i^{+k} and c_i^{-l} are as defined in (4.1.2).

The variables x_i^{+k} and x_i^{-l} are the deviations of x_i from x_i^{*j-1} in incremental steps of λ_i^j, where it is understood that if x_i^{*j-1} is at its lower bound, then the variables x_i^{-l} (corresponding to decrements from x_i^{*j-1}) would not be present in $(\mathrm{LPAP})^j$.

Similar comments hold for x_i^{+k} if x_i^{*j-1} is at its upper bound. The number of variables x_i^{+k}, x_i^{-l} vary depending on λ_i^j and also the value of x_i^{*j-1} in the interval $[l_i, u_i]$ (see Figure 2).

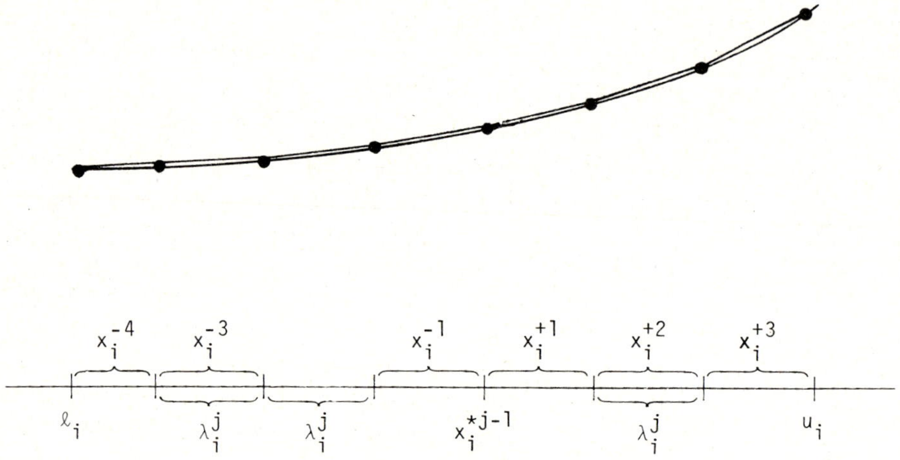

Fig. 2. Correspondence between segments and variables.

In (1.1), the variable x_i corresponds to flow on an arc in the network, while in $(\text{LPAP})^j$ the variables x_i^{+k}, x_i^{-l} correspond to flows on parallel arcs between the same pair of nodes. Clearly, the computational efficiency of (6.5) depends on how efficiently the subproblems $(\text{LPAP})^j$ are solved. As j (iteration number) increases, λ^j gets smaller and smaller and the number of variables x_i^{+k} and x_i^{-l} increases. It would seem that an enormous number of function evaluations would be required to set up the approximation and because of their size, the sub-problems $(\text{LPAP})^j$ would be unmanageable. It is, however, not necessary to generate a global approximation of f_i. It suffices to *work locally around* x_i^{*j-1} and generate additional arcs only as needed, while simultaneously discarding arcs. It is therefore sufficient to work, at anytime, with at most two arcs x_i^+ and x_i^- for each of the variables x_i.

Strategy for modifying the current approximation

Because of the convexity of f_i, we have

$$\ldots, c_i^{-2} \leq c_i^{-1} \leq c_i^{+1} \leq c_i^{+2}, \ldots. \tag{6.8.1}$$

(Note, however, that in LPAP^j, the cost coefficient of x_i^{-l} is $-c_i^{-l}$ and not c_i^{-l}.) It is clear from (6.8.1) that if x_i^{+1} prices out unfavorably, then so do all x_i^{+k}. Similar comments hold for x_i^{-1} and x_i^{-l}. We can, therefore, initially start with two variables x_i^+ and x_i^- with cost coefficients c_i^{+1} and $-c_i^{-1}$. It is again easy to see that if x_i^+ prices out favorably, then x_i^- cannot and vice versa. In the process of solving LPAP^j, suppose variable x_i^+ (x_i^-) reaches its temporary bound λ_i^j and becomes nonbasic. We can then modify the current value of x_i and then use x_i^+ (x_i^-) itself, as if it were a new variable corresponding to the immediately succeeding (preceding) segment in the approximation of $\hat{f}_i$. An additional functional evaluation of f_i is thus made at this stage and the cost coefficients are modified. If, however, x_i^+ (x_i^-) reaches its temporary bound λ_i^j, but remains basic, then the succeeding and preceding arcs do not price favorably and nothing need be done.

Pivoting in this fashion, we would either reach a stage where x_i^+ (x_i^-) no longer reaches its temporary bound λ_i^j, or the current value of x_i reaches its original upper bound $u_i(l_i)$. We can declare the current value as an optimal solution of LPAP^j if none of the variables x_i^+ and x_i^- $(i=1,2,\ldots n)$ price out favorably. The strategy thus is *column generation* combined with *column dropping.* Since the column generated is the same as the column dropped, from a computational viewpoint *columns are neither added nor dropped.*

In Algorithm 6.5, the grid size is reduced from iteration to iteration by a factor of α. It has been our experience that if α is set to 0.25, we need not move more than two or three segments from the current iterate. (Experimentation with $\alpha = 0.5$ has shown that the overall computing time does not change significantly, though the number of iterations to ε-optimality increases.) Thus, though the problems (LPAP)^j look formidable, they can be solved efficiently by the above computational strategy.

In (LPAP)^j for Algorithm 6.5, there are parallel arcs between every pair of nodes. We have experimented with pricing on multiple arcs successively. To be specific, suppose arc x_i^- reaches its lower bound, thus increasing the current value of x_i. Then, we can price out x_i^+ and select it as the entering variable if it prices favorably. Similarly, when an additional arc is added, it is possible to select this arc as the entering variable if it prices favorably. It has been our experience that such strategies do not help. In fact, experimentation has shown that such strategies increase the total number of pivots by 15% to 20%. In solving (LPAP)^j, we use a very simple unsophisticated pricing strategy as used in the primal network code RNET. It is possible that a suitable pricing strategy would improve the computational efficiency significantly.

6.9. *Computing lower bounds on the optimal value*

If the Lagrangian relaxation (5.2.1) can be solved in closed form, then it gives us an easy means for computing a lower bound on the optimal value. This, however, may not be possible in some problems.

Lemma 4.2.2 gives lower and upper bounds on the optimal value when a locally approximating problem $P(\hat{l}, \hat{m}, \hat{u})$ is solved. In the implicit grid method, after solving LPAP^j, we have available a local 2-segment approximation for which also x^{*j} (the optimal solution of LPAP^j) is optimal. Lemma 4.2.2 requires that x^{*j} be *not artificially bounded.* Note that in the implicit grid method nonbasic variables are never artifically bounded. If in x^{*j}, a basic variable x_i is at its temporary bound λ_i^j, then adding an additional (succeeding or preceding) segment does not affect the optimality of x^{*j}. Thus, a 2-segment approximation is on hand, such that x^{*j} is optimal for this approximation and further x^{*j} is not artifically bounded.

In order to compute the lower bound, we need to estimate $\hat{e}_i$, as given by (4.2.1). If the functions f_i are differentiable, then we can compute the linear support to f_i at x^{*j} and estimate $\hat{e}_i$ as the maximum deviation between this linear support and the 2-segment approximation. Computationally, this requires only $\nabla f(x^{*j})$ and no further function evaluations are needed.

For all the test problems given in Section 7, we have computed the lower bounds by both Lagrangian relaxation and this method. Close to an optimal solution, the two lower bounds agreed up to 6 or 7 figures. We thus have a lower bounding technique which would be highly useful in situations where Lagrangian relaxations are difficult to solve. Meyer (1980) describes yet another approach that uses only function values, but is less accurate than the methods used here.

7. Computational results

This section describes the test problems solved in this study, and presents computational experience with both the Algorithms 5.1 and 6.5. Algorithm 6.5 was implemented as per the strategies described in Section 6.8. Algorithm 5.1 was tested with the local approximations consisting of 2, 4 and 6 segments. The slopes of all the segments were computed and stored explicitly. It is, however, possible to implement Algorithm 5.1 also by generating additional segments only as needed. In all cases, the starting feasible solution was obtained in the first iteration, wherein a linearized subproblem (LPAP[1]) was solved with a full artificial basis.

The linearized subproblems were solved by a suitably modified version of RNET (Grigoriadis and Hsu, 1979). RNET was modified so as to maintain the cost coefficients and dual variables in double precision. All the test problems were run with the same partial pricing frequency. (The RNET pricing frequency was set to 4.0.) The test problems were run using the FORTRAN 77 compiler (with optimization) on the VAX 11/780 running under the UNIX operating system. The execution times reported are in CPU seconds exclusive of input and include output time. (The output was only two lines per iteration.)

7.1. *Quadratic data fitting problems*

Except for the first problem (with 100 arcs), these problems were generated from actual data on input–output tables of the West German economy. The data was obtained from Bachem and Korte (1981). The problem is to adjust the input–output coefficients of the previous year given the row and column sums for the current year, with an additional constraint that if an input–output coefficient in the previous year is zero, it must be zero for the current year as well. The problem can be formulated as

$$\min_{x_{ij}} \sum_i \sum_j (x_{ij} - t_{ij})^2$$

$$\text{s.t.} \quad \sum_{i=1}^{k} x_{ij} = d_j, \quad i = 1, \ldots, k,$$

$$\sum_{j=1}^{k} x_{ij} = s_i, \quad i = 1, \ldots, k,$$

$$0 \leq x_{ij} \leq u_{ij}, \quad i, j = 1, \ldots, k,$$

with the understanding that some arcs in this transportation problem may not be present. The upper bound u_{ij} for the ijth arc is determined as $u_{ij} = \min\{t_{ij} + 1000, 100 \times t_{ij}\}$, where t_{ij} is the target flow on the ijth arc. The numerical values of t_{ij} range from 1 to 6500. For these problems, computation was terminated if the lower and upper bounds on the optimal value agree up to 7 figures. (See Table 1.)

7.2. Reservoir control problems

These are the problems described in Bertsekas (1976). The problem is to schedule water release from a reservoir subject to upper and lower bounds on the total volume of water in the reservoir. These problems can be formulated as separable nonlinear network flow problems. We have tested two sets (four problems in each set) of these problems; the first with quadratic objective function terms and the second with exponential objective function terms. In all cases, computation was terminated if the lower and upper bounds on the optimal value agree up to 6 figures. Computational results with quadratic objectives are presented in Table 2 and the computational results with exponential objective functions are presented in Table 3.

7.3. Water distribution problems

The third class of problems are pipe flow problems as described by Collins et al. (1978). The objective function consists of a few linear terms and many nonlinear terms (a few integrals and the rest of the form $c_i|x_i|^{2.85}$). We have tested three problems of this type. These problems were obtained from Dembo and Klincewicz (1978). For all the 3 problems, computation was terminated if the lower and upper bounds agree up to 7 figures and the computational results are as shown in Table 4.

8. Conclusions

The numerical results of the preceding section indicate that the implicit grid strategy produces in 6–8 iterations highly accurate solutions for the broad set of

Table 1

Quadratic data fitting problems

Problem		Algorithm 6.5 (implicit grid)	Algorithm 5.1			Best known lower bound
			2-Segments	4-Segments	6-Segments	
100 Arcs	Obj. Value	6600.0001	6600.0001	6600.0002	6600.0001	6599.9999
20 Nodes	CPU Time(sec)	3.233	5.967	5.200	7.149	
	# Iterations	7	13	9	9	
	# Pivots	1026	1850	1640	2053	
144 Arcs	Obj. Value	7302243.43	7302243.70	7302243.77	7302243.37	7302243.20
24 Nodes	CPU Time(sec)	7.117	9.683	12.734	14.407	
	# Iterations	7	14	13	12	
	# Pivots	2193	2707	3760	4116	
144 Arcs	Obj. Value	2714896.11	2714898.11	2714896.52	2714896.17	2714895.87
24 Nodes	CPU Time(sec)	7.867	10.183	12.982	17.100	
	# Iterations	7	13	13	13	
	# Pivots	2445	2900	3636	4536	
2202 Arcs	Obj. Value	5634201.59	5634201.78			5634201.37
109 Nodes	CPU Time(sec)	258.700	312.367			
	# Iterations	7	15	not run	not run	
	# Pivots	53 498	53 477			
2238 Arcs	Obj. Value	12022985.31	12022985.79			12022984.64
109 Nodes	CPU Time(sec)	239.834	320.350	not run	not run	
	# Iterations	7	15			
	# Pivots	46 672	53 148			

Table 2

Reservoir control problems (Quadratic objective terms)

Problem		Algorithm 6.5 (Implicit Grid)	Algorithm 5.1			Best known lower bound
			2-Segments	4-Segments	6-Segments	
23 Arcs 13 Nodes	Obj. Value	−1975.6491	−1975.6491	−1975.6491	−1975.6491	−1975.6492
	CPU Time(sec)	0.300	0.417	0.433	0.500	
	# Iterations	6	12	11	10	
	# Pivots	47	50	52	57	
103 Arcs 53 Nodes	Obj. Value	−8731.0245	−8731.0257	−8731.0257	−8731.0257	−8731.0261
	CPU Time(sec)	1.116	1.650	1.866	2.217	
	# Iterations	6	13	12	11	
	# Pivots	219	279	248	330	
207 Arcs 105 Nodes	Obj. Value	−17393.533	−17393.553	−17393.553	−17393.553	−17393.555
	CPU Time(sec)	2.467	3.384	3.817	4.300	
	# Iterations	6	12	11	10	
	# Pivots	492	530	569	554	
729 Arcs 366 Nodes	Obj. Value	−60750.487	−60750.488	−60750.488	−60750.488	−60750.491
	CPU Time(sec)	12.316	17.450	24.500	30.750	
	# Iterations	6	13	12	11	
	# Pivots	1993	2007	1912	2319	

Table 3

Reservoir control problems (Exponential objective terms)

Problem		Algorithm 6.5 (Implicit Grid)	Algorithm 5.1			Best known lower bound
			2-Segments	4-Segments	6-Segments	
23 Arcs	Obj. Value	12.6411	12.6412	12.6412	12.6412	12.6411
13 Nodes	CPU Time(sec)	0.400	0.667	0.683	0.700	
	# Iterations	6	13	10	10	
	# Pivots	40	50	44	49	
103 Arcs	Obj. Value	56.5602	56.5603	56.5602	56.5602	56.5601
53 Nodes	CPU Time(sec)	1.450	2.400	2.700	2.917	
	# Iterations	6	13	11	10	
	# Pivots	215	242	245	264	
207 Arcs	Obj. Value	124.758	124.758	124.758	124.758	124.7578
105 Nodes	CPU Time(sec)	3.050	5.433	5.800	5.933	
	# Iterations	6	16	12	10	
	# Pivots	456	541	532	507	
729 Arcs	Obj. Value	476.266	476.266	476.266	476.266	476.266
366 Nodes	CPU Time(sec)	16.650	25.316	37.650	40.717	
	# Iterations	7	15	12	12	
	# Pivots	1841	1910	1872	2125	

Table 4

Water distribution problems

Problem		Algorithm 6.5 (Implicit Grid)	Algorithm 5.1			Best known lower bound
			2-Segments	4-Segments	6-Segments	
	Obj. Value	−32392.731	−32392.730	−32392.730	−32392.730	−32392.731
46 Arcs	CPU Time(sec)	2.300	3.883	6.383	5.967	
30 Nodes	# Iterations	7	15	13	11	
	# Pivots	414	347	580	589	
	Obj. Value	−48199.858	−48199.858	−48199.859	−48199.859	−48199.859
196 Arcs	CPU Time(sec)	11.633	20.567	31.184	32.167	
150 Nodes	# Iterations	7	17	17	14	
	# Pivots	1823	1719	2888	2753	
	Obj. Value	−206175.21	−206175.21	−206175.22	−206175.22	−206175.24
906 Arcs	CPU Time(sec)	73.834	172.584	163.617	195.400	
666 Nodes	# Iterations	8	23[a]	15	14	
	# Pivots	12 166	15 755	15 036	18 185	

[a]Computation terminated after 23 iterations.

test problems ranging from 23 to more than 2000 variables. With a fixed grid strategy, 7–23 iterations are required to achieve comparable accuracy. Since the time per iteration for the implicit grid method is generally comparable to or only slightly higher than the simplest of the fixed grid approaches (which is generally the best of the fixed grid methods), overall computing times are consistently better for the implicit grid approach. Further significant improvements in computing times appear attainable through the reduction of the time spent in pivoting by exploitation of pricing and basis preservation strategies.

Appendix

In this appendix, we present the proofs of Theorems 6.6 and 6.7. First, we give the definitions of strongly convex and boundedly convex functions, which were introduced by Wolfe (1970).

Definition. Let $f(x)$ be a twice differentiable function from $\mathbb{R}^n \to \mathbb{R}^1$. $f(x)$ is *strongly convex* on $[l, u]$ with constant $q > 0$, iff q is a uniform lower bound on all the eigenvalues of the Hessian matrix $H(x) = [\partial^2 f(x)/\partial x_i\, \partial x_j]$ for $x \in [l, u]$. $f(x)$ is *boundedly convex* on $[l, u]$ with constant Q, if it is convex on $[l, u]$ and Q is a uniform upper bound on all the eigenvalues of the Hessian matrix for $x \in [l, u]$.

Before proving Theorem 6.6, we would like to establish the following lemma.

Lemma. *Let Algorithm 6.5 generate a sequence of points $\{x^{*j}\}$ and let $\{\mu^{*j}\}$ be the corresponding sequence of dual multipliers associated with the constraints $Ax = b$. If the right and left derivatives $f'_{i+}(l_i)$ and $f'_{i-}(u_i)$ are finite for all $i = 1, 2, \ldots n$, then the sequence $\{\mu^{*j}\}$ is bounded.*

Proof. As shown in Section 6.8, each of the problems $\mathrm{AP}^j(x^{*j-1}, \lambda^j)$ can be converted to an $\mathrm{LP}(\mathrm{LPAP}^j)$. Given an optimal basis matrix of $(\mathrm{LPAP})^j$, we define a basis matrix of (1.1) as follows.

If any of the variables x_i^{+k} or x_i^{-l} is basic in LPAP^j, then define x_i to be basic for (1.1). Let B^j be the corresponding basis submatrix of A. Then, the optimal multipliers can be expressed as $\mu^{*j} = c_B^{j^{\mathrm{T}}} B^{j^{-1}}$, where $c^j = (c_1^j, c_2^j, \ldots, c_n^j)^{\mathrm{T}}$ and each of c_i^j is the slope of a segment of $\hat{f}_i^j$.

For $l_i \leq y < z \leq u_i$, by the convexity of f_i, we have

$$f'_{i+}(l_i) \leq f'_{i+}(y) \leq \frac{f_i(z) - f_i(y)}{z - y} = \frac{f_i(y) - f_i(z)}{y - z} \leq f'_{i-}(z) \leq f'_{i-}(u).$$

Thus c_B^j remains bounded for all j as $j \to \infty$. Since

$$\|\mu^{*j}\| \leq \|c_B^j\|\, \|B^{j^{-1}}\|,$$

and there are only finitely many basis submatrices of A, the sequence $\{\mu^{*j}\}$ remains bounded. □

Proof of Theorem 6.6. For convenience of notation, we rewrite the constraint set $S := \{x \mid Ax = b, l \leq x \leq u\}$ as $S' := \{x \mid h(x) = 0, x \in X\}$ where $h : \mathbb{R}^n \to \mathbb{R}^m$ is affine and $X = \{x \mid l \leq x \leq u\}$.

We have that x^{*j} and μ^{*j} are optimal primal and dual solutions of $\mathrm{AP}^j(x^{*j-1}, \lambda^j)$, and hence $\{x^{*j}, \mu^{*j}\}$ satisfy the saddle point necessary optimality conditions:

$$\hat{f}^j(x^{*j}) + \langle \mu, h(x^{*j}) \rangle \leq \hat{f}^j(x^{*j}) + \langle \mu^{*j}, h(x^{*j}) \rangle \leq \hat{f}^j(x) + \langle \mu^{*j}, h(x) \rangle \tag{A.1}$$

where the first inequality holds for all $\mu \in \mathbb{R}^m$ and the second inequality holds for all $x \in X$. Since $f(x) \leq \hat{f}^j(x)$ and $\hat{f}^j(x) \leq f(x) + \varepsilon^j \ \forall x \in X$, from the second inequality of (A.1), we have

$$f(x^{*j}) + \langle \mu^{*j}, h(x^{*j}) \rangle \leq f(x) + \varepsilon^j + \langle \mu^{*j}, h(x) \rangle \quad \text{for all } x \in X. \tag{A.2}$$

The sequence $\{x^{*j}\}$ is always bounded and by the previous lemma, $\{\mu^{*j}\}$ is also bounded. Let $\{x^{**}, \mu^{**}\}$ be an accumulation point of $\{x^{*j}, \mu^{*j}\}$.

By Proposition 6.2, $\varepsilon^j \to 0$ as $j \to \infty$. Thus, for any fixed $x \in X$, as $j \to \infty$, by the continuity of f and h,

$$f(x^{**}) + \langle \mu^{**}, h(x^{**}) \rangle \leq f(x) + \langle \mu^{**}, h(x) \rangle. \tag{A.3}$$

Since the LHS of (A.2) is independent of x, (A.3) holds for all $x \in X$.

By the continuity of h, $h(x^{**}) = 0$ and we have for all $\mu \in \mathbb{R}^m$,

$$f(x^{**}) + \langle \mu, h(x^{**}) \rangle \leq f(x^{**}) + \langle \mu^{**}, h(x^{**}) \rangle. \tag{A.4}$$

(A.3) and (A.4) together constitute the saddle point sufficient optimality conditions for (1.1). □

In order to prove the convergence rate result as stated in Theorem 6.7, we need the following lemma, (see, e.g., Thakur, 1978), which gives the maximum error of approximation in approximating a boundedly convex function (or a function with a Lipschitz continuous derivative) by a piecewise-linear function.

Lemma. *Let $g : [a, b] \to \mathbb{R}$ be twice differentiable on $[a, b]$. Let $\hat{g}(x)$ be a piecewise-linear approximation of $g(x)$ in $[a, b]$, with a subdivision interval length λ. If $g(x)$ satisfies the condition*

(i) $|g''(x)| \leq Q \ \forall x \in [a, b]$, *or*

(ii) $|g'(x) - g'(y)| \leq Q|x - y| \ \forall x, y \in [a, b]$.

Then,

$$\max_{x \in [a,b]} |g(x) - \hat{g}(x)| \leq Q\lambda^2/8.$$

Proof of Theorem 6.7. By the strong convexity of f, we have

$$f(y)-f(x)-\nabla f(x)^{\mathrm{T}}(y-x)\geq\frac{q}{2}\|y-x\|^2 \quad \forall x, y\in[l, u].$$

Let $\{x^{*j}\}$ be the sequence of points generated by Algorithm 6.5. Let x^{**} be an optimal solution of (1.1). From the above inequality, we get

$$f(x^{*j})-f(x^{**})-\nabla f(x^{**})^{\mathrm{T}}(x^{*j}-x^{**})\geq\frac{q}{2}\|x^{*j}-x^{**}\|^2.$$

Since x^{**} is optimal for (1.1) and x^{*j} is feasible for (1.1), by the minimum principle, we have that

$$\nabla f(x^{**})^{\mathrm{T}}(x^{*j}-x^{**})\geq 0.$$

Therefore,

$$f(x^{*j})-f(x^{**})\geq\frac{q}{2}\|x^{*j}-x^{**}\|^2$$

or

$$\|x^{*j}-x^{**}\|^2\leq\frac{2}{q}\{f(x^{*j})-f(x^{**})\}.$$

Combining the above lemma with Proposition 6.3, we have

$$f(x^{*j})-f(x^{**})\leq\sum_{i=1}^{n}\varepsilon_i^j\leq\frac{Q\|\lambda^j\|^2}{8},$$

where λ^j is the subdivision interval length in the jth iteration. Thus,

$$\|x^{*j}-x^{**}\|^2\leq\frac{2}{q}\,\frac{Q\|\lambda^j\|^2}{8}.$$

Since $\lambda^j=\alpha\lambda^{j-1}$, $j=1, 2, \ldots$, we have

$$\|x^{*j}-x^{**}\|\leq\left[\left(\frac{Q}{4q}\right)^{1/2}\|\lambda_0\|\right]\alpha^j, \quad j=1, 2, \ldots,$$

where $0<\alpha<1$. Thus, the sequence $\{x^{*j}\}$ converges to x^{**} linearly at rate α from the very first iteration.

References

A. Bachem and B. Korte, "Mathematical programming and the estimation of input-output matrices", in: G. Castellani and P. Mazzolani, eds., *Mathematical Programming and its Economic Applications* (Angeli, Milano, 1981).

M.S. Bazaraa and C.M. Shetty, *Nonlinear programming, theory and algorithms* (John Wiley & Sons, NY, 1979).

P. Beck, L. Lasdon and M. Enqquist, "A reduced gradient algorithm for nonlinear network flow problems", *ACM Transactions on Mathematical Software* 9 (1983) 57-70.

D.P. Bertsekas, "On the Goldstein-Levitin-Poljak gradient projection method", *IEEE Transactions on Automatic Control* 21 (1976) 174-184.

M. Collins, L. Cooper, R. Helgason, J. Kennington and L. LeBlanc, "Solving the pipe network analysis problems using optimization techniques", *Management Science* 24 (1978) 747-760.

R.W. Cottle and S.G. Duvall, "A Lagrangian relaxation algorithm for the constrained matrix problem", Technical Report SOL 82-10, Stanford University (Stanford, CA, 1982).

R. Dembo and J. Klincewicz, "An approximate second order algorithm for network flow problems with convex separable costs", Series #21, School of Business, Yale University (New Haven, CT, 1978).

M.D. Grigoriadis and Tau Hsu, "RNET: the Rutgers minimum cost network flow subroutine", *SIGMAP Bulletin* (1979) 17-18.

P.A. Jensen and W.J. Barnes, *Network flow programming* (John Wiley & Sons, NY, 1980).

C.Y. Kao and R.R. Meyer, "Secant approximation methods for convex optimization", *Mathematical Programming Study* 14 (1981) 143-162.

R.R. Meyer, "Computational aspects of two-segment separable programming", Technical Report #382, Computer Sciences Department, University of Wisconsin (Madison, WI, 1980).

R.T. Rockafeller, "Optimization in networks", Lecture notes, Department of Mathematics, University of Washington (Seattle, WA, 1976).

R.T. Rockafeller, "Monotropic programming", *Nonlinear programming 4* (Academic Press, NY, 1981).

L.S. Thakur, "Error analysis for convex separable programs", *SIAM Journal of Applied Mathematics* 34 (1978) 704-714.

P. Wolfe, "Convergence theory in nonlinear programming", in: J. Abadie, ed., *Integer and nonlinear programming* (North-Holland, Amsterdam, 1970) 1-36.

Mathematical Programming Study 22 (1984) 206-216
North-Holland

NORMAL SOLUTIONS OF LINEAR PROGRAMS

O. L. MANGASARIAN
University of Wisconsin-Madison, USA

Received 21 March 1983
Revised manuscript received 15 February 1984

The solvability of a linear program is characterized in terms of the existence of a fixed projection on the feasible region, of all sufficiently large positive multiples of the gradient of the objective function. This projection turns out to be the normal solution obtained by projecting the origin on the optimal solution set. By seeking the solution with least 2-norm which minimizes the 1-norm infeasibility measure of a system of linear inequalities or of the optimality conditions of a linear program, one is led to a simple minimization problem of a convex quadratic function on the nonnegative orthant which is guaranteed to be solvable by a successive overrelaxation (SOR) method. This normal solution is an exact solution if the original system is solvable, otherwise it is an error-minimizing solution. New computational results are given to indicate that SOR methods can solve very large sparse linear program that cannot be handled by an ordinary linear programming package.

AMS (MOS) Subject Classifications: 90C05, 90C06, 90C20, 15A39.

Key words: Linear Programming, Linear Inequalities, Large Scale Systems, Quadratic Programming.

1. Introduction

A *normal solution* to a linear program is an exact solution with some least norm property if the linear program is solvable, otherwise it is an *approximate* solution with some least norm property also. By an *approximate* solution we mean a point which minimizes a measure of satisfaction of the optimality conditions of the linear program. By considering normal solutions we are led to:

(i) Iterative successive overrelaxation (SOR) methods capable of solving very large linear programs.

(ii) Approximate solutions to poorly posed or unsolvable linear programs.

(iii) A stable solution or approximate solution, to a linear program, endowed with a least norm property.

For solvable linear programs our normal solution is essentially equivalent to that of Tikhonov and Arsenin [16] which they obtain by solving an *asymptotic* problem [16, Theorem 1, p. 226], whereas our solution is obtained by solving a simpler exact problem, problem (2.2) for any $\varepsilon \in (0, \bar{\varepsilon}]$ for some $\bar{\varepsilon} > 0$ (Theorem 2.1). Tikhonov and Arsenin's weaker asymptotic result comes about because they square the

Sponsored by the United States Army under Contract No. DAAG29-80-C-0041. This material is based on work sponsored by National Science Foundation Grant MCS-8200632.

objective function of the linear program in their regularization problem [16, p. 226] and thereby lose an essential exact feature of our problem (2.2). Tikhonov and Arsenin also do not consider the important case of possibly unsolvable linear programs (Section 4), nor do they give explicit computational methods for solving their asymptotic problem.

We now outline our principal results and their relation to other work. In Section 2 we consider normal solutions of solvable linear programs. In Theorem 2.1 we give a complete characterization of the solvability of a linear program in terms of a 2-norm projection on the feasible region of a sufficiently large but finite positive multiple of the gradient of the objective function. This projection turns out to be fixed and equal to the unique 2-norm projection of the origin on the *optimal solution set.* Part of Theorem 2.1, a(i), follows readily from [13, Theorem 1], while its converse, part a(ii), which is essential for a comprehensive justification of the linear programming SOR Algorithm 2.3, has not been available before. Theorem 2.2, which follows from Theorem 2.1 and quadratic programming duality, characterizes the solvability of a linear program in terms of the solvability of a convex quadratic function minimization on the nonnegative orthant (2.7) without any a priori assumptions regarding the solvability of the linear program (2.1) as was the case in [11, 12]. In addition, Theorem 2.2 gives the complete basis for the linear programming SOR Algorithm 2.3 and its convergence (Theorem 2.4) thereby sharpening earlier convergence results [11, 12].

In Section 3 we turn our attention to a system of possibly inconsistent linear inequalities (3.1) and reduce it to the problem of finding the unique least 2-norm solution of the problem of minimizing the 1-norm infeasibility measure of (3.1). The principal advantage of this approach is that it leads to the SOR Algorithm 3.1 which, unlike most other iterative procedures [1, 15, 3] which require an a priori consistency assumption, will converge no matter whether the original system (3.1) is consistent or not. In either case Algorithm 3.1 will give an exact or approximate solution with least 2-norm (Theorem 3.2). Among the potential useful applications of this approach is in the image reconstruction techniques of tomography [7, 8] which require the solution of enormous sparse systems of linear equations with nonnegative variables. Most current iterative techniques for the tomography problem [7, 8, 2] need an a priori assumption regarding the consistency of the original system. In contrast our Algorithm 3.1 needs no such assumption.

In Section 4 we consider possibly unsolvable linear programs and reduce their solution to finding the least 2-norm primal–dual solution which minimizes the 1-norm of the optimality conditions of the given linear program. This approach leads to an SOR algorithm that is guaranteed to work whether the original linear program is solvable or not. In either case it will give an exact or approximate solution with least 2-norm.

Finally in Section 5 we give some numerical comparisons for one version of our linear programming SOR algorithm with the XMP version [14] of the revised simplex method for medium and large size sparse linear programs. These comparisons indicate that SOR methods can solve very large sparse linear programs that *cannot be solved* by an ordinary linear programming package.

We briefly describe the notation used. All matrices and vectors are real. For the $m\times n$ matrix A we denote row i by A_i, column j by $A_{\cdot j}$ and the element in row i and column j by A_{ij}. For x in the real n-dimensional Euclidean space $\mathbb{R}^n$, x_i denotes element i for $i=1,\ldots,n$, and x_+ denotes the vector with components $(x_+)_i=\max\{x_i,0\}$, $i=1,\ldots,n$. Vectors are either row or column vectors depending on the context. For x and y in $\mathbb{R}^n$, xy denotes the scalar product $\sum_{i=1}^n x_iy_i$, while $\|x\|_t$ for $1\leq t\leq\infty$ denotes the t-norm $(\sum_{i=1}^n |x_i|^t)^{1/t}$, $\|x\|$ denotes an arbitrary but fixed norm on $\mathbb{R}^n$ and $\|A\|$ denotes the subordinate matrix norm $\max_{\|x\|=1}\|Ax\|$ for an $m\times n$ matrix. The vector e will denote a vector of ones in any real Euclidean space. $\mathbb{R}^n_+$ will denote the nonnegative orthant $\{x\mid x\in\mathbb{R}^n, x\geq 0\}$. For a point c in $\mathbb{R}^n$, a closed set X in $\mathbb{R}^n$ and a number $t\in[1,\infty]$ the t-norm projection $p_t(c,X)$ of the point c on X is defined by

$$\|c-p_t(c,X)\|_t=\min_{x\in X}\|c-x\|_t$$

For a function $f:\mathbb{R}^n\to\mathbb{R}$ which is twice differentiable on $\mathbb{R}^n$, $\nabla f(x)$ denotes the n-dimensional gradient vector at x with components $\nabla_{x_i}f(x)$, $i=1,\ldots,n$, and $\nabla^2 f(x)$ denotes the $n\times n$ Hessian matrix at x with elements $(\nabla^2 f(x))_{ij}$, $i,j=1,\ldots,n$.

2. Normal solutions of solvable linear programs

We consider here the linear program

$$\text{maximize} \quad cx \qquad \text{subject to} \quad x\in X:=\{x\mid x\in\mathbb{R}^n, Ax\leq b, x\geq 0\} \tag{2.1}$$

where b and c are given vectors in $\mathbb{R}^m$ and $\mathbb{R}^n$ respectively and A is a given $m\times n$ real matrix. Let $\bar{X}$ denote the (possibly empty) optimal solution set of (2.1). We shall assume throughout this section that this linear program is feasible, that is X is nonempty. We begin with the following fundamental and geometrically plausible result.

2.1. Theorem. *Let the linear program* (2.1) *be feasible. Then*

a. (i) $\max_{x\in X} cx$ *has a solution* $\Rightarrow \exists\bar{\varepsilon}>0$: $p_2\left(\frac{c}{\varepsilon}, X\right)=p_2(0,\bar{X})$ *for all* $\varepsilon\in(0,\bar{\varepsilon}]$,

(ii) $\left.\begin{array}{l}\max_{x\in X} cx \textit{ has a solution}\\ \textit{and } \bar{x}=p_2(0,\bar{X})\end{array}\right\} \Leftarrow \exists\bar{\varepsilon}>0, \bar{x}$: $p_2\left(\frac{c}{\varepsilon}, X\right)=\bar{x}$ *for all* $\varepsilon\in(0,\bar{\varepsilon}]$,

where $p_2(x,X)$ *denotes the* 2*-norm projection of* x *on* X.

b. $$\sup_{x\in X} cx=\infty \quad \Leftrightarrow \quad \left\|p_2\left(\frac{c}{\varepsilon}, X\right)\right\| \to \infty \quad as\ \varepsilon \to 0+.$$

Proof. a(i): By noting that $p_2(c/\varepsilon, X)$ is a solution of either of the equivalent problems

$$\min_{x\in X} \left\|x-\frac{c}{\varepsilon}\right\|_2^2 \quad \Leftrightarrow \quad \min_{x\in X} -cx+\frac{\varepsilon}{2}\|x\|_2^2 \tag{2.2}$$

the implication of a(i) follows from Theorem 1 of [13].

a(ii): Since $\bar{x}=p_2(c/\varepsilon, X)$ for all $\varepsilon\in(0, \bar{\varepsilon}]$, there exists a $(u(\varepsilon), v(\varepsilon))\in\mathbb{R}^{m+n}$ such that $(\bar{x}, u(\varepsilon), v(\varepsilon))$ satisfies the Karush–Kuhn–Tucker conditions [9] for (2.2), that is

$$\begin{gathered}\varepsilon\bar{x}-c+A^{\mathrm{T}}u(\varepsilon)-v(\varepsilon)=0, \quad v(\varepsilon)\bar{x}=0, \quad A\bar{x}\leqslant b, \quad u(\varepsilon)(A\bar{x}-b)=0, \\ (\bar{x}, u(\varepsilon), v(\varepsilon))\geqslant 0, \forall\varepsilon\in(0, \bar{\varepsilon}]\end{gathered} \tag{2.3}$$

By the fundamental theorem for the existence of basic feasible solutions for linear equations with nonnegative variables [6, Theorem 2.11], and the complementarity conditions $u(\varepsilon)(A\bar{x}-b)=0$, $v(\varepsilon)\bar{x}=0$, it follows that there exist $(u(\varepsilon), v(\varepsilon))$ satisfying (2.3) such that all elements of $(u(\varepsilon), v(\varepsilon))$ not corresponding to some subset of k linearly independent columns of $[A^{\mathrm{T}}-I]$ are zero. Since the rank of $[A^{\mathrm{T}}-\mathrm{I}]$ is n, it follows that we can take $k=n$ and denote by $B(\varepsilon)$ as this 'basis' matrix of n linearly independent columns of $[A^{\mathrm{T}}-I]$. Hence it follows for such a 'basic' solution $(u(\varepsilon), v(\varepsilon))$ satisfying (2.3) that

$$\|u(\varepsilon)\ \ v(\varepsilon)\|\leqslant(\|c\|+\bar{\varepsilon}\|\bar{x}\|)\|B(\varepsilon)^{-1}\| \quad \forall\varepsilon\in(0, \bar{\varepsilon}].$$

Since $[A^{\mathrm{T}}-I]$ contains a finite number of basis matrices it follows that for some basis matrix B, $\|B(\varepsilon)^{-1}\|\leqslant\|B^{-1}\|$ for all $\varepsilon\in(0, \bar{\varepsilon}]$ and consequently

$$\|u(\varepsilon)\ \ v(\varepsilon)\|\leqslant(\|c\|+\bar{\varepsilon}\|\bar{x}\|)\|B^{-1}\| \quad \forall\varepsilon\in(0, \bar{\varepsilon}]. \tag{2.4}$$

Now let $\{\varepsilon^i\}$ be a sequence of positive numbers in $(0, \bar{\varepsilon}]$ converging to 0. Then there exists a sequence $\{(u(\varepsilon^i), v(\varepsilon^i))\}$ satisfying (2.3) and (2.4) and hence it is bounded and has an accumulation point $(\bar{u}, \bar{v})$ satisfying

$$-c+A^{\mathrm{T}}\bar{u}-\bar{v}=0, \quad \bar{v}\bar{x}=0, \quad A\bar{x}\leqslant b, \quad \bar{u}(A\bar{x}-b)=0, \quad (\bar{x}, \bar{u}, \bar{v})\geqslant 0. \tag{2.5}$$

These are the Karush–Kuhn–Tucker conditions for the linear program (2.1) and hence $\bar{x}$ solves (2.1). Since $(\bar{x}, u(\varepsilon), v(\varepsilon))$ also satisfies (2.3) which are also the Karush–Kuhn–tucker conditions for

$$\min\ \tfrac{1}{2}\|x\|_2^2 \qquad \text{subject to} \quad Ax\leqslant b, \quad x\geqslant 0, \quad cx\geqslant c\bar{x}$$

with optimal $x=\bar{x}$ and optimal multiplier vector of $(u(\varepsilon)/\varepsilon, v(\varepsilon)\varepsilon, 1/\varepsilon)$, it follows that $\bar{x}=p_2(0, \bar{X})$.

(b)($\Leftarrow$): If not then the linear program (2.1) has a solution and by part a(i) of this theorem $\exists\bar{\varepsilon}>0$: $p_2(c/\varepsilon, X)=p_2(0, \bar{X})$ for all $\varepsilon\in(0, \bar{\varepsilon}]$. This however contradicts the hypothesis that $\|p_2(c/\varepsilon, X)\|\to\infty$ as $\varepsilon\to 0+$.

(⇒): If not then, for a sequence of positive numbers $\{\varepsilon^i\}$ converging to zero, the sequence $\{\|p_2(c/\varepsilon^i, X)\|\}$ is bounded. By defining $x(\varepsilon^i) := p_2(c/\varepsilon^i, X)$ we get that $x(\varepsilon^i)$ and some $(u(\varepsilon^i), v(\varepsilon^i)) \in \mathbb{R}^{m+n}$ satisfy the Karush–Kuhn–Tucker conditions for $\min_{x\in X} \|x - c/\varepsilon^i\|_2^2$ for $i = 1, 2, \ldots$, that is

$$\varepsilon^i x(\varepsilon^i) - c + A^{\mathrm{T}} u(\varepsilon^i) - v(\varepsilon^i) = 0, \quad v(\varepsilon^i) x(\varepsilon^i) = 0,$$
$$Ax(\varepsilon^i) \le b, \quad u(\varepsilon^i)(Ax(\varepsilon^i) - b) = 0, \quad (x(\varepsilon^i), u(\varepsilon^i), v(\varepsilon^i)) \ge 0. \tag{2.6}$$

By the same argument as in the proof of part a(ii) of this theorem we can show that the sequence $\{(u(\varepsilon^i), v(\varepsilon^i))\}$ satisfying (2.6) can be taken as bounded since $\{x(\varepsilon^i)\}$ is also bounded. Thus the sequence $\{(x(\varepsilon^i), u(\varepsilon^i), v(\varepsilon^i))\}$ is bounded and has an accumulation point $(\bar{x}, \bar{u}, \bar{v})$ satisfying (2.5). Hence $\bar{x}$ solves (2.1) which contradicts the hypothesis that $\sup_{x\in X} cx = \infty$. □

By noting that the quadratic programming dual [9] to (2.2) is

$$\underset{(u,\, v)\in\mathbb{R}_+^{m+n}}{\text{minimize}} \quad \tfrac{1}{2}\|A^{\mathrm{T}}u - v - c\|_2^2 + \varepsilon bu \tag{2.7}$$

where the primal and dual variables x and (u, v) are related by

$$x = \frac{1}{\varepsilon}(-A^{\mathrm{T}}u + v + c), \tag{2.8}$$

the following theorem is a direct consequence of Theorem 2.1.

2.2. Theorem. *The linear program* (2.1) *is solvable if and only if there exists an* $\bar{\varepsilon} > 0$ *such that for each* $\varepsilon \in (0, \bar{\varepsilon}]$ *the quadratic program* (2.7) *has a solution* $(u(\varepsilon), v(\varepsilon))$ *and such that the vector* $\bar{x}$ *defined by*

$$\bar{x} := \frac{1}{\varepsilon}(-A^{\mathrm{T}}u(\varepsilon) + v(\varepsilon) + c), \quad \varepsilon \in (0, \bar{\varepsilon}], \tag{2.9}$$

is independent of ε, *in which case* $\bar{x} = p_2(0, \bar{X})$.

If we define the objective function of (2.7) by

$$f(z) := \tfrac{1}{2}\|A^{\mathrm{T}}u - v - c\|_2^2 + \varepsilon bu, \quad z := \begin{pmatrix} u \\ v \end{pmatrix} \tag{2.10}$$

then we can prescribe an SOR procedure for solving (2.7) which in view of Theorem 2.2 solves the linear program (2.1). The SOR procedure is essentially a gradient projection algorithm of the following type

$$z_j^{i+1} = (z_j^i - \omega(\nabla^2 f(z^i))_{jj}^{-1} \nabla_{z_j} f(z_1^{i+1}, \ldots, z_{j-1}^{i+1}, z_j^i, \ldots, z_{m+n}^i))_+,$$
$$0 < \omega < 2, \; j = 1, \ldots, m+n. \tag{2.11}$$

More specifically [12] the following SOR algorithm for solving the linear program (2.1) follows directly from (2.11) and (2.10).

2.3. LPSOR(a, b, c) Algorithm. Choose $(u^0, v^0) \in \mathbb{R}^{m+n}$, $\omega \in (0, 2)$ and $\varepsilon > 0$. Having (u^i, v^i) determine (u^{i+1}, v^{i+1}) as follows:

$$u_j^{i+1} = \left(u_j^i - \frac{\omega}{\|A_j\|_2^2}\left(A_j\left(\sum_{\substack{l=1 \\ \text{for } j>1}}^{j-1} (A^T)_{\cdot l} u_l^{i+1} + \sum_{l=j}^{m} (A^T)_{\cdot l} u_l^i - v^i - c\right) + \varepsilon b_j\right)\right)_+$$

$$\text{if } A_j \neq 0, \quad u_j^{i+1} = 0 \text{ if } A_j = 0, \quad j = 1, \ldots, m,$$

$$v^{i+1} = (v^i - \omega(-A^T u^{i+1} + v^i + c))_+.$$

Note that Algorithm 2.3 is sparsity-preserving for it works with the *rows* of A only and the product AA^T need not be computed.

The following convergence theorem which follows from Theorem 2.2 above and [12, Theorem 2] sharpens previous LPSOR convergence theorems [11, Theorem 3.2] and [12, Theorem 4].

2.4. LPSOR(A, b, c) Convergence Theorem. (a) *The linear program* (2.1) *has a solution if and only if there exists a real positive number* $\bar{\varepsilon}$ *such that for each* $\varepsilon \in (0, \bar{\varepsilon}]$, *each accumulation point* $(u(\varepsilon), v(\varepsilon))$ *of the sequence* $\{(u^i, v^i)\}$ *generated by the* LPSOR(A, b, c) *Algorithm* 2.3 *solves* (2.7) *and the corresponding* $\bar{x}$ *determined by* (2.9) *is independent of* ε, *in which case* $\bar{x} = p_2(0, \bar{X})$.

(b) *If the linear program* (2.1) *has a solution and its constraints satisfy the Slater constraint qualification, that is* $Ax < b$ *for some* $x \geq 0$, *then the sequence* $\{(u^i, v^i)\}$ *of the* LPSOR(A, b, c) *Algorithm* 2.3 *is bounded and has an accumulation for each* $\varepsilon \in (0, \bar{\varepsilon}]$ *for some* $\bar{\varepsilon} > 0$.

Computational results for the Algorithm 2.3 are given in Section 5.

3. Normal solutions of possibly inconsistent linear inequalities

We consider in this section the possibly inconsistent system of linear inequalities

$$Ax \leq b, \quad x \geq 0, \tag{3.1}$$

where A is a given $m \times n$ matrix and b is a given vector in $\mathbb{R}^m$. If we try to 'solve' the above system by an SOR [10] procedure applied to the obvious 2-norm minimization problem

$$\min_{x \geq 0} \|(Ax - b)_+\|_2^2 = \min_{(x, y) \geq 0} \|Ax + y - b\|_2^2 := \min_{(x, y) \geq 0} \theta(x, y) \tag{3.2}$$

one needs the condition

$$\nabla\theta(x, y) = \begin{bmatrix} A^T(Ax + y - b) \\ Ax + y - b \end{bmatrix} > 0 \quad \text{for some } (x, y) \in \mathbb{R}^{n+m} \tag{3.3}$$

to guarantee boundedness of the SOR iterates [10, Theorem 2.2], which by the Gordan Theorem [9, Theorem 2.4.5] is equivalent to the condition that

$$Ax \leq 0, \quad 0 \neq x \geq 0, \quad \text{has no solution} \tag{3.4}$$

Unfortunately this condition is *not* satisfied in general, as is the case when the feasible region is nonempty and unbounded. To avoid this difficulty we use the SOR procedure of Section 2 to find the 2-norm projection of the origin in $\mathbb{R}^{n+m}$ on the nonempty solution set of the linear program

$$\min_{(x,\, y) \in \mathbb{R}_+^{n+m}} \{ey \mid Ax - y \leq b\} \tag{3.5}$$

which is the equivalent of the problem of minimizing the 1-norm feasibility of (3.1)

$$\min_{x \in \mathbb{R}_+^n} \|(Ax - b)_+\|_1. \tag{3.6}$$

The key feature of this approach is that the *SOR procedure will work no matter whether the system* (3.1) *is consistent or not.* In either case the SOR procedure will obtain the unique solution $(\bar{x}, \bar{y})$ of (3.5) with least 2-norm. In terms of the original inequalities (3.1), $\bar{x}$ is the unique solution of (3.6) which minimizes $\|x, (Ax - b)_+\|_2$. Needless to say, if (3.1) is consistent then $\bar{x}$ is the unique 2-norm projection of the origin in R^n on the nonempty feasible region determined by (3.1). To obtain an SOR procedure for solving (3.5) we take the dual of the quadratic perturbation of (3.5)

$$\min_{(x,\, y) \in \mathbb{R}_+^{n+m}} \left\{ ey + \frac{\varepsilon}{2} \|x, y\|_2^2 \,\middle|\, Ax - y \leq b \right\} \tag{3.7}$$

which turns out to be [9]

$$\min_{(u,\, v,\, w) \in \mathbb{R}_+^{m+n+m}} \tfrac{1}{2}\|A^{\mathrm{T}}u - v\|_2^2 + \tfrac{1}{2}\|u + w - e\|_2^2 + \varepsilon bu := \min_{(u,\, v,\, w) \in \mathbb{R}_+^{m+n+m}} \psi(u, v, w) \tag{3.8}$$

with (x, y) related to (u, v, w) by

$$x = \frac{1}{\varepsilon}(-A^{\mathrm{T}}u + v), \qquad y = \frac{1}{\varepsilon}(u + w - e). \tag{3.9}$$

Since

$$\nabla\psi(u, v, w) = \begin{bmatrix} A(A^{\mathrm{T}}u - v) + u + w - e + \varepsilon b \\ -(A^{\mathrm{T}}u - v) \\ u + w - e \end{bmatrix},$$

$$\nabla^2\psi(u, v, w) = \begin{bmatrix} AA^{\mathrm{T}} + I & -A & I \\ -A^{\mathrm{T}} & I & 0 \\ I & 0 & I \end{bmatrix}, \tag{3.10}$$

it follows that $\nabla\psi(0, e, \lambda e) > 0$ for sufficiently large λ and consequently the iterates of the SOR algorithm of [10, Algorithm 3.2, Remark 3.2] applied to 3.8 will have an accumulation for *all* positive values of ε. In particular we have the following algorithm and convergence theorem.

3.1. LISOR(A, b) Algorithm. Choose $(u^0, v^0, w^0) \in \mathbb{R}_+^{m+n+m}$, $\omega \in (0, 2)$ and $\varepsilon > 0$. Having (u^i, v^i, w^i) determine $(u^{i+1}, v^{i+1}, w^{i+1})$ as follows:

$$u_j^{i+1} = \left(u_j^i - \frac{\omega}{1 + \|A_j\|_2^2}\left(A_j\left(\sum_{\substack{l=1\\ \text{for } j>1}}^{j-1} (A^T)_{\cdot l} u_l^{i+1} + \sum_{l=j}^{m} (A^T)_{\cdot l} u_l^i - v^i\right) + u_j^i + w_j^i - 1 + \varepsilon b_j\right)\right)_+,$$

$$j = 1, \ldots, m, \quad v^{i+1} = (v^i + \omega(A^T u^{i+1} - v^i))_+,$$

$$w^{i+1} = (w^i - \omega(u^{i+1} + w^i - e))_+$$

3.2. LISOR(A, b) Convergence Theorem. *For each $\varepsilon > 0$ the iterates (u^i, v^i, w^i) of the LISOR(A, b) algorithm are bounded and have an accumulation point $(u(\varepsilon), v(\varepsilon), w(\varepsilon))$. For all $\varepsilon \in (0, \bar{\varepsilon}]$ for some $\bar{\varepsilon} > 0$, the point $(\bar{x}, \bar{y})$*

$$\bar{x} := \frac{1}{\varepsilon}(-A^T u(\varepsilon) + v(\varepsilon)), \qquad \bar{y} := \frac{1}{\varepsilon}(u(\varepsilon) + w(\varepsilon) - e) \tag{3.11}$$

is independent of ε and is the unique solution of (3.5) *with least* 2*-norm, and $\bar{x}$ is the unique solution* (3.6) *with least $\|x, (Ax - b)_+\|_2$.*

Proof. That the iterates (u^i, v^i, w^i) have an accumulation point which solves (3.8) for all $\varepsilon > 0$ follows from Theorem 2.2 of [10]. That for $\varepsilon \in (0, \bar{\varepsilon}]$ for some $\bar{\varepsilon} > 0$, $(\bar{x}, \bar{y})$ defined by (3.11) is the unique solution of (3.5) with least 2-norm follows from the duality equivalence of (3.7) and (3.8) and from Theorem 2.1. Since problems (3.5) and (3.6) are equivalent and $\bar{y} = (A\bar{x} - b)_+$ for a solution of (3.5), it follows that $\bar{x}$ is the unique solution of (3.6) with least value of $\|x, (Ax - b)_+\|_2$. □

We note here Eremin's algorithm [5] which is one of the few iterative algorithms capable of handling inconsistent inequalities. Eremin gives no computational experience and the presence in his algorithm of a positive stepsize λ_i satisfying $\lambda_i \to 0$ and $\sum_{i=1}^{\infty} \lambda_i = \infty$ may cause slow convergence.

An interesting application of the above method is to the problem of image reconstruction techniques [7, 8] where the fundamental problem is to solve the system

$$Bx = d, \quad x \geq 0, \tag{3.12}$$

where typically the $m \times n$ matrix B may be of order 28000×6000 with less than 1% of nonzero elements [8]. Iterative methods are well suited for such large sparse problems. Unfortunately such methods often require assumptions that are rarely verifiable. Typically such methods assume a priori that the system (3.12) has a solution [7, 8]. In contrast our proposed LISOR method requires no assumptions whatsoever when applied to the equivalent problem

$$Bx \leq d, \quad -Bx \leq -d, \qquad x \geq 0. \tag{3.13}$$

In particular LISOR$(\binom{B}{-B}, \binom{d}{-d})$ will lead to the unique solution $\bar{x}$ of $\min_{x \in \mathbb{R}^n_+} \|Bx - d\|_1$ with least $\|x, Bx - d\|_2$.

4. Normal solutions of possibly unsolvable linear programs

We consider here again the linear program (2.1) but make no assumptions whatsoever regarding its feasibility or solvability. The idea here is to apply the LISOR Algorithm 3.1 to the equivalent linear complementarity problem for (2.1) [4]

$$Mz + q \geq 0, \quad qz \leq 0, \qquad z \geq 0, \tag{4.1}$$

where

$$z = \begin{pmatrix} x \\ u \end{pmatrix} \in \mathbb{R}^{n+m}, \qquad M = \begin{pmatrix} 0 & A^T \\ -A & 0 \end{pmatrix}, \qquad q = \begin{pmatrix} -c \\ b \end{pmatrix} \tag{4.2}$$

and u is the dual variable. Direct application of the convergence Theorem 3.2 to the LISOR $(\binom{-M}{q}, \binom{q}{0})$ algorithm gives that the iterates of LISOR $(\binom{-M}{q}, \binom{q}{0})$ are bounded for all $\varepsilon > 0$ and that for all $\varepsilon \in (0, \bar{\varepsilon}]$ for some $\bar{\varepsilon} > 0$ they lead to a $\bar{z}$ which is independent of ε and such that $\bar{z}$ is the unique solution of

$$\min_{z \in \mathbb{R}^{n+m}_+} \|(-Mz - q, qz)_+\|_1 \tag{4.3}$$

with least $\|z, (-Mz - q, qz)_+\|_2$. Hence $\bar{z} = \binom{\bar{x}}{\bar{u}}$, is an exact solution of the primal–dual pair with least 2-norm if the linear program (2.1) is solvable. Else, it is the unique solution of (4.3) with least 2-norm for the vector in $\mathbb{R}^{2(n+m)+1}$ composed of the primal-dual variable z, the primal–dual infeasibility $(-Mz - q)_+$, and the primal–dual objective function inequality $(qz)_+$.

5. Computational results

Computational experiments have been carried so far on the LPSOR Algorithm 2.3 only. Results on medium-sized problems were given in [11]. We give below new computational results for randomly generated large sparse problems carried on the VAX 11/780 with double accuracy floating point addition time of 4.6 μs and multiplication time of 6.0 μs. Comparisons were made with Marston's XMP revised simplex linear programming code [14]. The results shown in Table 1 are all for a matrix A with essentially a 'tridiagonal' structure and fully dense last column and row. The XMP accuracy was to within 12-figure accuracy of the true objective function when it managed to obtain a solution. The accuracy of the LPSOR was measured by the ∞-norm of the primal infeasibility of the numerical solution and the relative deviation of the computed maximum value from the true maximum. The table indicates that for the accuracy obtained, the LPSOR method becomes competitive with the simplex method as the problem size gets larger and that for very large problems, SOR methods may be the only viable methods of solution.

Table 1

Comparison of the Revised Simplex Code XMP and LPSOR for Solving 2.1: m = no. of inequality constraints, n = no. of nonnegative variables

		XMP		LPSOR		
m	n	Iteration No.	Hr:Min:Sec	Iteration No.	Hr:Min:Sec	Relative Accur.
100	200	123	0:00:11	180	0:00:17	10^{-6}
500	1 000	746	0:03:12	520	0:05:11	10^{-9}
1 000	1 000	2 309	0:42:02	1 640	0:26:12	10^{-4}
2 500	10 000	Could not solve[a]		480	0:37:25	10^{-4}
5 000	20 000	Could not solve[b]		660	1:17:53	10^{-4}

[a] Program was killed after more than 3 hours of CPU time.

[b] Program used virtual memory spaced much larger than physical memory, so it ran inefficiently and had to be killed within 10 minutes of CPU time which corresponded to over 8 hours of real elapsed time.

Acknowledgement

I wish to acknowledge the original and tireless programming effort of my former student David P. Anderson in obtaining all the computational results of this paper.

References

[1] S. Agmon, "The relaxation method for linear inequalities", *Canadian Journal of Mathematics* 6 (1954) 382–392.
[2] Y. Censor, "Row-action methods for huge and sparse systems and their applications", *SIAM Review* 23 (1981) 444–466.
[3] Y. Censor and T. Elfving: "New Methods for linear inequalities", *Linear algebra and Its Applications* 42 (1982) 199–211.
[4] R. W. Cottle and G. B. Dantzig: "Complementary pivot theory of mathematical programming", *Linear Algebra and Its Applications* 1 (1968) 103–125.
[5] I. I. Eremin, "Iteration method for Čebyšev approximations for sets of incompatible linear inequalities", *Soviet Mathematics Doklady* 3 (1962) 570–572.
[6] D. Gale, *The theory of linear economic models* (McGraw-Hill, New York, 1960).
[7] G. T. Herman, Image reconstruction from projections: The fundamentals of computerized tomography (Academic Press, New York, 1980).
[8] G. T. Herman, A. Lent and P. H. Lutz, "Relaxation methods for image reconstruction", *Communications of the ACM* 21 (1978) 152–158.
[9] O. L. Mangasarian, Nonlinear programming (McGraw-Hill, New York, 1969).

[10] O. L. Mangasarian, "Solution of symmetric linear complementarity problems by iterative methods", *Journal of Optimization Theory and Applications* 22 (1977) 465–485.

[11] O. L. Mangasarian, "Iterative solution of linear programs", *SIAM Journal on Numerical Analysis* 18 (1981) 606–614.

[12] O. L. Mangasarian, "Sparsity-preserving SOR algorithms for separable quadratic and linear programming problems", *Computers and Operations Research* 11 (1984) 105–112.

[13] O. L. Mangasarian and R. R. Meyer, "Nonlinear perturbation of linear programs", *SIAM Journal on Control and Optimization* 17 (1979) 745–757.

[14] R. E. Marsten, "The design of the XMP linear programming library", MIS Technical Report 80–2, Management Information Systems, The University of Arizona, Tuscon, Arizona.

[15] T. S. Motzkin and I. J. Schoenberg, "The relaxation method for linear inequalities", *Canadian Journal of Mathematics* 6 (1954) 393–404.

[16] A. N. Tikhonov and V. Y. Arsenin, Solutions of ill-posed problems (Halsted Press, Wiley, New York, 1977).

Mathematical Programming Study 22 (1984) 217-230
North-Holland

LOCAL STRUCTURE OF FEASIBLE SETS IN NONLINEAR PROGRAMMING, PART II: NONDEGENERACY

Stephen M. ROBINSON
University of Wisconsin-Madison, Madison, WI 53706, USA

Received 12 June 1983
Revised manuscript received 29 March 1984

In this paper we examine the local structure of the feasible set of a nonlinear programming problem under the condition of nondegeneracy. We introduce this condition, examine its relationships to known properties of optimization problems, and show that when it holds at a given point the portion of the feasible set near that point is diffeomorphic to a simple convex set (often polyhedral). Moreover, this diffeomorphic relation is stable under small changes in the problem functions.

Key words: Nonlinear Programming, Feasible Sets.

0. Introduction

In this paper we study the system of nonlinear constraints

$$h(x)=0, \quad x\in C\cap \Omega, \tag{0.1}$$

where h is a C^r function from an open subset Ω of $\mathbb{R}^n$ to $\mathbb{R}^m$ ($r\geq 1$), and C is a nonempty closed convex set in $\mathbb{R}^n$. We shall be concerned with properties of the set $\mathcal{F}$ of points satisfying (0.1) under a certain regularity condition that we term nondegeneracy. This condeition is defined in Section 1, where its relationships to standard constraint qualifications used in nonlinear programming are also explored.

In Section 2 we introduce perturbations into the function h, and we show that under nondegeneracy the portions of the feasible sets of (0.1), and of all problems 'close' to (0.1), that are near x_0 are mutually C^r diffeomorphic. Geometrically, this means that if we do not perturb h very much then we are essentially working with the same problem, up to a C^r transformation of the coordinates. Moreover, this common feasible set has, with the correct choice of coordinates, a very simple structure: it is just the intersection of the part of C near x_0 with the tangent space to $h^{-1}(0)$ through x_0. The coordinates through which this simplification is made are easy to compute, and this means that nonlinear optimization problems whose feasible sets look like (0.1) can actually be regarded, locally, as problems of optimization on a fixed convex set which will be polyhedral if C is polyhedral.

Sponsored by the U.S. National Science Foundation under Grant No. MCS 8200632. Preliminary research for this paper was done at the Centre de Recherche de Mathématiques de la Décision, Université Paris-IX Dauphine, with travel support from C.N.R.S., and the writing was completed at the International Institute for Applied Systems Analysis, Laxenburg, Austria. The author thanks all of these agencies for their support of this work.

The main point of Section 3 is to apply this point of view to establish some properties of local minimizers of such problems. We show that all such local minimizers must lie on a certain face of C, so that the optimization problem in question may be further simplified by restricting one's attention to that face of C. Finally, we show that when the problem is thus restricted the local minimizer in question inherits from the original problem the property of nondegeneracy, so that all of the results established under the nondegeneracy hypothesis will remain valid for the restricted problem.

1. Nondegeneracy

In [5] we introduced a fundamental decomposition of $\mathbb{R}^n$ around a feasible point for (0.1), and we used it to prove various results about the feasible set $\mathcal{F} = C \cap h^{-1}(0)$. Since we shall use the same decomposition here, we first summarize its properties.

Given a point $x_0 \in \mathcal{F}$, denote $h'(x_0)$ by D. Let M be the subspace of $\mathbb{R}^n$ parallel to the affine hull A of the set C, and assume $D(M) = \mathbb{R}^m$. Denote $M \cap \ker D$ by K, and let L and J be subspaces complementary to K in M and in $\ker D$ respectively. Then one has $\mathbb{R}^n = J \oplus K \oplus L$, and one can define the projectors P_J, P_K, and P_L onto the spaces indicated along, in each case, the other two spaces. One has $P_J + P_K + P_L = I$, and the product of any two of these three projectors is zero. We shall denote by P_0 the projector $P_J + P_K$ onto $\ker D$ along L.

With these definitions the restriction of D to L is a bijection onto $\mathbb{R}^m$, and this bijection has an inverse D^-. By a slight abuse of notation we can regard D^- as a linear operator from $\mathbb{R}^m$ to $\mathbb{R}^n$ whose image is L and for which $DD^- = I$ and $D^-D = P_L$. One way to compute D^- is to let E be any bijection from $\mathbb{R}^m$ to L, then set $D^- = E(DE)^{-1}$; the inverse of DE exists because we assumed $D(M) = \mathbb{R}^m$.

The condition of nondegeneracy that we introduce here deals with the relationship of the spaces just described to the convex set C appearing in (0.1). Geometrically, it says that the space L can be chosen so that it is complementary to the tangent space ($\ker D$) to $h^{-1}(0)$ at x_0, while at the same time for any x in C near x_0, movement along L near x does not cause one to leave C.

Definition 1.1. Let x_0 satisfy (0.1). We say x_0 is a *nondegenerate feasible point* for (0.1) if $D(M) = \mathbb{R}^m$ and there exist an open neighborhood N of the origin in $\mathbb{R}^n$ and a subspace L, complementary to K in M, such that

$$C \cap (x_0 + N) + L \cap N \subset C. \tag{1.1}$$

It might appear that instead of (1.1) one could impose the simpler condition that $x_0 + L \cap N \subset C$. However, this condition is not equivalent to (1.1), and in fact we shall need (1.1). To see that the two conditions are not equivalent, consider the closed convex cone C in $\mathbb{R}^3$ defined by

$$C := \{(x_1, x_2, x_3) | x_1 \geq 0, x_3 \geq 0, x_2^2 \leq 2x_1x_3\}.$$

Evidently aff $C = \mathbb{R}^3$. If we let $x_0 := (0, 0, 1)$ and $K := \mathbb{R}^2 \times \{0\}$, then we can let $L := \{0\}^2 \times \mathbb{R}$, and for an appropriate neighborhood N of the origin in $\mathbb{R}^3$ we have $x_0 + L \cap N \subset C$. However, for any positive ε we have $(\frac{1}{2}\varepsilon^2, \varepsilon, 1) \in C$, but for all small positive δ, $(\frac{1}{2}\varepsilon^2, \varepsilon, 1 - \delta) \notin C$. Hence the condition (1.1) is not satisfied.

It is not difficult to show that if L is complementary in M to $M \cap \ker h'(x_0)$, then for x near x_0, L is also complementary to $M \cap \ker h'(x)$. It follows easily that if x_0 is a nondegenerate feasible point for (0.1) then so is each feasible point sufficiently close to x_0.

If x_0 is a nondegenerate feasible point for (0.1) then it is necessarily also a regular point in the sense of [5]. To see this, note that $L \oplus K = M$, so that $D(L) = D(M) = \mathbb{R}^m$. Therefore, for the neighborhood N of (1.1) it must be true that $D(L \cap N)$ is a neighborhood of the origin in $\mathbb{R}^m$. We can rewrite (1.1) as

$$(C - x_0) \cap N + L \cap N \subset C - x_0,$$

and by applying D to both sides we see that $D(C - x_0)$ is a neighborhood of 0 in $\mathbb{R}^m$. But this is one of the equivalent ways of expressing regularity [5, Proposition 1.1].

There are several ways in which (1.1) might be satisfied. One obvious way is for C to be a cylinder of the form $Q + C'$, where Q is a subspace and C' is convex. If $Q + K = M$, then the requirement is satisfied. Of course, nondegeneracy may hold also under much less stringent conditions, since (1.1) deals only with local behavior. In case C is polyhedral, as it is in many practical applications, then the *tangent cone* to C at x_0, $T_C(x_0)$, coincides with $C - x_0$ near the origin. In that case we can obtain a simple characterization of nondegeneracy in terms of $T_C(x_0)$ and its polar $N_C(x_0)$, the *normal cone* to C at x_0. The statement of this characterization uses the idea of the *lineality space* of a cone Z: this space, written lin Z, is the largest subspace in Z (i.e., $Z \cap (-Z)$).

Proposition 1.2. *Suppose C is polyhedral and x_0 is feasible for* (0.1). *Then each of the following is necessary and sufficient for x_0 to be nondegenerate*:

a. $D[\text{lin } T_C(x_0)] = \mathbb{R}^m$.

b. $\{u \mid D^* u \in \text{aff } N_C(x_0)\} = \{0\}$.

Proof. The two conditions are equivalent since

$$D[\text{lin } T_C(x_0)] = D[\{\text{aff } N_C(x_0)\}^\perp] = \{u \mid D^* u \in \text{aff } N_C(x_0)\}^\perp.$$

To show that (a) implies nondegeneracy, we first observe that lin $T_C(x_0) \subset M$, so under (a) we have $D(M) = \mathbb{R}^m$. Next, we note that to establish (1.1) it is enough to show that

$$\text{lin } T_C(x_0) + K = M. \tag{1.2}$$

Indeed, if (1.2) holds we can take L to be any subspace of lin $T_C(x_0)$ that is complementary to K in M. Then since C is polyhedral, $T_C(x_0)$ coincides with $C - x_0$ near the origin, so that (1.1) will hold for an appropriate N.

To establish (1.2) we need only show that lin $T_C(x_0)+K\supset M$, since the reverse inclusion is trivial. Let $m\in M$; since $D[\text{lin } T_C(x_0)]=\mathbb{R}^m$ there is some $l\in \text{lin } T_C(x_0)$ with $Dl=Dm$. But then $m-l\in K$, so $m=l+(m-l)\in \text{lin } T_C(x_0)+K$, which proves (1.2).

We next show that nondegeneracy implies (a). We are given that $D(M)=\mathbb{R}^m$ and that a subspace L exists, complementary to K in M, for which (1.1) holds for some neighborhood N. In particular, it follows from (1.1) that $L\cap N\subset C-x_0$. If we write cone A for the cone generated by a set A, we then have

$$L=\text{cone } L\cap N\subset \text{cone}(C-x_0)\subset \text{cl cone}(C-x_0)=T_C(x_0),$$

so $L\subset \text{lin } T_C(x_0)$. But we observed earlier that if nondegeneracy holds then $D(L)=\mathbb{R}^m$, so

$$\mathbb{R}^m=D(L)\subset D[\text{lin } T_C(x_0)],$$

and this completes the proof.

At this point, it may be helpful to review some special cases in order to put the idea of nondegeneracy into perspective. First, let us consider the linear programming problem in which $C=\mathbb{R}^n_+$ and $h(x)=Ax-b$. If x_0 is a basic feasible point corresponding to a partition of A into $[B, N]$, where B is a basis for $\mathbb{R}^m$, then we have

$$T_C(x_0)=\bigtimes_{i=1}^{n} Q_i,$$

where

$$Q_i=\begin{cases}\mathbb{R} & \text{if } (x_0)_i>0,\\ \mathbb{R}_+ & \text{otherwise.}\end{cases}$$

However, since x_0 is a basic feasible point the only possible indices i for which $(x_0)_i$ could be positive are $1,\ldots,m$, and therefore lin $T_C(x_0)$ is contained in the subspace of the first m coordinates. But for nondegeneracy as expressed in part (a) of Proposition 1.2 to hold, we must have dim lin $T_C(x_0)=m$: that is, lin $T_C(x_0)$ must be exactly the subspace of the first m coordinates. However, this in turn implies that $(x_0)_1,\ldots,(x_0)_m$ must all be positive, so we see that in this case nondegeneracy in the sense used here reduces to the standard concept of nondegeneracy in linear programming.

For another example, consider a system of nonlinear constraints of the type often seen in the literature:

$$g_1(z)\leq 0,\qquad g_2(z)=0,\tag{1.3}$$

where g_1 and g_2 are differentiable functions from an open set $W\subset\mathbb{R}^k$ into $\mathbb{R}^p$ and $\mathbb{R}^q$ respectively. To rewrite (1.3) in the form (0.1), let $n=k+p$ and $m=p+q$; write $x=(z, s)$ with

$$h(x)=h(z, s)=\begin{bmatrix} g_1(z)+s\\ g_2(z)\end{bmatrix}$$

and take $C=\mathbb{R}^k\times\mathbb{R}^p_+$. If z_0 is feasible for (1.3), then with $s_0=-g_1(z_0)$ the point $x_0=(z_0, s_0)$ satisfies $h(x_0)=0$, $x_0\in C$. We have

$$D=h'(x_0)=\begin{bmatrix} g_1'(z_0) & I \\ g_2'(z_0) & 0 \end{bmatrix} \quad \text{and} \quad \operatorname{lin} T_C(x_0)=\mathbb{R}^k\times\left(\mathop{\times}_{i=1}^{p} Q_i\right),$$

with

$$Q_i=\begin{cases}\{0\} & \text{if } i\in\mathscr{A},\\ \mathbb{R} & \text{if } i\notin\mathscr{A},\end{cases}$$

where $\mathscr{A}$ is the set $\{i\in\{1,\ldots,p\}\mid g_1(z_0)_i=0\}$ of indices corresponding to active inequality constraints. Therefore D will carry $\operatorname{lin} T_C(x_0)$ onto $\mathbb{R}^{p+q}$ if and only if the set $\{g_1'(z_0)_i \mid i\in\mathscr{A}\}\cup\{g_2'(z_0)_j \mid 1\le j\le q\}$ has rank equal to the sum of q and the cardinality of $\mathscr{A}$: that is, if and only if the gradients of the constraints active at z_0 are linearly independent. Thus, here again nondegeneracy corresponds to a well known property for the special case under consideration.

One might reasonably ask whether, in the case of a polyhedral set C, a representation like (1.3) could be used instead of (0.1). That is, could we simply represent the set C by a system of linear inequality and/or equality constraints, adjoin those constraints to the equation $h(x)=0$, and thereby obtain an equivalent system of the form (1.3)? Of course, the answer is yes insofar as this procedure maintains the same feasible set, but it is no in the sense that in the new system a given point may be degenerate whereas it was nondegenerate in the representation (0.1).

For an example to illustrate this difficulty, consider the system

$$\langle a, x\rangle=0, \quad x\in C, \tag{1.4}$$

where a is a vector in $\mathbb{R}^4$ with $a\neq 0$ and C is the polyhedral convex set defined by

$$C=\{(x_1, x_2, x_3, x_4) \mid x_3\ge\max\{|x_1|, |x_2|\}\}.$$

The origin is feasible for (1.4), and we have

$$\operatorname{lin} T_C(0)=\{0\}^3\times\mathbb{R}\subset\mathbb{R}^4.$$

Evidently the image of $\operatorname{lin} T_C(0)$ under the linear operator $\langle a,\cdot\rangle$ is $\mathbb{R}$, so by Theorem 1.2 the origin is a nondegenerate feasible point for (1.4). However, if we try to represent C by a system of linear inequalities we obtain the system

$$\begin{bmatrix} 1 & 0 & -1 & 0 \\ -1 & 0 & -1 & 0 \\ 0 & 1 & -1 & 0 \\ 0 & -1 & -1 & 0 \end{bmatrix} x\le 0, \quad \langle a, x\rangle=0. \tag{1.5}$$

It is not hard to show that (1.5) is a minimal representation of the feasible set in the sense of Telgen [7]. Thus, any representation must have at least five linear constraints, and all five must pass through the origin since the solution set of (1.5) is a cone. But then their gradients (in $\mathbb{R}^4$) cannot be linearly independent, so that the origin is a degenerate feasible point.

It follows that the representation (0.1) is indeed more general, at least for purposes of representing nondegeneracy, than is (1.3): that is, there exist feasible sets which can be represented in the form (0.1), whose nondegenerate points become degenerate under *any* equivalent representation of the form (1.3). This may indicate that (0.1) is a better choice than (1.3) for use as a 'canonical' representation for the feasible set of a nonlinear programming problem.

2. Local structure of the feasible set

Having the idea of nondegeneracy as introduced in Section 1, we can now proceed to study the structure of the feasible set near a nondegenerate point x_0. In fact, we are interested not only in the feasible set of the given problem but also in the feasible sets of all problems close to it in an appropriate sense. To make this idea precise we introduce a parameter p lying in some open subset Π of a real Banach space, and a function $h: \Omega \times \Pi \to \mathbb{R}^m$. The set we shall study is the set of $x \in \Omega$ satisfying

$$h(x, p) = 0, \quad x \in C, \tag{2.1)(p}$$

and we denote the set of all such x by $\mathcal{F}(p)$. To link this notation to the analysis of Section 1 we specify a point $p_0 \in \Pi$ and identify the 'unperturbed' function $h(x, p_0)$ with the $h(x)$ of Section 1. In all of what follows we assume that $h(\cdot, \cdot)$ is C^r $(r \geq 1)$.

By applying the implicit-function theorem to the function

$$F(x, y, p) := D^- h(x, p) + (I - D^- D)[x - (x_0 + y)]$$

using the hypothesis that $D(M) = \mathbb{R}^m$, one can prove the following theorem.

Theorem 2.1. *If $D(M) = \mathbb{R}^m$, there exist neighborhoods U of the origin in $\mathbb{R}^n$, V of p_0 in Π, and W of x_0 in $\mathbb{R}^n$, and a unique function x: $U \times V \to W$, such that x is continuous at $(0, p_0)$ and*

(i) $x(0, p_0) = x_0$,

and, for each $y \in U$ and $p \in V$,

(ii) $x(y, p) \in A$ *and* $h[x(y, p), p] = 0$,

(iii) $P_0[x(y, p) - x_0] = P_0 y$.

Further, x is C^r on $U \times V$.

We omit the proof since it is almost identical to the proof given in [5]. The only difference is in the presence of the parameter p, which causes no essential change in the argument; of course, one has to appeal to a version of the implicit-function theorem for Banach spaces (e.g., [1, Theorem 1(4.XVII)]). In many applications the space of perturbations may be finite-dimensional, but we have made it a Banach space here in order to cover situations in which it may be, e.g., a function space.

The main use we shall make of Theorem 2.1 will be a proof that if x_0 is a nondegenerate feasible point of $(2.1)(p_0)$, then for each p near p_0 the portion of

$\mathcal{F}(p)$ near x_0 is C^r diffeomorphic to the portion of the fixed convex set $\Lambda := K \cap (C - x_0)$ that is near the origin. Thus, under nondegeneracy the local structure of the sets $\mathcal{F}(p)$ near x_0 does not change. The next theorem formulates this result.

Theorem 2.2. *Assume x_0 is a nondegenerate feasible point of (2.1)(p_0). Then there exist open neighborhoods $U_* \subset U$, $V_* \subset V$, and $W_* \subset W$, such that if for each $p \in V_*$ one defines*

$$\psi_p := x(\cdot, p) | U_* \cap \Lambda \quad \text{and} \quad \theta_p := P_0[(\cdot) - x_0] | W_* \cap \mathcal{F}(p),$$

then ψ_p is a C^r diffeomorphism of $U_* \cap \Lambda$ onto $W_* \cap \mathcal{F}(p)$, and θ_p is its inverse.

Proof. Recall that since x_0 is a nondegenerate feasible point there is a neighborhood N of the origin in $\mathbb{R}^n$ such that

$$(C - x_0) \cap N + L \cap N \subset C - x_0. \tag{2.2}$$

Shrink the neighborhoods U, V, and W of Theorem 2.1 to open neighborhoods U_*, V_*, and W' such that if $y \in K \cap U_*$, $p \in V_*$ and $z \in W'$ then $y \in N$, and such that U_* is open and convex with

(a) $x(y, p) \in W'$ and $P_L[x(y, p) - x_0] \in N$, and

(b) $z - x_0 \in N$ and $-P_L(z - x_0) \in N$.

Finally, let $W_* = W' \cap [x_0 + P_0^{-1}(U_*)]$.

Choose any $p \in V_*$; we first show that $\psi_p : U_* \cap \Lambda \to W_* \cap \mathcal{F}(p)$ and that $\theta_p * \psi_p$ is the identity of $U_* \cap \Lambda$. Suppose $y \in U_* \cap \Lambda$. Then as $P_0[x(y, p) - x_0] = P_0 y = y$, we have

$$\begin{aligned} x(y, p) - x_0 &= (P_0 + P_L)[x(y, p) - x_0] \\ &= y + P_L[x(y, p) - x_0] \in (C - x_0) \cap N + L \cap N \subset C - x_0. \end{aligned}$$

However, we already know $h[x(y, p)] = 0$, and we know $x(y, p) \in W'$ and $P_0[x(y, p) - x_0] = y \in U_*$, so $x(y, p) \in W_*$. Thus $x(y, p) \in W_* \cap \mathcal{F}(p)$, so $\psi_p : U_* \cap \Lambda \to W_* \cap \mathcal{F}(p)$. If we apply θ_p we find that $(\theta_p \circ \psi_p)(y) = \theta_p[x(y, p)] = P_0[x(y, p) - x_0] = y$, so $\theta_p \circ \psi_p$ is the identity of $U_* \cap \Lambda$.

Now suppose that $z \in W_* \cap \mathcal{F}(p)$. Then $\theta_p(z) = P_0[z - x_0] = (I - P_L)(z - x_0) \in (C - x_0) \cap N + L \cap N \subset C - x_0$. However, since $z \in C$ we have $z - x_0 \in M$, so $P_0(z - x_0) \in K$, and since $z \in W_*$ we know $P_0(z - x_0) \in U_*$. Thus $\theta_p : W_* \cap \mathcal{F}(p) \to U_* \cap \Lambda$. Now if we set $y = \theta_p(z)$ we have $z \in M + x_0 = A$, $h(z, p) = 0$ (since $z \in \mathcal{F}(p)$), and $P_0(z - x_0) = \theta_p(z) = y$. By the local uniqueness assertion of Theorem 2.1 we then have $z = x(y, p) = (\psi_p \circ \theta_p)(z)$, so that $\psi_p \circ \theta_p$ is the identity of $W_* \cap \mathcal{F}(p)$. This completes the proof.

Evidently, if p_1 and p_2 belong to V_* then $\psi_{p_2} \circ \theta_{p_1}$ is a C^r diffeomorphism of $W_* \cap \mathcal{F}(p_1)$ onto $W_* \cap \mathcal{F}(p_2)$ whose inverse is $\psi_{p_1} \circ \theta_{p_2}$. This provides a significant contrast to the situation of [5], in which only regularity (as opposed to nondegeneracy) held. For example, consider the regular system

$$(1 \quad -1 \quad p)x = 0, \quad x \in \mathbb{R}^2_+ \times \mathbb{R},$$

where $p_0 = 0$. The feasible set $\mathcal{F}(p_0)$ is the half-plane $\{(\alpha, \alpha, \beta) | \alpha \geq 0, \beta \in \mathbb{R}\}$, of dimension 2 in $\mathbb{R}^3$. The origin is a degenerate feasible point. If p is now changed from 0 to small positive or negative values, then the new feasible set $\mathcal{F}(p)$ acquires a vertex at the origin and therefore cannot be diffeomorphic to $\mathcal{F}(p_0)$.

We can obtain a more detailed picture of the structure of $\mathcal{F}(p)$ near x_0 if we examine the faces of Λ that meet U_*. Every point of Λ is the relative interior of exactly one face of Λ [6, Theorem 18.2]. The intersection of such a relative interior with the open set U_* is again a relatively open convex set, so for each $p \in V_*$ its image under ψ_p is a C^r manifold whose dimension equals that of the face in question. The collection of all such manifolds must form a partition of $W_* \cap \mathcal{F}(p)$, so we have the following theorem.

Theorem 2.3. *Assume the notation and hypotheses of Theorem 2.2. For each $p \in V_*$, the set $W_* \cap \mathcal{F}(p)$ is a union of mutually disjoint C^r manifolds, each being the image under ψ_p of $U_* \cap \operatorname{ri} F$ for a distinct face F of Λ. If C is polyhedral then for each p there are only finitely many such manifolds.*

Proof. Everything has been proved except for the last statement, which follows from the fact that a polyhedral convex set has only finitely many faces [6, Theorem 19.1].

3. Application to nonlinear programming

Here we apply the results of Sections 1 and 2 to study the local solutions of the nonlinear optimization problem

$$\min_x \; f(x, p) \qquad \text{subject to} \quad h(x, p) = 0, \; x \in C, \qquad (3.1)(p)$$

where f is a C^r function from $\Omega \times \Pi$ into $\mathbb{R}$, and where h and C are as previously defined. Note that if x_0 is a nondegenerate feasible point for $(3.1)(p_0)$, then Theorem 2.2 says that for $p \in V^*$ the sets

$$W^* \cap C \cap \{x | h(x, p) = 0\}, \qquad U^* \cap (C - x_0) \cap K$$

are C^r diffeomorphic copies of each other. Hence, if we are interested only in the behavior of $(3.1)(p)$ near x_0, we might just as well study the problem

$$\min_y \; \phi(y, p) \qquad \text{subject to} \quad y \in \Lambda, \qquad (3.2)(p)$$

where we have set

$$\phi(y, p) = f[x(y, p), p] = [f(\cdot, p) \circ \psi_p](y).$$

Note that the feasible set of (3.2)(p) is a convex set, independent of p, which will be polyhedral if C is polyhedral. For the rest of this section we make the blanket assumption that C is polyhedral and x_0 is a nondegenerate feasible point of (3.1)(p_0).

We shall exploit the relationship between (3.1)(p) and (3.2)(p) in several different ways in what follows. Our first result is a simple proof of the well known first order optimality condition. It will also yield geometric insight that we can apply to find other results.

Proposition 3.1. *If x_0 is a local minimizer of* (3.1)(p_0), *then*

$$f_x(x_0, p_0)P_0 \in -N_C(x_0). \tag{3.3}$$

Proof. Since x_0 is a local minimizer of (3.1)(p_0) the origin is a local minimizer of (3.2)(p_0). Suppose $d \in T_C(x_0)$; since $L \subset \text{lin } T_C(x_0)$ and $P_0 d = d - P_L d$, we have $P_0 d \in T_C(x_0)$. However, since $d \in T_C(x_0) \subset M$, we have $P_J d = 0$, and thus $P_0 d = P_K d \in K$. Hence $P_0 d \in K \cap T_C(x_0)$, and for small $\lambda > 0$ we have $\lambda P_0 D \in K \cap (C - x_0)$ since $(C - x_0)$ agrees with $T_C(x_0)$ near the origin. But then $\lambda P_0 d$ is feasible for (3.2)(p_0) and therefore $\phi(\lambda P_0 d, p_0) \geq \phi(0, p_0)$ for small λ. Hence we must have $\phi_y(0, p_0)P_0 d \geq 0$, but by the chain rule $\phi_y(0, p_0) = f_x(x_0, p_0)P_0$, and since $P_0^2 = P_0$ we have $f_x(x_0, p_0)P_0 d \geq 0$ for all $d \in T_C(x_0)$. But this means $f_x(x_0, p_0)P_0 \in -N_C(x_0)$ and this completes the proof.

One may think that (3.3) is not the usual first order optimality condition. To see that it really is, recall that

$$P_0 = I - h_x(x_0, p_0)^- h_x(x_0, p_0)$$

and define

$$\lambda(x_0, p_0) := -[h_x(x_0, p_0)^-]^* f_x(x_0, p_0). \tag{3.4}$$

Then (3.3) becomes

$$f_x(x_0, p_0) + h_x(x_0, p_0)^* \lambda(x_0, p_0) \in -N_C(x_0), \tag{3.5}$$

which is the familiar multiplier rule.

It is worth noting that under the nondegeneracy hypothesis the multipliers appearing in (3.5) are unique. To see this, suppose that λ_1 and λ_2 satisfy (3.5) when substituted for $\lambda(x_0, p_0)$ there. Then

$$h_x(x_0, p_0)^*(\lambda_1 - \lambda_2) \in \text{aff } N_C(x_0),$$

and by part (b) of Proposition 1.2 we then have $\lambda_1 = \lambda_2$.

Now denote $\phi_y(0, p_0)$ by g_0. The necessary optimality condition in Proposition 3.1 shows that, if x_0 is a local minimizer of (3.1)(p_0), then x_0 belongs to the set

$$\partial \psi_C^*(-g_0) = \{x \in C \mid \langle -g_0, x \rangle = \sup_{c \in C} \langle -g_0, c \rangle\}.$$

It is easy to show that this set is a face of C; we shall denote it by F_0. It is also easy to show that $(F_0 - x_0) \cap K$ is a face of Λ, which we shall denote by Φ_0. What may be less obvious is that if we apply ψ_p to Φ_0, we recover, locally, exactly the intersection of F_0 with $h^{-1}(0)$. This is the content of the next result.

Proposition 3.2. *If x_0 is a local minimizer of* (3.1)(p_0), *then for any $p \in V_*$,*

$$\psi_p[U_* \cap \Phi_0] = W^* \cap F_0 \cap \mathcal{F}(p).$$

Proof. Suppose $y \in U_* \cap \Phi_0$; let $x = \psi_p(y) = x(y, p)$. By Theorem 2.2, $x \in W_* \cap \mathcal{F}(p)$. By Theorem 2.1, $P_0(x - x_0) = P_0 y$. By the definition of Φ_0, we have $y \in K \cap (F_0 - x_0)$; thus,

$$0 = \langle g_0, y\rangle = \phi_y(0, p_0)y = f_x(x_0, p_0)P_0 y = f_x(x_0, p_0)P_0(x - x_0) = \langle g_0, x - x_0\rangle,$$

so $x \in F_0$.

For the opposite inclusion, suppose that $x \in W^* \cap F_0 \cap \mathcal{F}(p)$. Let $y = \theta_p(x) = P_0(x - x_0)$; then we know that $x = \psi_p(y)$ and $y \in U_* \cap \Lambda$ by Theorem 2.2; thus in particular $y \in K$ and $x_0 + y \in C$. However, since $P_0^2 = P_0$ we have

$$0 = \langle g_0, x - x_0\rangle = f_x(x_0, p_0)P_0(x - x_0) = f_x(x_0, p_0)P_0[P_0(x - x_0)] = \langle g_0, y\rangle,$$

and thus $x_0 + y \in F_0$, which completes the proof.

It will turn out that the face Φ_0 and, therefore, its image $F_0 \cap h^{-1}(0)$, contain all of the local minimizers of (3.2)(p) and (3.1)(p) respectively. In order to establish this fact, we show first that the local minimizers of (3.1)(p) and (3.2)(p) are in $1-1$ correspondence for each p.

Proposition 3.3. *For any $p \in V_*$ and any $y \in U_* \cap \Lambda$, y is a local minimizer of* (3.2)(p) *if and only if $\psi_p(y)$ is a local minimizer of* (3.1)(p).

Proof. Obvious, since ψ_p and its inverse θ_p are diffeomorphisms and $\phi(\cdot, p) = f(\cdot, p) \circ \psi_p$.

The next result shows that the local minimizers of (3.2)(p) and of the more tightly constrained problem

$$\min \quad \phi(y, p) \qquad \text{subject to} \quad y \in \Phi_0, \tag{3.6)(p}$$

are identical. From now on, we assume that $r \geq 2$.

Theorem 3.4. *There exist neighborhoods U_1 of the origin in $\mathbb{R}^n$ and V_1 of p_0 in Π such that for any $(y, p) \in U_1 \times V_1$, y is a local minimizer of* (3.2)(p) *if and only if y is a local minimizer of* (3.6)(p).

Proof. We shall require the following lemma, proved in the appendix.

Lemma 3.5. *Let C be a polyhedral convex set in $\mathbb{R}^n$. Let $z_0 \in \mathbb{R}^n$ and define $Z_0 := \partial\psi_C^*(z_0)$. Then there is a neighborhood U of z_0 such that if $z \in U$ then $\partial\psi_C^*(z) = \partial\psi_{Z_0}^*(z)$.*

This lemma says, geometrically, that the face of C picked out by z is the same as the face of Z_0 picked out by z, as long as z is close to z_0. A related result was proved in [3]. The lemma obviously makes essential use of the property of polyhedrality.

We also need the following lemma, proved by Mangasarian and Meyer [2, Lemma A.1]. In [2] this is proved for the maximum norm, but since norms on finite-dimensional spaces are equivalent, it holds also for the Euclidean norm. We restate it in the notation of this paper.

Lemma 3.6. *Let C, z_0 and Z_0 be as in Lemma 3.5. Assume $Z_0 \neq \emptyset$, and for $x \in \mathbb{R}^n$ let $z(x)$ be the projection of x on Z_0. Then there is a constant $\gamma > 0$ such that for each $x \in C$,*

$$\langle z_0, x - z(x)\rangle \leq -\gamma\|x - z(x)\|.$$

For the proof of Theorem 3.4, note first that if y is a local minimizer of (3.2)(p) that is close to 0, and if p is close to p_0, then $y \in \partial\psi_\Lambda^*[-\phi_y(y, p)]$. Using Lemma 3.5 we conclude that $y \in \partial\psi_{\Phi_0}^*[-\phi_y(y, p)]$ and, in particular, that $y \in \Phi_0$. But as y is a local minimizer of (3.2)(p), it is evidently also a local minimizer of the more tightly constrained problem (3.6)(p). For the converse, we can deduce from our hypothesis about f_{xx} that there exist a ball $2\rho B$ of radius 2ρ about the origin in $\mathbb{R}^n$, a neighborhood V_1 of p_0 in Π, and a constant λ, such that if $y_1, y_2 \in 2\rho B$ and $p \in V_1$ then

$$\phi(y_2, p) = \phi(y_1, p) + \phi_y(y_1, p)(y_2 - y_1) + \alpha(y_1, y_2, p),$$

with $|\alpha(y_1, y_2, p)| \leq \frac{1}{2}\lambda\|y_1 - y_2\|^2$. Further, if ρ and V_1 are taken to be small enough, then $\rho \leq \lambda^{-1}\gamma$ and if $y \in 2\rho B$ and $p \in V_1$ then $y \in U_*$, $p \in V_*$, and $\|\phi_y(y, p) - \phi_y(0, p_0)\| < \frac{1}{2}\gamma$, where γ comes from Lemma 3.6 with $z_0 := -\phi_y(0, p_0)$. Now let $U_1 := \rho B$. Let $(y_0, p) \in U_1 \times V_1$ and suppose y_0 is a local minimizer of $\phi(\cdot, p)$ on Φ, so that for some positive η, if $y \in \Phi$ with $\|y_0 - y\| < \eta$ then $\phi(y, p) \geq \phi(y_0, p)$. Choose any $y \in \Lambda$ with $\|y - y_0\| < \min\{\eta, \rho\}$. Let y_Φ be the projection of y on Φ; then as $y_0 \in \Phi$ we have $\langle y - y_\Phi, y_0 - y_\Phi\rangle \leq 0$, so that $\|y_0 - y_\Phi\|^2 + \|y - y_\Phi\|^2 \leq \|y - y_0\|^2$. Hence $\|y\| \leq \|y - y_0\| + \|y_0\| < 2\rho$ and similarly $\|y_\Phi\| < 2\rho$. Now we have

$$\phi(y, p) = \phi(y_\Phi, p) + \phi'(y_\Phi, p)(y - y_\Phi) + \alpha,$$

where $|\alpha| \leq \frac{1}{2}\lambda\|y - y_\Phi\|^2$. Therefore

$$\begin{aligned}\phi(y, p) - \phi(y_0, p) &= [\phi(y, p) - \phi(y_\Phi, p)] + [\phi(y_\Phi, p) - \phi(y_0, p)] \\ &\geq \phi(y, p) - \phi(y_\Phi, p) = \phi_y(y_\Phi, p)(y - y_\Phi) + \alpha,\end{aligned}$$

since $\phi(y_\Phi, p) \geq \phi(y_0, p)$ because $y_\Phi \in \Phi$ with $\|y_\Phi - y_0\| \leq \|y - y_0\| < \eta$. Applying Lemma 3.6 with $z_0 = -\phi_y(0, p_0)$, we find that

$$\phi_y(0, p_0)(y - y_\Phi) \geq \gamma\|y - y_\Phi\|.$$

But also, since $y_\Phi \in 2\rho B$ and $p \in V_1$,

$$\|\phi_y(y_\Phi, p) - \phi_y(0, p_0)\| < \tfrac{1}{2}\gamma.$$

Hence we conclude that

$$\phi_y(y_\Phi, p)(y - y_\Phi) \geq \tfrac{1}{2}\gamma\|y - y_\Phi\|.$$

From (3.7) we then have

$$\phi(y, p) - \phi(y_0, p) \geq \phi_y(y_\Phi, p)(y - y_\Phi) + \alpha \geq \tfrac{1}{2}\gamma\|y - y_\Phi\| - \tfrac{1}{2}\lambda\|y - y_\Phi\|^2 \geq 0,$$

since $\|y - y_\Phi\| \leq \rho \leq \lambda^{-1}\gamma$. It follows that y_0 is a local minimizer of (3.2)(p), which completes the proof of Theorem 3.4.

We have thus shown that, for the purpose of optimization, when considering (3.2)(p) one may as well restrict one's attention to the face Φ_0: i.e., to the problem (3.6)(p). (This explains, incidentally, the prominent appearance of Φ_0 in the second-order optimality conditions.) The immediate corollary to this is that in considering (3.1)(p) one may look only at the image of Φ_0 under ψ_p; as we showed in Proposition 3.2, this will, locally, be the restriction of $\mathscr{F}(p)$ to the face F_0 of C picked out by $f_x(x_0, p_0)P_0$. Thus we could consider, instead of (3.1)(p), the more tightly constrained problem

$$\min_x \; f(x, p) \qquad \text{subject to} \quad h(x, p) = 0, \; x \in F_0. \tag{3.8)(p}$$

By Proposition 3.2 the local minima (near x_0) of (3.8)(p) will be the images under ψ_p of those of (3.6)(p). The latter, in turn, are by Theorem 3.4 identical to those of (3.2)(p), which finally are the images under θ_p of those of (3.1)(p) by Proposition 3.3. We conclude that, for any p near p_0, the local minimizers of (3.1)(p) and of (3.8)(p) are the same.

It is worth pointing out here that if F is any face of C that contains x_0, then the nondegeneracy hypothesis actually implies that x_0 is a nondegenerate feasible point for the system

$$h(x, p_0) = 0, \quad x \in F, \tag{3.9}$$

in which C has been replaced by F. To see this, let U be an open ball about the origin in $\mathbb{R}^n$, small enough so that the neighborhood N in the definition of nondegeneracy contains U. If $f - x_0 \in (F - x_0) \cap U$, then

$$(f - x_0) + L \cap U \subset (F - x_0) \cap U + L \cap U \subset C - x_0.$$

However, $f + L \cap U$ is then a relatively open convex subset of C, which meets F. Thus, by [6, Theorem 18.1] we have $f + L \cap U \subset F$. But this implies

$$(F - x_0) \cap U + L \cap U \subset F - x_0,$$

so that x_0 is a nondegenerate feasible point for (3.9). In particular, if in place of F we put the face F_0, we find that x_0, as a feasible point of (3.8)(p_0), inherits the nondegeneracy that it had as a feasible point of (3.1)(p_0).

Appendix

The purpose of this appendix is to prove the key lemma required in the proof of Theorem 3.4. For convenience, we restate the lemma here.

Lemma 3.5. *Let C be a polyhedral convex set in $\mathbb{R}^n$. Let $z_0 \in \mathbb{R}^n$ and define $Z_0 := \partial\psi_C^*(z_0)$. Then there is a neighborhood U of z_0 such that if $z \in U$ then $\partial\psi_C^*(z) = \partial\psi_{Z_0}^*(z)$.*

Proof. We first show that for z near z_0 we have $\partial\psi_C^*(z) \subset \partial\psi_{Z_0}^*(z)$. Indeed, if $y \in \partial\psi_C^*(z)$ then for each $c \in C$, $\langle z, c - y\rangle \leqslant 0$. This holds in particular if $c \in Z_0$, so if $y \in Z_0$ then $y \in \partial\psi_{Z_0}^*(z)$. Thus we have to show that there is a neighborhood of z_0 such that for all z in that neighborhood we have $\partial\psi_C^*(z) \subset Z_0$. If this were not so, there would be a sequence $\{z_n\}$ converging to z_0, such that for each n there exists $y_n \in \partial\psi_C^*(z_n)\backslash Z_0$. Each set $\partial\psi_C^*(z_n)$ is a face of C, but the set of such faces is finite since C is polyhedral [6, Theorem 19.1]. Therefore some face G of C recurs infinitely often among the $\partial\psi_C^*(z_n)$. Let $g \in G$; then for infinitely many n the pair (z_n, g) belongs to the (closed) graph of $\partial\psi_C^*$. But then (z_0, g) belongs to this graph too, so $g \in \partial\psi_C^*(z_0)$ and thus in fact $G \subset \partial\psi_C^*(z_0) \subset Z_0$. However, this contradicts the existence of the y_n, so we see that $\partial\psi_C^*(z) \subset \partial\psi_{Z_0}^*(z)$ for all z in some neighborhood, say U_1, of z_0.

To prove $\partial\psi_{Z_0}^*(z) \subset \partial\psi_C^*(z)$ for z near z_0, we first obtain a formula for the inverse of $\partial\psi_{Z_0}^*$: i.e., for N_{Z_0}. If $z \in Z_0$ then, near 0, $C - z$ and $Z_0 - z$ agree with $T_C(z)$ and $T_{Z_0}(z)$ respectively. Also, if we denote by L the halfline $\{-\lambda z_0 \mid \lambda \geqslant 0\}$, then we know that

$$Z_0 - z = (C - z) \cap L^0.$$

Thus for some neighborhood V of the origin,

$$V \cap T_{Z_0}(z) = V \cap (Z_0 - z) = V \cap (C - z) \cap L^0 = V \cap T_C(z) \cap L^0.$$

But since $T_{Z_0}(z)$ and $T_C(z) \cap L^0$ are cones, we then have $T_{Z_0}(z) = T_C(z) \cap L^0$ and hence, by polyhedrality, $N_{Z_0}(z) = N_C(z) + L$. Thus, for each $z \in C$ we have

$$N_{Z_0}(z) = \begin{cases} N_C(z) + L & \text{if } z \in Z_0, \\ \emptyset & \text{if } z \notin Z_0. \end{cases} \tag{A.1}$$

Having this representation for N_{Z_0}, we next show that, for small elements of $N_{Z_0}(z)$, only small elements of L need be used in the representation (A.1). Recall that N_C is constant on the relative interior of any convex subset of C, and in particular this holds for faces of C. But the relative interiors of the faces partition C [6, Theorem 18.2], and thus N_C takes only finitely many distinct values. Each of these is a cone in $\mathbb{R}^n$: call them $K_1, \ldots, K_I$. For each i let G_i be the convex polyhedral multifunction taking $t \in \mathbb{R}_+$ to $K_i - tz_0$. Then $K_i + L = G_i(\mathbb{R}_+)$, so by [4, Proposition 2] there is some $\alpha_i > 0$ with

$$B \cap (K_i + L) = B \cap G_i(\mathbb{R}_+) \subseteq G_i([0, \alpha_i]).$$

It follows that

$$(\alpha_i^{-1}B)\cap(K_i+L)\subset G_i([0,1]),$$

and if we let $\alpha := \max_{i=1}^{I}\alpha_i$, then $\alpha>0$ and

$$(\alpha^{-1}B)\cap(K_i+L)\subset G_i([0,1]),\quad i=1,\ldots,I. \tag{A.2}$$

Hence in order to represent elements of K_i+L having norm not greater than α^{-1}, we do not need to use elements of L having length more than 1.

Now define a neighborhood U of z_0 by letting $U=U_1\cap(Z_0+\alpha^{-1}B)$. Let $z\in U$; then either $\partial\psi^*_{Z_0}(z)$ is empty or it contains some point, say y. In the latter case $z\in N_{Z_0}(y)$, so if $w\in Z_0$ then $\langle z, w-y\rangle\leq 0$. But then also $\langle z-z_0, w-y\rangle\leq 0$ since $\langle z_0, w-y\rangle=0$ because w and y belong to Z_0. Thus $z-z_0\in N_{Z_0}(y)$, so by (A.1) and (A.2) we have for some i, $z-z_0\in G_i[0,1]$. Thus there is some $t\in[0,1]$ with

$$z-z_0\in N_C(y)-tz_0. \tag{A.3}$$

However, we also know that $(1-t)z_0\in N_C(y)$, since $N_C(y)$ is a cone and $y\in\partial\psi^*_C(z_0)$. Combining this with (A.3) we find that $z\in N_C(y)$, so that $y\in\partial\psi^*_C(z)$. It follows that $\partial\psi^*_{Z_0}(z)\subset\partial\psi^*_C(z)$, and this completes the proof of Lemma 3.5.

References

[1] L.V. Kantorovich and G.P. Akilov, *Functional analysis in normed spaces* (Macmillan, Inc., New York, 1959).

[2] O.L. Mangasarian and R.R. Meyer, "Nonlinear perturbation of linear programs", *SIAM Journal on Control and Optimization* 17 (1979) 745–752.

[3] S.M. Robinson, "Strongly regular generalized equations", *Mathematics of Operations Research* 5 (1980) 43–62.

[4] S.M. Robinson, "Some continuity properties of polyhedral multifunctions", *Mathematical Programming Study* 14 (1981) 206–214.

[5] S.M. Robinson, "Local structure of feasible sets in nonlinear programming, Part I: Regularity", in: V. Pereyra and A. Reinoza, eds., *Numerical methods: Proceedings, Caracas 1982* (Springer-Verlag, Berlin, 1983) pp. 240–251.

[6] R.T. Rockafellar, *Convex analysis* (Princeton University Press, Princeton, NJ, 1970).

[7] J. Telgen, "Minimal representation of convex polyhedral sets", *Journal of Optimization Theory and Applications* 38 (1982) 1–24.

Mathematical Programming Study 22 (1984) 231-236
North-Holland

PERFORMANCE OF APPROXIMATE ALGORITHMS FOR GLOBAL MINIMIZATION

J.B. ROSEN
Computer Science Department, University of Minnesota, Minneapolis, MN 55455, USA

Received 25 March 1983
Revised manuscript received 29 March 1984

The performance of a class of algorithms for solving global minimization problems is analyzed. Problems which may have a large number of variables appearing only linearly (in addition to the nonlinear variables) are considered. The analysis is based on finding an ε-approximate solution, in the sense that the function value found is known to be no more than $\varepsilon\Delta\psi_1$ greater than the global minimum, where $\varepsilon\Delta\psi_1$ is a known scale factor and the tolerance ε is specified. Each algorithm considered is characterized by two quantities α and ρ. A bound on the computing effort F_T required to obtain an ε-approximate solution is given in terms of the problem size, α, ρ and ε. In particular, it is shown that F_T increases no more than linearly with the number of linear variables, and that provided $\alpha\rho \leq 1$, the value of F_T is a linear function of $(1/\varepsilon)$.

Key words: Global Minimization, Linear Constraints, Approximate Solution, Computational Analysis.

We consider the following linearly constrained, concave global minimization problem:

$$\operatorname*{global\,min}_{(x,y)\in\Omega} \psi(x, y) \tag{1}$$

where

$$\psi(x, y) = \phi(x) + d^{\mathrm{T}}y \tag{2}$$

and

$$\Omega = [(x, y) | A_1x + A_2y = b; x, y \geq 0]. \tag{3}$$

The function $\phi(x)$ is a differentiable, concave function, with $b \in \mathbb{R}^m$, $x \in \mathbb{R}^n$ and $y \in \mathbb{R}^k$, so that $\Omega \subset \mathbb{R}^{n+k}$. It is assumed that the polyhedron Ω is a nonempty, bounded polytope. The matrices A_1 and A_2 are $(m \times n)$ and $(m \times k)$ respectively, so that there are m linear equality constraints. The number of nonlinear variables is assumed to be relatively small ($n \leq 50$), but we wish to consider problems for which k may be much larger than n. It is well known that for a concave function every local minimum (and therefore the global minimum) is attained at a vertex of the polytope.

We wish to find the minimum value ψ^* of $\psi(x, y)$ on Ω and a vertex (x^*, y^*) at which it is attained. An ε-approximate solution will also be accepted, that is, any

This research was supported in part by the Computer Science Section of the National Science Foundation under Research Grant MCS 8101214.

feasible point $(x_\varepsilon^*, y_\varepsilon^*)$, such that

$$\psi(x_\varepsilon^*, y_\varepsilon^*) - \psi^* \leq \varepsilon \Delta\psi_1 \tag{4}$$

where ε is a specified tolerance, and $\Delta\psi_1$ is defined below.

This analysis is valid for any branch and bound algorithm in which the domain considered for further search at each iteration is either reduced in size, or is partitioned into two or more subdomains. In both cases the difference $\Delta\psi_i = \bar{\psi}_i - \underline{\psi}_i$, between the upper bound $\bar{\psi}_i$, and the lower bound $\underline{\psi}_i$ at the ith iteration, is reduced at each iteration. In the case where the domain is partitioned, every subdomain is eliminated whose lower bound exceeds the minimum known upper bound (given by the incumbent solution value). Examples of branch and bound algorithms which satisfy these conditions are given in [2, 5, 6, 7, 8, 9].

In each of these algorithms a value of $\Delta\psi_1$ is determined by some initial calculation. To simplify the presentation, we will consider algorithms for the case where ϕ is a quadratic, concave function [5, 6, 7, 9]. A similar analysis can be carried out for other related branch and bound algorithms for global minimization over a polytope. For this quadratic case the Hessian of ϕ is a negative semi-definite, symmetric matrix. Its eigenvalues (all real) and corresponding orthogonal eigenvectors, u_i, $i = 1, \ldots, n$, can easily be computed.

The initial value $\Delta\psi_1$ is then obtained by solving a multiple-cost-row linear program, with a total of $2n$ cost rows, where each cost row is the vector u_i or $-u_i$. The functions $\pm u_i^T x$ are minimized, subject to $(x, y) \in \Omega$. This gives at most $2n$ vertices of Ω. That vertex with the smallest function value gives the initial upper bound $\bar{\psi}_1$. The vertices obtained also permit the construction of a rectangular domain which contains Ω, and whose faces contain the vertices. This enclosing rectangular domain permits the calculation of the initial lower bound $\underline{\psi}_1$, so that $\Delta\psi_1 = \bar{\psi}_1 - \underline{\psi}_1$. Details are given in [6].

A tolerance ε is specified, and the algorithm terminates successfully when

$$\Delta\psi_l \leq \varepsilon \Delta\psi_1. \tag{5}$$

The upper bound $\bar{\psi}_i$ obtained at each iteration is given by the minimum known value of $\psi(x, y)$ corresponding to a feasible point $(\bar{x}_i, \bar{y}_i)$. Then at $i = l$, we have $\bar{\psi}_l - \Delta\psi_l \leq \psi^* \leq \bar{\psi}_l$, so that $\bar{\psi}_l = \psi(\bar{x}_l, \bar{y}_l)$ differs from ψ^* by at most $\Delta\psi_l \leq \varepsilon\Delta\psi_1$. Therefore, by the definition (4), $(\bar{x}_l, \bar{y}_l)$ is an ε-approximate minimum point and $\bar{\psi}_l$ is its ε-approximate optimal function value.

The algorithms considered are best represented in terms of a tree structure. The root of the tree (node 1, level 1) corresponds to the original feasible doman Ω. An upper and lower bound are computed, and their difference $\Delta\psi_1$ is known at node 1. Computing $\Delta\psi_1$ requires the solution of $2n$ linear programs (LP's). The domain Ω is then partitioned, and the active subdomains determined. An upper bound α on the number of active subdomains is known, where $\alpha = \alpha(n) \geq 1$, will depend on n, and the specific algorithm used, but not on k the number of linear variables. For example, in [5] each subdomain is bisected at each iteration, so that in the worst case (no pruning) the number of subdomains will double at each iteration; that is, $\alpha = 2$. Thus level 2 of the tree will contain no more than α nodes, where each node corresponds to an active subdomain of Ω. This partitioning process is continued, with each active subdomain yielding no more than α active partitions (nodes), at each successive subdivision. A tree of degree α and depth l is thus generated by the partitioning process. For the ith level of the tree, $i = 1, \ldots, l$, an upper and lower bound, and their difference $\Delta\psi_i$ are computed. These quantities will satisfy the relation

$$\Delta\psi_{i+1} \leqslant \rho\Delta\psi_i, \quad i = 1, \ldots, l-1, \tag{6}$$

where ρ, $0 < \rho < 1$, will depend on the specific algorithm used. A tree of this type is illustrated in Fig. 1 for $\alpha = 3$ and $l = 4$.

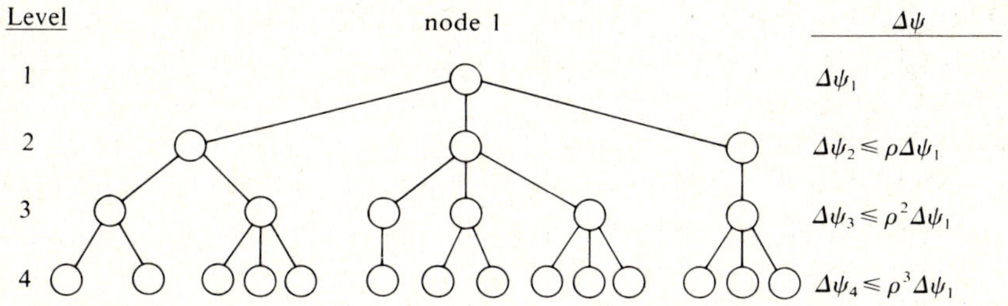

Fig. 1

A special case is given by an algorithm in which the bound difference $\Delta\psi_i$ is reduced at each level by a reduction in the size of the active feasible domain. Rather than partitioning the original domain into two or more active subdomains, the original domain is reduced at each iteration by adding one or more cutting planes, and solving one or more LP's. These cutting planes are determined so as to eliminate subdomains for which the lower bound exceeds the best known upper bound. This special case corresponds to $\alpha = 1$.

The analysis which follows shows how small ρ must be in order that the performance of an algorithm is satisfactory. It may be possible by a theoretical analysis to obtain an upper bound $\bar{\rho} < 1$, to the value of ρ for a given algorithm (see for example [5]). Alternatively, an average value of ρ may be estimated by appropriate computational testing of an algorithm.

In order to analyze the performance of the type of algorithm considered, we first obtain a relation giving the minimum value of l which satisfies (5), in terms of ρ and ε. Directly from (6) we have

$$\Delta\psi_l \leqslant \rho^{l-1}\Delta\psi_1. \tag{7}$$

The relation (5) is then satisfied if we choose l so that $\rho^{l-1} \leqslant \varepsilon$, or

$$l - 1 \geqslant \eta, \tag{8}$$

where

$$\eta = \eta(\varepsilon, \rho) \equiv \log\left(\frac{1}{\varepsilon}\right) \Big/ \log\left(\frac{1}{\rho}\right).$$

An integer l satisfying (8) is given by

$$l = l(\varepsilon, \rho) = 1 + \bar{\eta},$$

where $\bar{\eta}$ = smallest integer $\geqslant \eta$.

This gives

$$l \leqslant 2 + \eta. \tag{10}$$

The total computing effort required for each algorithm considered will depend primarily on the total number of nodes in the tree. More specifically, $2n$ LP's are solved for the first node, and ν LP's for each node in levels 2 through l of the tree, where ν is a small integer which will depend on the algorithm, and typically is $\nu = 1$ or 2. Thus if we denote by N the total number of nodes in the tree, then N_{LP}, the total number of LP's solved, is given by

$$N_{LP} = 2n + \nu(N - 1). \tag{11}$$

For $\alpha = 1$, we have one node per level, so that $N = l$. Directly from (10) we get

$$N \leqslant 2 + \eta = 2 + \left[\log\left(\frac{1}{\varepsilon}\right) \Big/ \log\left(\frac{1}{\rho}\right)\right]. \tag{12}$$

For $\alpha > 1$, an upper bound on N in terms of α and l is given by:

$$N \leqslant \sum_{i=0}^{l-1} \alpha^i = \frac{\alpha^l - 1}{\alpha - 1} \leqslant \frac{\alpha^l}{\alpha - 1}. \tag{13}$$

Then using (10) and (8) we get

$$\begin{aligned} \log N &\leqslant l \log \alpha - \log(\alpha - 1) \leqslant (2 + \eta) \log \alpha - \log(\alpha - 1) \\ &= \log\left[\frac{\alpha^2}{\alpha - 1}\right] + \eta \log \alpha = \log\left[\frac{\alpha^2}{\alpha - 1}\right] + \omega(\alpha, \rho) \log\left[\frac{1}{\varepsilon}\right], \end{aligned} \tag{14}$$

where

$$\omega(\alpha, \rho) \equiv \log \alpha / \log\left[\frac{1}{\rho}\right]. \tag{15}$$

Directly from (14) we obtain the more convenient bound on N:

$$N \leqslant \frac{\alpha^2}{\alpha - 1}\left[\frac{1}{\varepsilon}\right]^{\omega(\alpha, \rho)}. \tag{16}$$

For $\alpha > 1$, it is clear from (16) that N cannot increase faster than linearly in $[1/\varepsilon]$ provided $\omega \leqslant 1$. From (15) we see that $\rho\alpha \leqslant 1$ and $\omega \leqslant 1$ are equivalent. Thus to avoid exponential growth of the number of nodes with $[1/\varepsilon]$, only those algorithms for which $\alpha\rho \leq 1$ should be considered.

In order to complete the analysis of computational effort and its dependence on problem size, we now take account of the effort required to solve the LP's. For any specific problem (m, n, k fixed), all LP's solved are of the same size, with m constraints and $n+k$ variables. We also let δ be the fractional density of the constraint coefficient matrix $[A_1 A_2]$. We assume that the total computational effort measured in floating point operations (flops) is essentially that required to solve the total number, N_{LP} of LP's given by (11). We also use an average figure for the number, $F_{\text{LP}} = \text{flops/LP}$, based on accepted typical performance results [3]. For recent theoretical investigations of the average behavior of the simplex method see [1, 4]. These typical results satisfy

$$F_{\text{LP}} \leq 2m^2(n+k)\delta. \tag{17}$$

Thus if $F_{\text{T}} = $ total flops required to obtain an ε-approximate solution, we have

$$F_{\text{T}} = F_{\text{LP}} N_{\text{LP}} \leq 2m^2(n+k)\delta[2n + \nu(N-1)], \tag{18}$$

where

$$N \leq \begin{cases} 2 + \left[\log\left(\dfrac{1}{\varepsilon}\right) \Big/ \log\left(\dfrac{1}{\rho}\right)\right] & \text{if } \alpha = 1, \\ \dfrac{\alpha^2}{\alpha - 1}\left(\dfrac{1}{\varepsilon}\right)^{\omega(\alpha, \rho)} & \text{if } \alpha > 1. \end{cases}$$

This bound shows the explicit dependence of total solution time on the problem size and the tolerance ε. Dependence on the number of equations m, and the number of linear variables k, is only as shown explicitly, that is, m^2 and linear in k. The dependence on the number of nonlinear variables n is more complicated since $\omega = \omega(\alpha, \rho)$, and α will depend on n, as well as the specific algorithm used.

Several important conclusions follow from this performance analysis.

1. Average solution time dependence on the number of linear variables is no greater than linear. Thus problems with k much larger than n can be solved.

2. The average time dependence on the number of equations m is the same as that for linear programming; that is, $O(m^2)$ for n, k, δ and ε fixed.

3. The performance of a specific algorithm is determined primarily by the value of $\rho < 1$, for $\alpha = 1$, and by the value of $\omega(\alpha, \rho)$ for $\alpha > 1$. Specifically, for $\alpha \geq 1$, a value of $\alpha\rho \leq 1$ will result in total computing effort (and therefore computing time) being a linear function of $(1/\varepsilon)$. Generally speaking, only those algorithms for which $\alpha\rho \leq 1$ should be considered as practical computational methods. A rough measure of efficiency in comparing two algorithms for which $\alpha > 1$ is the corresponding value of $\omega(\alpha, \rho)$. The algorithm with the smaller ω should be more efficient.

References

[1] K.H. Borgwardt, "Some distribution independent results about the asymptotic order of the average number of pivot steps of the simplex method", *Mathematics of Operations Research* 7 (1982) 441–462.

[2] J.E. Falk and K.R. Hoffman, 'A successive underestimating method for concave minimization problems", *Mathematics of Operations Research* 1 (1976) 251–259.

[3] S.I. Gass, *Linear programming methods and applications* (McGraw-Hill, New York, 1975).

[4] M. Haimovich, "The simplex algorithm is very good!—On the expected number of pivot steps and related properties of random linear programs", Manuscript, Columbia University (New York, 1983).

[5] B. Kalantari, "Large scale concave quadratic minimization and extensions", Ph.D. Thesis, University of Minnesota (Minneapolis, MN, 1984).

[6] J.B. Rosen, "Global minimization of a linearly constrained concave function by partition of feasible domain", *Mathematics of Operations Research* 8 (1983) 215–230.

[7] J.B. Rosen, "Parametric global minimization for large-scale problems", Technical Report 83-11 (revised) Computer Science Department, University of Minnesota (Minneapolis, MN 1983).

[8] N.V. Thoai and H. Tuy, "Convergent algorithms for minimizing a concave function", *Mathematics of Operations Research* 5 (1980) 556–566.

[9] N. Zilverberg, "Global minimization for large scale linearly constrained systems", Ph.D. Thesis, University of Minnesota (Minneapolis, MN, 1983).

Mathematical Programming Study 22 (1984) 237–251
North-Holland

SOME OPTIMIZATION PROBLEMS WITH EXTREMAL OPERATIONS

Karel ZIMMERMANN

Karls-Universität Prag, Abt. Kybernetik und Operationsforschung, Malostr. Nam. 25, CS 11800 Praha, CSSR

Received 11 April 1983
Revised manuscript received 28 March 1984

We consider an algebraic structure $\mathscr{E}(\oplus, \otimes)$ satisfying the axioms (A1)–(A6) described below. These axioms are similar to those satisfied by finite elements of linear radicable blogs in the sense of [1]. Certain 'distance' function $d(x, y)$ is introduced for $x, y \in \mathscr{E}^m$ with nonzero components (formula (2.5) below). The problem of minimizing $d(x, \bar{x})$ subject to conditions $x = B \otimes u$, $u \in U = \{u \in \mathscr{E}^n \mid h \leq u \leq H\}$, where h, H, $\bar{x}$ and B are given, is solved via a finite polynomial threshold-type algorithm. The concretization for the case when $\mathscr{E}$ is the principal interpretation from [1] and motivation from machine time scheduling problems from [1, 3] are briefly discussed.

Key words: Nonconvex Optimization, Optimization in Ordered Algebraic Structures.

1. Introduction

In this article, we shall consider optimization problem on an algebraic structure, which was studied for instance in [4, 1, 2, 3]. In case that this structure is the principal interpretation of minimax algebras from [1], such problems arise in the context of certain machine time scheduling problems studied in [1, 4].

2. Necessary concepts and notations

We shall assume in the sequel that $\mathscr{E}$ is a set with operations $\oplus$, $\otimes$ satisfying the following axioms:

(A1) $\mathscr{E}$ is a commutative semigroup with respect to $\oplus$, $\otimes$;

(A2) $\alpha \otimes (\beta \oplus \gamma) = (\alpha \otimes \beta) \oplus (\alpha \otimes \gamma)$ $\forall \alpha, \beta, \gamma \in \mathscr{E}$;

(A3) $\exists \theta, \bar{1} \in \mathscr{E}$ such that $\alpha \oplus \theta = \alpha$, $\alpha \otimes \bar{1} = \alpha$, $\alpha \otimes \theta = \theta$ $\forall \alpha \in \mathscr{E}$;

(A4) $\alpha \oplus \beta = \alpha$ or β $\forall \alpha, \beta \in \mathscr{E}$;

(A5) $\forall \alpha \in \mathscr{E} \setminus \{\theta\}$ $\exists \alpha^{-1} \in \mathscr{E}$ such that $\alpha \otimes \alpha^{-1} = \bar{1}$;

(A6) $\forall \beta \in \mathscr{E}$ and integer $k \geq 1$ there is a unique $\alpha \in \mathscr{E}$ such that $\alpha^k = \beta$, where $\alpha^k \equiv \alpha \otimes \cdots \otimes \alpha$ (k-times).

Remark 2.1. If $\mathscr{E} \setminus \{\theta\}$ is the set of all finite elements of a radicable blog in the sense of [1], then (A1)–(A3), (A5) and (A6) are satisfied.

It can be shown that a binary relation $\leqslant$ defined by

$$\alpha \leqslant \beta \quad \Leftrightarrow \quad \alpha \oplus \beta = \beta \qquad \forall \alpha, \beta \in \mathscr{E} \tag{2.1}$$

is a total order over $\mathscr{E}$ (cf. [1, 2]). We remark that according to (A4), $\alpha \geqslant \theta\ \forall \alpha \in \mathscr{E}$. We shall set in the sequel:

$$\max_{j \in I} \alpha_j \equiv \sum_{j \in I}^{\oplus} \alpha_j \text{ if } I \text{ is finite} \tag{2.2}$$

and

$$\max_{j \in I} \alpha_j \equiv \theta \quad \text{if } I = \emptyset. \tag{2.3}$$

Further we shall define $\mathscr{E}_+$ as follows

$$\mathscr{E}_+ = \{\alpha \in \mathscr{E} \mid \alpha \neq \theta\}. \tag{2.4}$$

Let us remark that operator 'max' is defined in accordance with relation $\leqslant$ defined by (2.1). We set further $\alpha < \beta$, if $\alpha \leqslant \beta$ and $\alpha \neq \beta\ \forall \alpha, \beta \in \mathscr{E}$.

Let $\mathscr{E}^k$ denote the Cartesian product $\mathscr{E} \times \cdots \times \mathscr{E}$ of k sets $\mathscr{E}$ (for an arbitrary natural k). If B is an $(m \times n)$-matrix with elements $b_{ij} \in \mathscr{E}$ and $u = (u_1, \ldots, u_n) \in \mathscr{E}^n$, we define $B \otimes u$ as an element of $\mathscr{E}^m$ with components

$$(B \otimes u)_i = \sum_{j=1}^{n\,\oplus} (b_{ij} \otimes u_j) = \max_{1 \leqslant j \leqslant n} (b_{ij} \otimes u_j) \quad \forall i = 1, \ldots, m.$$

We shall set $x \leqslant y$ for $x, y \in \mathscr{E}^k$ if $x \leqslant y\ \forall i$.

Remark 2.2. It can be shown (cf. [1, 2]) that, for all $u, v \in \mathscr{E}^n$, $u \leqslant v$ implies $B \otimes u \leqslant B \otimes v$ and that for all $\alpha, \beta \in \mathscr{E}_+$, $\gamma \in \mathscr{E}$, $\alpha \leqslant \beta$ implies $\gamma \otimes \alpha \leqslant \gamma \otimes \beta$ and thus $\gamma \otimes \beta^{-1} \leqslant \gamma \otimes \alpha^{-1}$.

We shall define a function $d : \mathscr{E}_+^m \times \mathscr{E}_+^m \to \mathscr{E}_+$ as follows:

$$d(x, y) = \sum_{i=1}^{m\,\oplus} ((x_i \otimes y_i^{-1}) \oplus (x_i^{-1} \otimes y_i)) \quad \forall x, y \in \mathscr{E}_+^m$$

or using the simplified notation

$$d(x, y) = \max_{1 \leqslant i \leqslant m} \max((x_i \otimes y_i^{-1}), (x_i^{-1} \otimes y_i)). \tag{2.5}$$

Remark 2.3. The function d can be interpreted as a certain 'distance' function defined for elements from $\mathscr{E}_+^m$ on the basis of the seminorm introduced in [1].

In this article, we shall consider the following optimization problem on $\mathscr{E}_+^m$:

$$\begin{aligned} &\text{minimize} \quad d(x, \bar{x}) \\ &\text{subject to} \quad x = B \otimes u, \\ &\qquad\qquad\quad u \in U = \{u \mid h \leqslant u \leqslant H\}, \end{aligned} \tag{P}$$

where $h, H \in \mathscr{E}_+^n$, $\bar{x} \in \mathscr{E}_+^m$, $B = \|b_{ij}\|$, $b_{ij} \in \mathscr{E}_+ \ \forall i = 1, \ldots, m, j = 1, \ldots, n$ are given and the minimization is understood with respect to the relation $\leqslant$ defined by (2.1) so that, for optimal solution x^{opt} of (P), $d(x^{\text{opt}}, \bar{x}) \leqslant d(x, \bar{x})$ for all feasible solutions of (P).

Remark 2.4. If we interpret u as a control vector, then the set

$$D(U) \equiv \{x \in \mathscr{E}^m \mid \exists u \in U \text{ such that } x = B \otimes u\}$$

can be interpreted as the set of elements x attainable from U and we have to find an attainable point minimizing the 'distance' $d(x, \bar{x})$. The problems of this type can occur as certain modifications of the machine time scheduling problems discussed for instance in [1, 4] (see also Section 5 below).

Remark 2.5. It follows immediately from Remark 2.2 that $\hat{x} = B \otimes H$ is the greatest element of $D(U)$ in the sense that $x \leqslant \hat{x}$ for all $x \in D(U)$.

Remark 2.6. Different publications use different notation for describing the algebraic structure $\mathscr{E}$ introduced above. In this article, we use the notation from [1]. We shall describe in this remark briefly another notation (cf. [4]). Let $(G, +, \leqslant)$ be a totally ordered, commutative, radicable group with neutral element 0. Then the absolute value $|\alpha| = \max(\alpha, -\alpha)$ is well-defined. The problem (P) can be rewritten in the form:

$$\text{Find } \min_{h \leqslant u \leqslant H} \max_{1 \leqslant i \leqslant m} \left| \max_{1 \leqslant j \leqslant n} (b_{ij} + u_j) - b_i \right|. \qquad (P)'$$

If we define

$$(Bu)_i = \max_{1 \leqslant j \leqslant n} (b_{ij} + u_j) \quad \forall i = 1, \ldots, m$$

and

$$\|c\| := \max_{1 \leqslant i \leqslant m} |c_i| \quad \text{for all } c \in G^m,$$

then $(P)'$ can be reformulated as

$$\text{Find } \min_{h \leqslant u \leqslant H} \|Bu - \bar{x}\|.$$

The notation used in $(P)'$ and $(P)''$ seems to be easier to understand for those readers who are not actively working with algebraic structures. Let us remark still that another notation slightly different from both mentioned above is used in [3].

Remark 2.7. Since we consider only the data from $\mathscr{E}_+$, the axiom (A3) could be weakened for purposes of this article to (A3)' $\exists \bar{1} \in \mathscr{E}$ such that $\alpha \otimes \bar{1} = \alpha$ for all $a \in \mathscr{E}$.

We shall introduce now the following notations:

$$N=\{1,\dots,n\},\qquad S=\{1,\dots,m\},$$

$$x(u)\equiv B\otimes u\quad \forall u\in U,\qquad D(U)\equiv\{x(u)\mid u\in U\},$$

$$F^{(i)}(u)\equiv\{r\in N\mid b_{ir}\otimes u_r=\max_{j\in N}(b_{ij}\otimes u_j)\}\quad \forall u\in U,\ i\in S,$$

$$F(u)\equiv\{p\in S\mid (x_p(u)\otimes \bar{x}_p^{-1})\oplus(x_p(u)^{-1}\otimes\bar{x}_p)=d(x(u),\bar{x})\}\quad \forall u\in U,$$

$$S^+(u)\equiv\{p\in S\mid x_p(u)>\bar{x}_p\}\quad \forall u\in U,\qquad S^-(u)\equiv S\setminus S^+(u)\quad \forall u\in U,$$

$$T(u)\equiv\bigcup_{p\in F(u)}F^{(p)}(u)\quad \forall u\in U,$$

$$G(u)\equiv\{i\in S\mid F^{(i)}(u)\subseteq T(u)\}\quad \forall u\in U,$$

$$V(u)\equiv\{r\in N\mid u_r=h_r\}\cap T(u)\quad \forall u\in U,$$

$$W(u)\equiv F(u)\cap S^+(u),$$

$$T^+(u)\equiv\bigcup_{p\in W(u)}F^{(p)}(u)\qquad \forall u\in U.$$

In the sequel, we shall prove some properties of problem (P) and derive a finite algorithm for solving (P).

3. Some properties of problem (P)

Lemma 3.1. *Let u^*, $\tilde{u}\in U$, $x^*=B\otimes u^*$, $\tilde{x}=B\otimes\tilde{u}$; let one of the following conditions be fulfilled*:

(a) $\exists p_0\in F(u^*)\cap S^+(u^*)$ such that $\tilde{x}_{p_0}\geq x^*_{p_0}$,

(b) $\exists p_0\in F(u^*)\cap S^-(u^*)$ such that $\tilde{x}_{p_0}\leq x^*_{p_0}$.

Then

$$d(x^*,\bar{x})\leq d(\tilde{x},\bar{x}).\tag{3.1}$$

Proof. Let (a) be fulfilled. Then

$$x^*_{p_0}>\bar{x}_{p_0}\quad\Rightarrow\quad x^*_{p_0}\otimes\bar{x}_{p_0}^{-1}>\mathbf{\bar{1}}>x^{*-1}_{p_0}\otimes\bar{x}_{p_0}$$

so that

$$d(x^*,\bar{x})=\max(x^*_{p_0}\otimes\bar{x}_{p_0}^{-1},\ x^{*-1}_{p_0}\otimes\bar{x}_{p_0})=x^*_{p_0}\otimes\bar{x}_{p_0}^{-1}$$

and since $\tilde{x}_{p_0}\geq x^*_{p_0}$, we have (cf. Remark 2.2):

$$x^*_{p_0}\otimes\bar{x}_{p_0}^{-1}\leq\tilde{x}_{p_0}\otimes\bar{x}_{p_0}^{-1}\leq d(\tilde{x},x^*)$$

so that (3.1) holds. If (b) is satisfied we similarly obtain:

$$d(x^*,\bar{x})=x^{*-1}_{p_0}\otimes\bar{x}_{p_0}\leq\tilde{x}_{p_0}^{-1}\otimes\bar{x}_{p_0}\leq d(\tilde{x},x^*).\qquad\square$$

Corollary. *Let* $\hat{x} = B \otimes H$, $x = B \otimes u$, $u \in U$. *Let us assume that one of the following conditions is fulfilled:*

(a)′ $F(H) \cap S^{-}(H) \neq \emptyset$;

(b)′ $V(H) \neq \emptyset$.

Then

$$d(\hat{x}, \bar{x}) \leq d(x, \bar{x}). \tag{3.2}$$

Proof. Let (a)′ be fulfilled and let $p_0 \in F(H) \cap S^{-}(H)$. We further have (cf. Remark 2.2) $x_{p_0} \leq \hat{x}_{p_0}$ so that (b) Lemma 3.1 is satisfied and (3.2) holds.

Let (b)′ hold and let $h_{r_0} = H_{r_0}$, where $r_0 \in F^{(p_0)}(H)$, $p_0 \in F(H)$. It is in this case $h_{r_0} = u_{r_0} = H_{r_0}$ so that

$$x_{p_0} = \max_{j \in N}(b_{ij} \otimes u_j) \geq b_{p_0 r_0} \otimes u_{r_0} = b_{p_0 r_0} \otimes H_{r_0} = \hat{x}_{p_0}$$

and since $x_{p_0} \leq \hat{x}_{p_0}$ (Remark 2.5), we have $x_{p_0} = \hat{x}_{p_0}$.

If $p_0 \in S^{+}(H)$, then (3.2) holds according to Lemma 3.1(a), if $p_0 \in S^{-}(H)$, then (3.2) holds according to Lemma 3.1(b). □

Lemma 3.2. *Let us assume that* $p_1, p_2 \in S^{+}(u)$ *and let*

$$x_{p_1}(u) \otimes \bar{x}_{p_1}^{-1} = x_{p_2}(u) \otimes \bar{x}_{p_2}^{-1}. \tag{3.3}$$

Then

$$r \in F^{(p_1)}(u) \cap F^{(p_2)}(u) \quad \Rightarrow \quad b_{p_1 r}^{-1} \otimes \bar{x}_{p_1} = b_{p_2 r}^{-1} \otimes \bar{x}_{p_2}. \tag{3.4}$$

Proof. Under the given assumptions we have:

$$x_{p_1}(u) = b_{p_1 r} \otimes u_r, \qquad x_{p_2}(u) = b_{p_2 r} \otimes u_r$$

so that we obtain, from (3.3),

$$b_{p_1 r} \otimes u_r \otimes \bar{x}_{p_1}^{-1} = b_{p_2 r} \otimes u_r \otimes \bar{x}_{p_2}^{-1} / \otimes u_r^{-1},$$

$$b_{p_1 r} \otimes \bar{x}_{p_1}^{-1} = b_{p_2 r} \otimes \bar{x}_{p_2}^{-1}$$

and thus

$$b_{p_1 r}^{-1} \otimes \bar{x}_{p_1} = b_{p_2 r}^{-1} \otimes \bar{x}_{p_2}. \quad \square$$

Corollary. *If* $r \in F^{(p)}(u)$, *then*

$$b_{pr}^{-1} \otimes \bar{x}_p = \alpha_r \quad \forall p \in W(u).$$

Proof. The assertion follows from the fact that

$$d(x(u), \bar{x}) = x_p(u) \otimes \bar{x}_p^{-1} \quad \forall p \in W(u)$$

so that (3.3) holds for any pair $p_1, p_2 \in W(u)$. □

Lemma 3.3. *Let $r_1, r_2 \in F^{(s)}(u)$, $u_{r_1} = t \otimes \beta_{r_1}$, $u_{r_2} = t \otimes \beta_{r_2}$, where $t > \theta$, $\beta_{r_1}, \beta_{r_2} \in \mathscr{E}$ are given. Then*

$$b_{sr_1} \otimes \beta_{r_1} = b_{sr_2} \otimes \beta_{r_2}. \tag{3.5}$$

Proof. Under the given assumptions we have

$$x_s(u) = b_{sr_1} \otimes u_{r_1} = b_{sr_2} \otimes u_{r_2}$$

so that

$$b_{sr_1} \otimes t \otimes \beta_{r_1} = b_{sr_2} \otimes t \otimes \beta_{r_2} - \otimes t^{-1}$$

$$b_{sr_1} \otimes \beta_{r_1} = b_{sr_2} \otimes \beta_{r_2}. \qquad \square$$

Corollary. *Under the given assumptions of Lemma 3.3 it follows that*

$$x_s(u) = v_s \otimes t,$$

where $v_s = b_{sr} \otimes \beta_r \ \forall r \in F^{(s)}(u)$.

4. The description of the algorithm for solving problem (*P*)

Let us suppose that starting with $u^{(0)} = H$ and $x^{(0)} = B \otimes H$, we have already obtained $u^{(1)}, \ldots, u^{(k-1)}$, $x^{(i)} = B \otimes u^{(i)}$, $i = 1, \ldots, k-1$, such that

$$d(x^{(i)}, \bar{x}) < d(x^{(i-1)}, \bar{x}) \quad \forall i = 1, \ldots, k-1.$$

We shall show that, if $F(u^{(k-1)}) \cap S^-(u^{(k-1)}) = \emptyset$ and $V(u^{(k-1)}) = \emptyset$, we can construct an element $u^k \in U$ such that, for $x^{(k)} = B \otimes u^{(k)}$,

$$d(x^{(k)}, \bar{x}) < d(x^{(k-1)}, \bar{x})$$

holds.

Further we shall show that, if $F(u^{(k-1)}) \cap S^-(u^{(k-1)}) \neq \emptyset$ or $V(u^{(k-1)}) \neq \emptyset$, then $x^{(k-1)} = B \otimes u^{(k-1)}$ is an optimal solution of (P).

Let us assume at first that both $F(u^{(k-1)}) \cap S^-(u^{(k-1)})$ and $V(u^{(k-1)})$ is empty and derive the formulae for the simultaneous parametric decrease of variables u_r, $r \in T(u^{(k-1)})$ resulting in a new 'control' $u^{(k)} \in U$ and new 'state' $x^{(k)} = B \otimes u^{(k)}$ with the properties:

$$x^{(k)} \leq x^{(k-1)}, \quad x^{(k)} \neq x^{(k-1)}, \quad d(x^{(k)}, \bar{x}) < d(x^{(k-1)}, \bar{x}).$$

Since $F(u^{(k-1)}) \cap S^-(u^{(k-1)}) = \emptyset$, the set $F(u^{(k-1)})$ is a subset of $S^+(u^{(k-1)})$ and thus similarly to the proof of Lemma 3.1 we can obtain:

$$d(x^{(k-1)}, \bar{x}) = x_p^{(k-1)} \otimes \bar{x}_p^{-1} \qquad \forall p \in F(u^{(k-1)}).$$

Let us now set

$$t = x_p(u) \otimes \bar{x}_p^{-1} \qquad \forall p \in F(u^{(k-1)}), \quad u \in U, \qquad u \leq u^{(k-1)}. \tag{4.1}$$

Then

$$t^{(k-1)} \equiv x_p(u^{(k-1)}) \otimes \bar{x}_p^{-1} = d(x^{(k-1)}, \bar{x}).$$

We shall decrease parameter t starting from the value $t = t^{(k-1)}$ via decreasing u_r, $r \in T(u^{(k-1)})$.

The decrease of t will be stopped if one of the following four cases occurs:

(A) Some variable (or variables) x_i, $i \in G(u^{(k-1)})$ does not decrease any more;

(B) The objective function $d(x, \bar{x})$ does not decrease any more;

(C) Some variable (or variables) x_i, $i \in G(u^{(k-1)})$ attains the value $\bar{x}_i$;

(D) Some variable (or variables) u_r attains the value h_r.

It follows from (4.1) that

$$x_p(u) = t \otimes \bar{x}_p \qquad \forall p \in F(u^{(k-1)})$$

so that

$$b_{pr} \otimes u_r = t \otimes \bar{x}_p \qquad \forall p \in F(u^{(k-1)}), \quad r \in F^{(p)}(u^{(k-1)})$$

and thus

$$u_r = t \otimes \bar{x}_p \otimes b_{pr}^{-1} \qquad \forall p \in F(u^{(k-1)}), \quad r \in F^{(p)}(u^{(k-1)}).$$

Since, according to the Corollary of Lemma 3.2, we can set

$$\bar{x}_p \otimes b_{pr}^{-1} = \alpha_r \qquad \forall p \in F(u^{(k-1)}) \quad \text{such that } r \in F^{(p)}(u^{(k-1)}),$$

we obtain

$$u_r = u_r(t) \equiv \alpha_r \otimes t \quad \forall r \in T(u^{(k-1)}). \tag{4.2}$$

Further we have for each $i \in G(u^{(k-1)})$:

$$x_i(u(t)) = b_{ir} \otimes u_r(t) = b_{ir} \otimes \alpha_r \otimes t \qquad \forall r \in F^{(i)}(u(t)).$$

Since, according to the Corollary of Lemma 3.3, we can set

$$b_{ir} \otimes \alpha_r = v_i \qquad \forall r \in F^{(i)}(u(t)),$$

we have for

$$t \geqslant \max_{j \in N \setminus T(u^{(k-1)})} (b_{ij} \otimes u_j^{(k-1)})$$

and $i \in G(u^{(k-1)})$:

$$x_i(u(t)) = v_i \otimes t. \tag{4.3}$$

We shall now derive the formulae for the values of t, at which cases (A), (B), (C), (D) occur.

Case (A)

The decrease of x_i, $i \in G(u^{(k-1)})$ can stop only if $x_i(u(t))$ attains the value

$$\alpha_i^{(k-1)} = \max_{j \in N \setminus T(u^{(k-1)})} (b_{ij} \otimes u_j^{(k-1)})$$

i.e. if

$$x_i(u(t)) = \alpha_i^{(k-1)}$$

or, according to (4.3),

$$v_i \otimes t = \alpha_i^{(k-1)}$$

and, since under our assumptions $v_i \neq \theta$,

$$t = \alpha_i^{(k-1)} \otimes v_i^{-1}.$$

The value $\tau_A^{(k-1)}$, at which for the first time at least one variable x_i, $i \in G(u^{(k-1)})$ does not decrease any more satisfies therefore the relation

$$\tau_A^{(k-1)} = \max_{i \in G(u^{(k-)})} (\alpha_i^{(k-1)} \otimes v_i^{-1}). \tag{4.4}$$

Case (B)

The value of $d(x, \bar{x})$ does not decrease any more, if at least one of the following cases occurs:

(i) $t = \alpha_p^{(k-1)} \otimes v_p^{-1}$ for some $p \in F(u^{(k-1)})$ so that some variable x_p, $p \in F(u^{(k-1)})$ decreases no more (compare Case (A));

(ii) $$t = \alpha_0^{(k-1)} \equiv \max_{i \in S \setminus G(u^{(k-1)})} [(x_i^{(k-1)} \otimes \bar{x}_i^{-1}) \oplus ((x_i^{(k-1)}) \otimes \bar{x}_i)];$$

(iii) parameter t (i.e. the current value of the objective function) and some variable (or variables) x_s, $s \in G(u^{(k-1)}) \cap S^-(u^{(k-1)})$, attain value $t_s^{(k-1)}$, $x_s(t_s^{(k-1)})$, at which

$$(x_s(t_s^{(k-1)}) \otimes \bar{x}_s^{-1}) \oplus (x_s(t_s^{(k-1)})^{-1} \otimes \bar{x}_s) = t_s^{(k-1)}$$

so that $d(x(u(t)), \bar{x}) > t_s^{(k-1)} = d(x(u(t_s^{(k-1)})), \bar{x})$ for $t < t_s^{(k-1)}$ (the detailed explanation see below).

Since the (i) was analysed already in Case (A) and (ii) gives already the corresponding value of t, it remains to investigate only (iii).

Since $s \in G(u^{(k-1)}) \otimes S^-(u^{(k-1)})$, it is $x_s(t) < \bar{x}_s$ for all $t \leq t^{(k-1)}$ so that similarly as in the proof of Lemma 3.1 we have, for $t \leq t^{(k-1)}$,

$$(x_s(t) \otimes x_s^{-1}) \oplus (\bar{x}_s(t)^{-1} \otimes \bar{x}_s) = x_s(t)^{-1} \otimes \bar{x}_s$$

so that $t_s^{(k-1)}$ is the solution of the equation

$$x_s(t)^{-1} \otimes \bar{x}_s = t$$

or, in accordance with (4.3),

$$(v_s \otimes t)^{-1} \otimes \bar{x}_s = t$$

so that

$$t^2 = \bar{x}_s \otimes v_s^{-1},$$

where we set $t^2 = t \otimes t$. The equation (4.5) is uniquely soluble according to (A6) so that we can assume that $t_s^{(k-1)}$ is its solution and set

$$\bar{t}^{(k-1)} = \max_{s \in G(u^{(k-1)}) \cap S^-(u^{(k-1)})} t_s^{(k-1)}.$$

Summarizing (i), (ii) and (iii), we obtain that the value $\tau_B^{(k-1)}$, at which for the first time Case (B) occurs is given by

$$\tau_B^{(k-1)} = \max\{ \max_{p \in F(u^{(k-1)})} \alpha_p^{(k-1)} \otimes v_p^{-1}, \alpha_0^{(k-1)}, \bar{t}^{(k-1)} \}. \tag{4.6}$$

Case (C)

The value of t at which some variable x_i, $i \in G(u^{(k-1)})$, attains the value $\bar{x}_i$ must solve the equation

$$x_i(u(t)) = \bar{x}_i$$

or according to (4.3)

$$v_i \otimes t = \bar{x}_i$$

so that

$$t = \bar{x}_i \otimes v_i^{-1}.$$

The value $\tau_C^{(k-1)}$ of t, at which at least one variable x_i, $i \in G(u^{(k-1)})$ attains the value $\bar{x}_i$ is therefore given by the relation:

$$\tau_C^{(k-1)} = \max_{i \in G(u^{(k-1)}) \cap S^+(u^{(k-1)})} (\bar{x}_i \otimes v_i^{-1}). \tag{4.7}$$

Case (D)

To determine the value of t, at which $u_r(t) = h_r$ for some $r \in T(u^{(k-1)})$, we have to solve the equation

$$u_r(t) = h_r$$

or according to (4.2)

$$\alpha_r \otimes t = h_r \quad \text{so that } t = h_r \otimes \alpha_r^{-1}.$$

The value $\tau_D^{(k-1)}$ of t, at which for the first time at least one variable u_r, $r \in T(u^{(k-1)})$ attains the value h_r is thus given by the relation

$$\tau_D^{(k-1)} = \max_{r \in T(u^{(k-1)})} (h_r \otimes \alpha_r^{-1}). \tag{4.8}$$

After having determined values $\tau_A^{(k-1)}$, $\tau_B^{(k-1)}$, $\tau_C^{(k-1)}$, $\tau_D^{(k-1)}$, we set:

$$\begin{aligned} t^{(k)} &= \max\{\tau_A^{(k-1)}, \tau_B^{(k-1)}, \tau_C^{(k-1)}, \tau_D^{(k-1)}\}, \\ u_r^{(k)} &= \alpha_r \otimes t^{(k)} \quad \forall r \in T(u^{(k-1)}), \\ u_r^{(k)} &= u_r^{(k-1)} \quad \forall r \in N \setminus T(u^{(k-1)}), \\ x^{(k)} &= B \otimes u^{(k)}, \end{aligned} \tag{4.9}$$

It is then $t^{(k)} = d(x^{(k)}, \bar{x}) < d(x^{(k-1)}, \bar{x}) = t^{(k-1)}$, $x^{(k)} \leq x^{(k-1)}$, $x^{(k)} \neq x^{(k-1)}$. We shall prove now that if either $F(u^{(k-1)} \cap S^-(u^{(k-1)}) \neq \emptyset$ or $V(u^{(k-1)}) \neq \emptyset$, then $x^{(k-1)} = B \otimes u^{(k-1)}$ is an optimal solution of (P). For this purpose, we shall need the following lemma.

Lemma 4.1. *Let $u^{(k)}$ be obtained according to* (4.9) *and $x^{(k)} \neq \bar{x}$. Then*

$$\begin{aligned} &\text{(a)} \quad T(u^{(k-1)}) \subseteq T^+(u^{(k)}) \subseteq T(u^{(k)}), \\ &\text{(b)} \quad u_r^{(k)} = H_r \text{ for all } r \in N \setminus T(u^{(k-1)}). \end{aligned} \tag{4.10}$$

Proof. It follows immediately from the investigation of cases (A), (B), (C), (D) above that:

$$T(u^{(k-1)}) \subseteq T(u^{(k)}) \quad \text{if } t^{(k)} = \tau_A^{(k-1)} \quad \text{or} \quad \tau_B^{(k-1)}$$

$$T(u^{(k-1)}) = T(u^{(k)}) \quad \text{if } t^{(k)} = \tau_C^{(k-1)} \quad \text{or} \quad \tau_D^{(k-1)}.$$

Since we assumed that $F(u^{(k-1)}) \cap S^-(u^{(k-1)}) = \emptyset$ and $x^{(k)} \neq \bar{x}$, we have

$$T(u^{(k-1)}) = T^+(u^{(k-1)}) = T^+(u^{(k)}) \subseteq T(u^{(k)}),$$

because $W(u^{(k-1)}) = W(u^{(k)})$, so that (4.10) holds.

To prove (b), let us remark that it follows from (4.10):

$$T(u^{(j)}) \subseteq T(u^{(k-1)}) \quad \text{for all } j,\ 0 \leq j \leq k-1$$

so that if $r \in N \setminus T(u^{(k-1)})$, then $r \notin T(u^{(j)})$ for all j, $0 \leq j \leq k-1$ and, according to (4.9),

$$u_r^{(k)} = u_r^{(k-1)} = \cdots = u_r^{(0)} = H_r \quad \forall r \in N \setminus T(u^{(k-1)}). \qquad \square$$

Theorem 4.1. *Let $u^{(k-1)}$ be obtained according to the process described above (i.e. $u^{(k-1)} = H$ if $k-1 = 0$, $u^{(k-1)}$ is defined according to* (4.9) *otherwise). Let us assume that either*

$$V(u^{(k-1)}) \neq \emptyset \tag{4.11}$$

or

$$F(u^{(k-1)}) \cap S^-(u^{(k-1)}) \neq \emptyset. \tag{4.12}$$

Let $\tilde{u}$ be an aribtrary element of U and $\tilde{x} = B \otimes \tilde{u}$. Then

$$d(x^{(k-1)}, \bar{x}) \leq d(\tilde{x}, \bar{x}) \tag{4.13}$$

(i.e. $x^{(k-1)}$ is an optimal solution of (P)).

Proof. If either (a) or (b) from Lemma 3.1 is satisfied for $u^* = u^{(k-1)}$, $x^* = x^{(k-1)}$, then (4.13) holds. If $k-1 = 0$, the optimality of $x^{(0)} = B \times H$ follows from the Corollary of Lemma 3.1. Therefore we only have to investigate the case in which $k-1 > 0$ and neither (a) nor (b) from Lemma 3.1 holds. Further we can exclude from our considerations the case $x^{(k-1)} = \bar{x}$, since the optimality of $x^{(k-1)}$ follows in this case immediately from the properties of d. We shall assume therefore that $x^{(k-1)} \neq \bar{x}$ and that

$$\tilde{x}_p < x_p^{(k-1)} \quad \forall p \in F(u^{(k-1)}) \cap S^+(u^{(k-1)}) \tag{4.14}$$

and in the same time

$$\tilde{x}_p > x_p^{(k-1)} \quad \forall p \in F(u^{(k-1)}) \cap S^-(u^{(k-1)}). \tag{4.15}$$

We shall show that if $x^{(k-1)} \neq \bar{x}$ and (4.14) and (4.15) hold for some $\tilde{x} = B \otimes \tilde{u}$, then $\tilde{u} \notin U$.

It follows from (4.14) for all $p \in W(u^{(k-1)})$ that

$$\tilde{x}_p = \max_{j \in N}(b_{pj} \otimes \tilde{u}_j) < x_p^{(k-1)} = \max_{j \in N}(b_{pj} \otimes u_j^{(k-1)}) = b_{pr} \otimes u_r^{(k-1)},$$

where $r \in F^{(p)}(u^{(k-1)})$; we therefore obtain

$$b_{pj} \otimes \tilde{u}_j < b_{pr} \otimes u_r^{(k-1)} \quad \forall j \in N, \qquad r \in F^{(p)}(u^{(k-1)})$$

so that for all $p \in W(u^{(k-1)})$

$$b_{pr} \otimes \tilde{u}_r < b_{pr} \otimes u_r^{(k-1)} \quad \forall r \in F^{(p)}(u^{(k-1)})$$

and thus (since $b_{pr} \neq \theta$)

$$\tilde{u}_r < u_r^{(k-1)} \quad \forall r \in T^+(u^{(k-1)}). \tag{4.16}$$

Let us assume now that $F(u^{(k-1)}) \cap S^-(u^{(k-1)}) = \emptyset$, so that (4.11) holds. In this case $F(u^{(k-1)}) = W(u^{(k-1)})$ so that $W(u^{(k-1)}) \neq \emptyset$ and

$$T(u^{(k-1)}) = T^+(u^{(k-1)}) \tag{4.17}$$

and therefore

$$\tilde{u}_r < u_r^{(k-1)} \quad \forall r \in T(u^{(k-1)}). \tag{4.18}$$

Let r_0 be an element of $V(u^{(k-1)})$. It follows then from (4.18) that

$$\tilde{u}_{r_0} < u_{r_0}^{(k-1)} = h_{r_0}$$

and thus $\tilde{u} \notin U$.

Let us assume now that (4.12) holds and let $p_0 \in F(u^{(k-1)}) \cap S^-(u^{(k-1)})$ so that according to (4.15) $\tilde{x}_{p_0} > x_{p_0}^{(k-1)}$. Let r_0 be an element of $F^{(p_0)}(u^{(k-1)})$. It is then

$$x_{p_0}^{(k-1)} = b_{p_0 r_0} \otimes u_{r_0}^{(k-1)} \quad \text{and} \quad r_0 \in T(u^{(k-1)}).$$

It follows from $\tilde{x}_{p_0} > x_{p_0}^{(k-1)}$ that there exists an index r_1 such that

$$\tilde{x}_{p_0} = b_{p_0 r_1} \otimes \tilde{u}_{r_1} > b_{p_0 r_0} \otimes u_{r_0}^{(k-1)} \geqslant b_{p_0 r_1} \otimes u_{r_1}^{(k-1)}$$

and thus

$$\tilde{u}_{r_1} > u_{r_1}^{(k-1)}. \tag{4.19}$$

It follows further from the construction of $u^{(k-1)}$, $x^{(k-1)}$ and from $k-1>0$ that $F(u^{(k-2)}) \cap S^-(u^{(k-2)}) = \emptyset$. Therefore according to Lemma 4.1(a)

$$T(u^{(k-2)}) \subseteq T^+(u^{(k-1)}) \subseteq T(u^{(k-1)}). \tag{4.20}$$

It follows further from (4.18) and (4.20) that $r_1 \in N \setminus T(u^{(k-2)})$ so that according to (4.9) for $k := k-1$ and Lemma 4.1(b) we obtain

$$u_{r_1}^{(k-1)} = u_{r_1}^{(k-2)} = H_{r_1}$$

and therefore according to (4.19)

$$\tilde{u}_{r_1} > u_{r_1}^{(k-1)} = H_{r_1}$$

so that again $\tilde{u} \notin U$. □

The algorithm for the solution of (P) can now be summarized as follows:

(1) Read n, m, B, h, H, $\bar{x}$; $N := \{1, \ldots, n\}$, $S := \{1, \ldots, m\}$, $k := 1$, $u^{(k-1)} := H$, $x^{(k-1)} := B \otimes u^{(k-1)}$;

(2) Determine sets $F^{(i)}(u^{(k-1)}) \forall i \in S$, $F(u^{(k-1)})$, $T(u^{(k-1)})$, $V(u^{(k-1)})$, $G(u^{(k-1)})$, $S^+(u^{(k-1)})$, $S^-(u^{(k-1)})$, $W(u^{(k-1)})$,

(3) If either $V(u^{(k-1)}) \neq \emptyset$ or $F(u^{(k-1)}) \cap S^-(u^{(k-1)}) \neq \emptyset$, then go to (7); otherwise go to (4).

(4) Determine $\tau_A^{(k-1)}$, $\tau_B^{(k-1)}$, $\tau_C^{(k-1)}$, $\tau_D^{(k-1)}$ according to (4.4), (4.6)–(4.8);

(5) Determine $t^{(k)}$, $u^{(k)}$, $x^{(k)}$ according to (4.9);

(6) $k := k+1$, go to (2).

(7) $x^{(k-1)}$ is an optimal solution of (P), END.

We shall derive now complexity bounds for this algorithm. At first we shall estimate the maximal value of k. Let us note that each time between two iterations $t^{(p)}$, $t^{(s)}$ with $p < s$, $t^{(p)} = \tau_A^{(p-1)} = \alpha_{i_0}^{(p-1)} \otimes v_{i_0}^{-1}$, $t^{(s)} = \tau^{(s-1)} = \alpha_{i_0}^{(s-1)} \otimes v_{i_0}^{-1}$, $i_0 \in F(u^{(p)})$, $T(u^{(p)}) \subset T(u^{(s)})$ maximally $m-1$ variables x_i, $i \in G(u^{(p)})$, $i \notin F(u^{(p)})$ can stop to decrease (sequentially one after the other), i.e. maximally $m-1$ iterations $t^{(k)} = \tau_A^{(k-1)} = \alpha_i^{(k-1)} \otimes v_i^{(k-1)}$ with $i \in G(u^{(k)})$, $i \notin F(u^{(k)}) = F(u^{(p)})$, $G(u^{(k)}) \subset G(u^{(k-1)})$, $T(u^{(k)}) = T(u^{(k-1)})$ can occur (i.e. $k = p+1, \ldots, p+m-1 = s-1$; cf. Example 5.1 below). This leads to the maximal number of such iterations $n(m-1)$. After that $m-1$ variables can (sequentially one after the other) attain the value $\bar{x}_i$, i.e. $m-1$ iterations $t^{(k)} = \tau_C^{(k-1)}$ can occur and the algorithm will finish after that in one step at some $x^{(q)}$, in which $V(u^{(q)}) \neq \emptyset$ (Case (D)) or $F(u^{(q)}) \cap S^-(u^{(q)}) \neq \emptyset$ (Case B(iii)). This leads us the total estimation

$$q \leq n(m-1) + (m-1) + 1 = n(m-1) + m. \tag{4.21}$$

Remark 4.1. Example 5.1 from the next section illustrates the situation, in which the bound given in (4.21) is really attained (in this example we have $m = n = 3$, so that $q \leq 3 \cdot 2 + 3 = 9$).

We shall derive now the complexity bounds for the performance of each iteration. We shall use the following notation: f is the number of $\otimes$-operations, g is the number of comparisons and r is the number of 'square root'-operations (i.e. the number of solutions of the equation (4.5)). The corresponding bounds are summarized in Table 4.1.

Table 4.1

	f	g	r
Case A	$(n-1)+2m$	$(n-1)+2m-2$	0
Case B(ii)	$2(m-1)$	$2(m-2)$	0
Case B(iii)	$m-1$	$m-2$	$m-1$
CaseC	m	m	m
Case D	n	n	0
Total	$2n+6m-4$	$2n+6m-8$	$m-1$

Remark 4.2. Let us note that Case B(i) was omitted, since it is contained in Case A and need no additional effort. Let us remark further that calculating τ_B needs still 3 additional comparisons (see (4.6) above), so that the total amount of comparisons for carrying out one iteration can be estimated by $(2n+6m-8)+3=2n+6m-5$.

Table 4.1, Remark 4.2 and estimation (4.21) immediately lead us to the conclusion that the algorithm suggested above is finite and polynomial in m, n (with respect to the number of $\otimes$-operations, comparisons and 'square root'-operations).

5. Some examples

Example 5.1. Let $\mathbb{R}_+$ be the set of all nonnegative numbers, set $\mathbb{R}_+^0=\mathbb{R}_+\setminus\{0\}$, $\mathscr{E}=\mathbb{R}_+$, $\alpha\oplus\beta=\max(\alpha,\beta)$, $\alpha\otimes\beta=\alpha\cdot\beta\ \forall\alpha,\beta\in\mathscr{E}$ so that $\theta=0$, $\bar{1}=1$, $\mathscr{E}_+=\mathbb{R}_+^0$ and $\mathscr{E}$ is the extremal algebra from [3]. Let further

$$\bar{x}^{\mathrm{T}}=(\tfrac{1}{4},1,1),\qquad B=\begin{pmatrix}1 & \frac{7}{10} & \frac{4}{10}\\ 1 & \frac{9}{10} & \frac{54}{70}\\ 1 & \frac{8}{10} & \frac{40}{70}\end{pmatrix},$$

$$h=(\tfrac{1}{2},\tfrac{1}{10},\tfrac{1}{10})^{\mathrm{T}},\qquad H=(10,10,10)^{\mathrm{T}},\qquad \text{so that } m=n=3.$$

The calculations according to the algorithm suggested in the preceding section are summarized in Table 5.1.

Let us note that in this example $F(u^{(k)})=\{1\}$ for all k, $0\leqslant k\leqslant 9$, so that Case (B) does not occur. Further we have: $S^{+}(u^{(k)})=\{1,2,3\}$, $S^{-}(u^{(k)})=\emptyset$ for $0\leqslant k\leqslant 6$,

Table 5.1

k	Case	$u^{(k)\mathrm{T}}$	$x^{(k)\mathrm{T}}$	$T(u^{(k)})$	$G(u^{(k)})$	$d(x^{(k)}, \bar{x})$
0	(A)	$(10, 10, 10)$	$(10, 10, 10)$	$\{1\}$	$\{1, 2, 3\}$	40
1	(A)	$(9, 10, 10)$	$(9, 9, 9)$	$\{1\}$	$\{1, 3\}$	36
2	(A)	$(8, 10, 10)$	$(8, 9, 8)$	$\{1\}$	$\{1\}$	32
3	(A)	$(7, 10, 10)$	$(7, 9, 8)$	$\{1, 2\}$	$\{1, 2, 3\}$	28
4	(A)	$(6, \frac{60}{7}, 10)$	$(6, \frac{54}{7}, \frac{48}{7})$	$\{1, 2\}$	$\{1, 3\}$	24
5	(A)	$(5, \frac{50}{7}, 10)$	$(5, \frac{54}{7}, \frac{40}{7})$	$\{1, 2\}$	$\{1\}$	20
6	(A)	$(4, \frac{40}{7}, 10)$	$(4, \frac{54}{7}, \frac{40}{7})$	$\{1, 2, 3\}$	$\{1, 2, 3\}$	16
7	(C)	$(\frac{7}{10}, 1, 7)$	$(\frac{7}{10}, \frac{27}{20}, 1)$	$\{1, 2, 3\}$	$\{1, 2, 3\}$	$\frac{28}{10}$
8	(C)	$(\frac{14}{27}, \frac{20}{27}, \frac{35}{27})$	$(\frac{14}{27}, 1, \frac{20}{27})$	$\{1, 2, 3\}$	$\{1, 2, 3\}$	$\frac{56}{27}$
9	(D)	$(\frac{1}{2}, \frac{20}{28}, \frac{10}{8})$	$(\frac{4}{8}, \frac{54}{56}, \frac{40}{56})$	$\{1, 2, 3\}$	$\{1, 2, 3\}$	2

$S^+(u^{(7)}) = \{1, 2\}$, $S^-(u^{(7)}) = \{3\}$, $S^+(u^{(8)}) = S^+(u^{(9)}) = \{1\}$, $S^-(u^{(8)}) = S^-(u^{(9)}) = \{1, 2\}$, $V(u^{(k)}) = \emptyset$ for $0 \leq k \leq 8$, $vV(u^{(9)}) = \{1\}$.

Example 5.2. Let $\mathbb{R}$ be the set of all real numbers, $\mathscr{E} = \mathbb{R} \cup \{-\infty\}$, $\alpha \oplus \beta = \max(\alpha, \beta)$, $\alpha \otimes \beta = \alpha + \beta \ \forall \alpha, \beta \in \mathscr{E}$, then $\mathscr{E}$ is so called principal interpretation from [1]. It can be shown that in this case $\theta = -\infty$, $\bar{1} = 0$, $\alpha^{-1} = -\alpha \ \forall \alpha \in \mathbb{R}$ so that the problem (P) can be reformulated as follows:

$$\begin{aligned} &\text{minimize} \quad d(x, \bar{x}) \equiv \max_{1 \leq i \leq m} |x_i - \bar{x}_i| \\ &\text{subject to} \quad x_i = \max_{1 \leq j \leq n} (b_{ij} + u_j), \quad i = 1, \ldots, m, \\ &\qquad\qquad\quad h_j \leq u_j \leq H_j, \quad j = 1, \ldots, n, \end{aligned} \tag{5.1}$$

where h_j, H_j, b_{ij}, $\bar{x}_i$ $\forall i, j$ are given elements of $\mathbb{R}$.

Remark 5.1. As a motivating example for problems of the form (5.1) can serve us the machine time scheduling problem from [1] or [4] (cf. [1, pp. 1–3] or [4, pp. 204–205]) with preassigned starting times in the cycle $r+1$, $\bar{x}$. We ask then for the vector of starting times $u = x(r-t)$ in one of the preceding cycles, $r-t$, which results in starting times $x(r+1)$ in cycle $r+1$ being, as close as possible to $\bar{x}$ in the sense of the Tshebyshef norm. The solution of this problem was obtained in [1, 4] under the additional constraint $x_i(r+1) \leq \bar{x}_i \ \forall i$.

Remark 5.2. It can be shown that the set of feasible solutions $x(u)$, $u \in U$ of problem (5.1) is in general nonconvex with respect to the usual set convexity in $\mathbb{R}^m$ but convex with respect to convexity introduced in [2, 3]. It can be shown further that $d(x(u), \bar{x})$ is in general a nonconvex function of u (with respect to the usual convexity of functions defined in $\mathbb{R}^n$).

References

[1] R.A. Cunninghame-Green, *Minimax algebra* (Springer, Berlin, Heidelberg, New York, 1979).

[2] L. Teschke and K. Zimmermann, "Extremale Algebren und verallgemeinerte Konvexitätsräume", *Ekonomicko-Matematický Obzor* 15 (1979) 74–84.

[3] N.N. Vorobev, Extremal algebra of positive matrices (in Russian), *Elektronische Informationsverarbeitung und Kybernetik* 3 (1967) 39–71.

[4] U. Zimmermann, *Linear and combinatorial optimization in ordered algebraic structures* (North-Holland, Amsterdam, 1981).